DIGITAL DESIGN

© 1991, 1984 by Prentice-Hall, Inc.
Upper Saddle River, New Jersey 07458

Printed in the United States of America

20 19 18 17 16 15 14 13 12

ISBN   0-13-212994-9

PRENTICE-HALL INTERNATIONAL (UK) LIMITED, LONDON
PRENTICE-HALL OF AUSTRALIA PTY. LIMITED, SYDNEY
PRENTICE-HALL CANADA INC., TORONTO
PRENTICE-HALL HISPANOAMERICANA, S.A., MEXICO
PRENTICE-HALL OF INDIA PRIVATE LIMITED, NEW DELHI
PRENTICE-HALL OF JAPAN, INC., TOKYO
PEARSON EDUCATION ASIA PTE. LTD., SINGAPORE
EDITORA PRENTICE-HALL DO BRASIL, LTDA., RIO DE JANEIRO

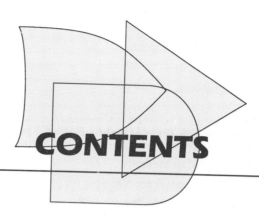

CONTENTS

3   SIMPLIFICATION OF BOOLEAN FUNCTIONS   *72*

4   COMBINATIONAL LOGIC   *114*

5   MSI AND PLD COMPONENTS   *152*

6   SYNCHRONOUS SEQUENTIAL LOGIC   *202*

7   REGISTERS, COUNTERS, AND THE MEMORY UNIT   *257*

8   ALGORITHMIC STATE MACHINES (ASM)   *307*

9   ASYNCHRONOUS SEQUENTIAL LOGIC   *341*

10   DIGITAL INTEGRATED CIRCUITS   399

11   LABORATORY EXPERIMENTS   436

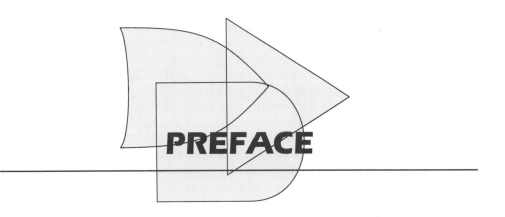

PREFACE

Digital Design is concerned with the design of digital electronic circuits. The subject is also known by other names such as logic design, digital logic, switching circuits, and digital systems. Digital circuits are employed in the design of systems such as digital computers, control systems, data communications, and many other applications that require electronic digital hardware. This book presents the basic tools for the design of digital circuits and provides methods and procedures suitable for a variety of digital design applications.

Many features of the second edition remain the same as those of the first edition. The material is still organized in the same manner. The first five chapters cover combinational circuits. The next three chapters deal with synchronous clocked sequential circuits. Asynchronous sequential circuits are introduced next. The last three chapters deal with various aspects of commercially available integrated circuits.

The second edition, however, offers several improvements over the first edition. Many sections have been rewritten to clarify the presentation. Chapters 1 through 7 and Chapter 10 have been revised by adding new up-to-date material and deleting obsolete subjects. New problems have been formulated for the first seven chapters. These replace the problem set from the first edition. Three new experiments have been added in Chapter 11. Chapter 12, a new chapter, presents the IEEE standard graphic symbols for logic elements.

The following is a brief description of the subjects that are covered in each chapter with an emphasis on the revisions that were made in the second edition.

Chapter 1 presents the various binary systems suitable for representing information in digital systems. The binary number system is explained and binary codes are illustrated. A new section has been added on signed binary numbers.

Chapter 2 introduces the basic postulates of Boolean algebra and shows the correlation between Boolean expressions and their corresponding logic diagrams. All possible logic operations for two variables are investigated and from that, the most useful logic gates used in the design of digital systems are determined. The characteristics of integrated circuit gates are mentioned in this chapter but a more detailed analysis of the electronic circuits of the gates is done in Chapter 11.

Chapter 3 covers the map and tabulation methods for simplifying Boolean expressions. The map method is also used to simplify digital circuits constructed with AND-OR, NAND, or NOR gates. All other possible two-level gate circuits are considered and their method of implementation is summarized in tabular form for easy reference.

Chapter 4 outlines the formal procedures for the analysis and design of combinational circuits. Some basic components used in the design of digital systems, such as adders and code converters, are introduced as design examples. The sections on multi-level NAND and NOR implementation have been revised to show a simpler procedure for converting AND-OR diagrams to NAND or NOR diagrams.

Chapter 5 presents various medium scale integration (MSI) circuits and programmable logic device (PLD) components. Frequently used digital logic functions such as parallel adders and subtractors, decoders, encoders, and multiplexers, are explained, and their use in the design of combinational circuits is illustrated with examples. In addition to the programmable read only memory (PROM) and programmable logic array (PLA) the book now shows the internal construction of the programmable array logic (PAL). These three PLD components are extensively used in the design and implementation of complex digital circuits.

Chapter 6 outlines the formal procedures for the analysis and design of clocked synchronous sequential circuits. The gate structure of several types of flip-flops is presented together with a discussion on the difference between pulse level and pulse transition triggering. Specific examples are used to show the derivation of the state table and state diagram when analyzing a sequential circuit. A number of design examples are presented with added emphasis on sequential circuits that use D-type flip-flops.

Chapter 7 presents various sequential digital components such as registers, shift registers, and counters. These digital components are the basic building blocks from which more complex digital systems are constructed. The sections on the random access memory (RAM) have been completely revised and a new section deals with the Hamming error correcting code.

Chapter 8 presents the algorithmic state machine (ASM) method of digital design. The ASM chart is a special flow chart suitable for describing both sequential and parallel operations with digital hardware. A number of design examples demonstrate the use of the ASM chart in the design of state machines.

Chapter 9 presents formal procedures for the analysis and design of asynchronous sequential circuits. Methods are outlined to show how an asynchronous sequential circuit can be implemented as a combinational circuit with feedback. An alternate implementation is also described that uses SR latches as the storage elements in an asynchronous sequential circuit.

Chapter 10 presents the most common integrated circuit digital logic families. The electronic circuits of the common gate in each family is analyzed using electrical circuit theory. A basic knowledge of electronic circuits is necessary to fully understand the material in this chapter. Two new sections are included in the second edition. One section shows how to evaluate the numerical values of four electrical characteristics of a gate. The other section introduces the CMOS transmission gate and gives a few examples of its usefulness in the construction of digital circuits.

Chapter 11 outlines 18 experiments that can be performed in the laboratory with hardware that is readily and inexpensively available commercially. These experiments use standard integrated circuits of the TTL type. The operation of the integrated circuits is explained by referring to diagrams in previous chapters where similar components are originally introduced. Each experiment is presented informally rather than in a step-by-step fashion so that the student is expected to produce the details of the circuit diagram and formulate a procedure for checking the operation of the circuit in the laboratory.

Chapter 12 presents the standard graphic symbols for logic functions recommended by ANSI/IEEE standard 91-1984. These graphic symbols have been developed for SSI and MSI components so that the user can recognize each function from the unique graphic symbol assigned to it. The best time to learn the standard symbols is while learning about digital systems. Chapter 12 shows the standard graphic symbols of all the integrated circuits used in the laboratory experiments of Chapter 11.

The various digital componets that are represented throughout the book are similar to commercial MSI circuits. However, the text does not mention specific integrated circuits except in Chapters 11 and 12. The practical application of digital design will be enhanced by doing the suggested experiments in Chapter 11 while studying the theory presented in the text.

Each chapter in the book has a list of references and a set of problems. Answers to most of the problems appear in the Appendix to aid the student and to help the independent reader. A *solutions manual* is available for the instructor from the publisher.

M. Morris Mano

1
Binary Systems

1-1 DIGITAL COMPUTERS AND DIGITAL SYSTEMS

Digital computers have made possible many scientific, industrial, and commercial advances that would have been unattainable otherwise. Our space program would have been impossible without real-time, continuous computer monitoring, and many business enterprises function efficiently only with the aid of automatic data processing. Computers are used in scientific calculations, commercial and business data processing, air traffic control, space guidance, the educational field, and many other areas. The most striking property of a digital computer is its generality. It can follow a sequence of instructions, called a *program,* that operates on given data. The user can specify and change programs and/or data according to the specific need. As a result of this flexibility, general-purpose digital computers can perform a wide variety of information-processing tasks.

The general-purpose digital computer is the best-known example of a digital system. Other examples include telephone switching exchanges, digital voltmeters, digital counters, electronic calculators, and digital displays. Characteristic of a digital system is its manipulation of *discrete elements* of information. Such discrete elements may be electric impulses, the decimal digits, the letters of an alphabet, arithmetic operations, punctuation marks, or any other set of meaningful symbols. The juxtaposition of discrete elements of information represents a quantity of information. For example, the letters *d, o,* and *g* form the word *dog.* The digits 237 form a number. Thus, a sequence of discrete elements forms a language, that is, a discipline that conveys information. Early digital computers were used mostly for numerical computations. In this case, the discrete elements used are the digits. From this application, the term *digital computer* has emerged. A more appropriate name for a digital computer would be a "discrete information-processing system."

Discrete elements of information are represented in a digital system by physical quantities called *signals*. Electrical signals such as voltages and currents are the most common. The signals in all present-day electronic digital systems have only two discrete values and are said to be *binary*. The digital-system designer is restricted to the use of binary signals because of the lower reliability of many-valued electronic circuits. In other words, a circuit with ten states, using one discrete voltage value for each state, can be designed, but it would possess a very low reliability of operation. In contrast, a transistor circuit that is either on or off has two possible signal values and can be constructed to be extremely reliable. Because of this physical restriction of components, and because human logic tends to be binary, digital systems that are constrained to take discrete values are further constrained to take binary values.

Discrete quantities of information arise either from the nature of the process or may be quantized from a continuous process. For example, a payroll schedule is an inherently discrete process that contains employee names, social security numbers, weekly salaries, income taxes, etc. An employee's paycheck is processed using discrete data values such as letters of the alphabet (names), digits (salary), and special symbols such as $. On the other hand, a research scientist may observe a continuous process but record only specific quantities in tabular form. The scientist is thus quantizing his continuous data. Each number in his table is a discrete element of information.

Many physical systems can be described mathematically by differential equations whose solutions as a function of time give the complete mathematical behavior of the process. An *analog computer* performs a direct *simulation* of a physical system. Each section of the computer is the analog of some particular portion of the process under study. The variables in the analog computer are represented by continuous signals, usually electric voltages that vary with time. The signal variables are considered analogous to those of the process and behave in the same manner. Thus, measurements of the analog voltage can be substituted for variables of the process. The term *analog signal* is sometimes substituted for *continuous signal* because "analog computer" has come to mean a computer that manipulates continuous variables.

To simulate a physical process in a digital computer, the quantities must be quantized. When the variables of the process are presented by real-time continuous signals, the latter are quantized by an analog-to-digital conversion device. A physical system whose behavior is described by mathematical equations is simulated in a digital computer by means of numerical methods. When the problem to be processed is inherently discrete, as in commercial applications, the digital computer manipulates the variables in their natural form.

A block diagram of the digital computer is shown in Fig. 1-1. The memory unit stores programs as well as input, output, and intermediate data. The processor unit performs arithmetic and other data-processing tasks as specified by a program. The control unit supervises the flow of information between the various units. The control unit retrieves the instructions, one by one, from the program that is stored in memory. For

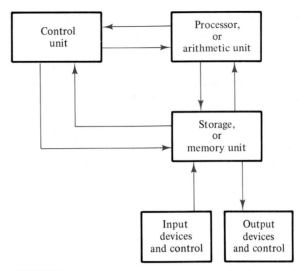

FIGURE 1-1

Block diagram of a digital computer

each instruction, the control unit informs the processor to execute the operation specified by the instruction. Both program and data are stored in memory. The control unit supervises the program instructions, and the processor manipulates the data as specified by the program.

The program and data prepared by the user are transferred into the memory unit by means of an input device such as a keyboard. An output device, such as a printer, receives the result of the computations and the printed results are presented to the user. The input and output devices are special digital systems driven by electromechanical parts and controlled by electronic digital circuits.

An electronic calculator is a digital system similar to a digital computer, with the input device being a keyboard and the output device a numerical display. Instructions are entered in the calculator by means of the function keys, such as plus and minus. Data are entered through the numeric keys. Results are displayed directly in numeric form. Some calculators come close to resembling a digital computer by having printing capabilities and programmable facilities. A digital computer, however, is a more powerful device than a calculator. A digital computer can accommodate many other input and output devices; it can perform not only arithmetic computations, but logical operations as well and can be programmed to make decisions based on internal and external conditions.

A digital computer is an interconnection of digital modules. To understand the operation of each digital module, it is necessary to have a basic knowledge of digital systems and their general behavior. The first four chapters of the book introduce the basic tools of digital design such as binary numbers and codes, Boolean algebra, and the basic building blocks from which electronic digital circuits are constructed. Chapters 5 and 7 present the basic components found in the processor unit of a digital computer.

The operational characteristics of the memory unit are explained at the end of Chapter 7. The design of the control unit is discussed in Chapter 8 using the basic principles of sequential circuits from Chapter 6.

It has already been mentioned that a digital computer manipulates discrete elements of information and that these elements are represented in the binary form. Operands used for calculations may be expressed in the binary number system. Other discrete elements, including the decimal digits, are represented in binary codes. Data processing is carried out by means of binary logic elements using binary signals. Quantities are stored in binary storage elements. The purpose of this chapter is to introduce the various binary concepts as a frame of reference for further detailed study in the succeeding chapters.

1-2   BINARY NUMBERS

A decimal number such as 7392 represents a quantity equal to 7 thousands plus 3 hundreds, plus 9 tens, plus 2 units. The thousands, hundreds, etc. are powers of 10 implied by the position of the coefficients. To be more exact, 7392 should be written as

$$7 \times 10^3 + 3 \times 10^2 + 9 \times 10^1 + 2 \times 10^0$$

However, the convention is to write only the coefficients and from their position deduce the necessary powers of 10. In general, a number with a decimal point is represented by a series of coefficients as follows:

$$a_5 a_4 a_3 a_2 a_1 a_0 . a_{-1} a_{-2} a_{-3}$$

The a_j coefficients are one of the ten digits (0, 1, 2, . . . , 9), and the subscript value j gives the place value and, hence, the power of 10 by which the coefficient must be multiplied.

$$10^5 a_5 + 10^4 a_4 + 10^3 a_3 + 10^2 a_2 + 10^1 a_1 + 10^0 a_0 + 10^{-1} a_{-1} + 10^{-2} a_{-2} + 10^{-3} a_{-3}$$

The decimal number system is said to be of *base,* or *radix,* 10 because it uses ten digits and the coefficients are multiplied by powers of 10. The *binary* system is a different number system. The coefficients of the binary numbers system have two possible values: 0 and 1. Each coefficient a_j is multiplied by 2^j. For example, the decimal equivalent of the binary number 11010.11 is 26.75, as shown from the multiplication of the coefficients by powers of 2:

$$1 \times 2^4 + 1 \times 2^3 + 0 \times 2^2 + 1 \times 2^1 + 0 \times 2^0 + 1 \times 2^{-1} + 1 \times 2^{-2} = 26.75$$

In general, a number expressed in base-r system has coefficients multiplied by powers of r:

$$a_n \cdot r^n + a_{n-1} \cdot r^{n-1} + \cdots + a_2 \cdot r^2 + a_1 \cdot r + a_0$$
$$+ a_{-1} \cdot r^{-1} + a_{-2} \cdot r^{-2} + \cdots + a_{-m} \cdot r^{-m}$$

The coefficients a_j range in value from 0 to $r - 1$. To distinguish between numbers of different bases, we enclose the coefficients in parentheses and write a subscript equal to the base used (except sometimes for decimal numbers, where the content makes it obvious that it is decimal). An example of a base-5 number is

$$(4021.2)_5 = 4 \times 5^3 + 0 \times 5^2 + 2 \times 5^1 + 1 \times 5^0 + 2 \times 5^{-1} = (511.4)_{10}$$

Note that coefficient values for base 5 can be only 0, 1, 2, 3, and 4.

It is customary to borrow the needed r digits for the coefficients from the decimal system when the base of the number is less than 10. The letters of the alphabet are used to supplement the ten decimal digits when the base of the number is greater than 10. For example, in the *hexadecimal* (base 16) number system, the first ten digits are borrowed from the decimal system. The letters A, B, C, D, E, and F are used for digits 10, 11, 12, 13, 14, and 15, respectively. An example of a hexadecimal number is

$$(B65F)_{16} = 11 \times 16^3 + 6 \times 16^2 + 5 \times 16 + 15 = (46687)_{10}$$

The first 16 numbers in the decimal, binary, octal, and hexadecimal systems are listed in Table 1-1.

TABLE 1-1
Numbers with Different Bases

Decimal (base 10)	Binary (base 2)	Octal (base 8)	Hexadecimal (base 16)
00	0000	00	0
01	0001	01	1
02	0010	02	2
03	0011	03	3
04	0100	04	4
05	0101	05	5
06	0110	06	6
07	0111	07	7
08	1000	10	8
09	1001	11	9
10	1010	12	A
11	1011	13	B
12	1100	14	C
13	1101	15	D
14	1110	16	E
15	1111	17	F

Arithmetic operations with numbers in base r follow the same rules as for decimal numbers. When other than the familiar base 10 is used, one must be careful to use only the r allowable digits. Examples of addition, subtraction, and multiplication of two binary numbers are as follows:

augend: 101101   minuend: 101101   multiplicand: 1011
addend: $+100111$ subtrahend: -100111 multiplier: $\times\ 101$
sum: 1010100   difference: 000110   1011
                                                                            0000
                                                                            1011
                                            product: 110111

The sum of two binary numbers is calculated by the same rules as in decimal, except that the digits of the sum in any significant position can be only 0 or 1. Any carry obtained in a given significant position is used by the pair of digits one significant position higher. The subtraction is slightly more complicated. The rules are still the same as in decimal, except that the borrow in a given significant position adds 2 to a minuend digit. (A borrow in the decimal system adds 10 to a minuend digit.) Multiplication is very simple. The multiplier digits are always 1 or 0. Therefore, the partial products are equal either to the multiplicand or to 0.

1-3 NUMBER BASE CONVERSIONS

A binary number can be converted to decimal by forming the sum of the powers of 2 of those coefficients whose value is 1. For example

$$(1010.011)_2 = 2^3 + 2^1 + 2^{-2} + 2^{-3} = (10.375)_{10}$$

The binary number has four 1's and the decimal equivalent is found from the sum of four powers of 2. Similarly, a number expressed in base r can be converted to its decimal equivalent by multiplying each coefficient with the corresponding power of r and adding. The following is an example of octal-to-decimal conversion:

$$(630.4)_8 = 6 \times 8^2 + 3 \times 8 + 4 \times 8^{-1} = (408.5)_{10}$$

The conversion from decimal to binary or to any other base-r system is more convenient if the number is separated into an *integer part* and a *fraction part* and the conversion of each part done separately. The conversion of an *integer* from decimal to binary is best explained by example.

Example 1-1

Convert decimal 41 to binary. First, 41 is divided by 2 to give an integer quotient of 20 and a remainder of $\frac{1}{2}$. The quotient is again divided by 2 to give a new quotient and remainder. This process is continued until the integer quotient becomes 0. The *coefficients* of the desired binary number are obtained from the *remainders* as follows:

Integer quotient			Remainder	Coefficient
$\dfrac{41}{2} =$	20	$+$	$\dfrac{1}{2}$	$a_0 = 1$
$\dfrac{20}{2} =$	10	$+$	0	$a_1 = 0$
$\dfrac{10}{2} =$	5	$+$	0	$a_2 = 0$
$\dfrac{5}{2} =$	2	$+$	$\dfrac{1}{2}$	$a_3 = 1$
$\dfrac{2}{2} =$	1	$+$	0	$a_4 = 0$
$\dfrac{1}{2} =$	0	$+$	$\dfrac{1}{2}$	$a_5 = 1$

answer: $(41)_{10} = (a_5 a_4 a_3 a_2 a_1 a_0)_2 = (101001)_2$

The arithmetic process can be manipulated more conveniently as follows:

Integer	Remainder
41	
20	1
10	0
5	0
2	1
1	0
0	1

$101001 = $ answer ∎

The conversion from decimal integers to any base-r system is similar to the example, except that division is done by r instead of 2.

Example 1-2   Convert decimal 153 to octal. The required base r is 8. First, 153 is divided by 8 to give an integer quotient of 19 and a remainder of 1. Then 19 is divided by 8 to give an integer quotient of 2 and a remainder of 3. Finally, 2 is divided by 8 to give a quotient of 0 and a remainder of 2. This process can be conveniently manipulated as follows:

$$
\begin{array}{c|c}
153 & \\
19 & 1 \\
2 & 3 \\
0 & 2
\end{array}
\quad = (231)_8 \qquad \blacksquare
$$

The conversion of a decimal *fraction* to binary is accomplished by a method similar to that used for integers. However, multiplication is used instead of division, and integers are accumulated instead of remainders. Again, the method is best explained by example.

**Example
1-3**

Convert $(0.6875)_{10}$ to binary. First, 0.6875 is multiplied by 2 to give an integer and a fraction. The new fraction is multiplied by 2 to give a new integer and a new fraction. This process is continued until the fraction becomes 0 or until the number of digits have sufficient accuracy. The coefficients of the binary number are obtained from the integers as follows:

	Integer		Fraction	Coefficient
$0.6875 \times 2 =$	1	+	0.3750	$a_{-1} = 1$
$0.3750 \times 2 =$	0	+	0.7500	$a_{-2} = 0$
$0.7500 \times 2 =$	1	+	0.5000	$a_{-3} = 1$
$0.5000 \times 2 =$	1	+	0.0000	$a_{-4} = 1$

Answer: $(0.6875)_{10} = (0.a_{-1}a_{-2}a_{-3}a_{-4})_2 = (0.1011)_2$ ∎

To convert a decimal fraction to a number expressed in base r, a similar procedure is used. Multiplication is by r instead of 2, and the coefficients found from the integers may range in value from 0 to $r - 1$ instead of 0 and 1.

**Example
1-4**

Convert $(0.513)_{10}$ to octal.

$$0.513 \times 8 = 4.104$$
$$0.104 \times 8 = 0.832$$
$$0.832 \times 8 = 6.656$$
$$0.656 \times 8 = 5.248$$
$$0.248 \times 8 = 1.984$$
$$0.984 \times 8 = 7.872$$

The answer, to seven significant figures, is obtained from the integer part of the products:

$$(0.513)_{10} = (0.406517 \ldots)_8$$ ∎

The conversion of decimal numbers with both integer and fraction parts is done by converting the integer and fraction separately and then combining the two answers. Using the results of Examples 1-1 and 1-3, we obtain

$$(41.6875)_{10} = (101001.1011)_2$$

From Examples 1-2 and 1-4, we have

$$(153.513)_{10} = (231.406517)_8$$

1-4 OCTAL AND HEXADECIMAL NUMBERS

The conversion from and to binary, octal, and hexadecimal plays an important part in digital computers. Since $2^3 = 8$ and $2^4 = 16$, each octal digit corresponds to three binary digits and each hexadecimal digit corresponds to four binary digits. The conversion from binary to octal is easily accomplished by partitioning the binary number into groups of three digits each, starting from the binary point and proceeding to the left and to the right. The corresponding octal digit is then assigned to each group. The following example illustrates the procedure:

$$(\underbrace{10}_{2} \ \underbrace{110}_{6} \ \underbrace{001}_{1} \ \underbrace{101}_{5} \ \underbrace{011}_{3} \ . \ \underbrace{111}_{7} \ \underbrace{100}_{4} \ \underbrace{000}_{0} \ \underbrace{110}_{6})_2 = (26153.7460)_8$$

Conversion from binary to hexadecimal is similar, except that the binary number is divided into groups of four digits:

$$(\underbrace{10}_{2} \ \underbrace{1100}_{C} \ \underbrace{0110}_{6} \ \underbrace{1011}_{B} \ . \ \underbrace{1111}_{F} \ \underbrace{0010}_{2})_2 = (2C6B.F2)_{16}$$

The corresponding hexadecimal (or octal) digit for each group of binary digits is easily remembered after studying the values listed in Table 1-1.

Conversion from octal or hexadecimal to binary is done by a procedure reverse to the above. Each octal digit is converted to its three-digit binary equivalent. Similarly, each hexadecimal digit is converted to its four-digit binary equivalent. This is illustrated in the following examples:

$$(673.124)_8 = (\underbrace{110}_{6} \ \underbrace{111}_{7} \ \underbrace{011}_{3} \ . \ \underbrace{001}_{1} \ \underbrace{010}_{2} \ \underbrace{100}_{4})_2$$

$$(306.D)_{16} = (\underbrace{0011}_{3} \ \underbrace{0000}_{0} \ \underbrace{0110}_{6} \ . \ \underbrace{1101}_{D})_2$$

Binary numbers are difficult to work with because they require three or four times as many digits as their decimal equivalent. For example, the binary number 111111111111 is equivalent to decimal 4095. However, digital computers use binary numbers and it is sometimes necessary for the human operator or user to communicate directly with the machine by means of binary numbers. One scheme that retains the binary system in the computer but reduces the number of digits the human must consider utilizes the relationship between the binary number system and the octal or hexadecimal system. By this method, the human thinks in terms of octal or hexadecimal numbers and performs the required conversion by inspection when direct communication with the machine is necessary. Thus the binary number 111111111111 has 12 digits and is expressed in octal as 7777 (four digits) or in hexadecimal as FFF (three digits). During communication between people (about binary numbers in the computer), the octal or hexadecimal representation is more desirable because it can be expressed more compactly with a third or a quarter of the number of digits required for the equivalent binary number. When the human communicates with the machine (through console switches or indicator lights or by means of programs written in *machine language*), the conversion from octal or hexadecimal to binary and vice versa is done by inspection by the human user.

1-5 COMPLEMENTS

Complements are used in digital computers for simplifying the subtraction operation and for logical manipulation. There are two types of complements for each base-r system: the radix complement and the diminished radix complement. The first is referred to as the r's complement and the second as the $(r - 1)$'s complement. When the value of the base r is substituted in the name, the two types are referred to as the 2's complement and 1's complement for binary numbers, and the 10's complement and 9's complement for decimal numbers.

Diminished Radix Complement

Given a number N in base r having n digits, the $(r - 1)$'s complement of N is defined as $(r^n - 1) - N$. For decimal numbers, $r = 10$ and $r - 1 = 9$, so the 9's complement of N is $(10^n - 1) - N$. Now, 10^n represents a number that consists of a single 1 followed by n 0's. $10^n - 1$ is a number represented by n 9's. For example, if $n = 4$, we have $10^4 = 10,000$ and $10^4 - 1 = 9999$. It follows that the 9's complement of a decimal number is obtained by subtracting each digit from 9. Some numerical examples follow.

The 9's complement of 546700 is $999999 - 546700 = 453299$.

The 9's complement of 012398 is $999999 - 012398 = 987601$.

For binary numbers, $r = 2$ and $r - 1 = 1$, so the 1's complement of N is $(2^n - 1) - N$. Again, 2^n is represented by a binary number that consists of a 1 followed by n 0's. $2^n - 1$ is a binary number represented by n 1's. For example, if $n = 4$, we have $2^4 = (10000)_2$ and $2^4 - 1 = (1111)_2$. Thus the 1's complement of a binary number is obtained by subtracting each digit from 1. However, when subtracting binary digits from 1, we can have either $1 - 0 = 1$ or $1 - 1 = 0$, which causes the bit to change from 0 to 1 or from 1 to 0. Therefore, the 1's complement of a binary number is formed by changing 1's to 0's and 0's to 1's. The following are some numerical examples.

The 1's complement of 1011000 is 0100111.

The 1's complement of 0101101 is 1010010.

The $(r - 1)$'s complement of octal or hexadecimal numbers is obtained by subtracting each digit from 7 or F (decimal 15), respectively.

Radix Complement

The r's complement of an n-digit number N in base r is defined as $r^n - N$ for $N \neq 0$ and 0 for $N = 0$. Comparing with the $(r - 1)$'s complement, we note that the r's complement is obtained by adding 1 to the $(r - 1)$'s complement since $r^n - N = [(r^n - 1) - N] + 1$. Thus, the 10's complement of decimal 2389 is $7610 + 1 = 7611$ and is obtained by adding 1 to the 9's-complement value. The 2's complement of binary 101100 is $010011 + 1 = 010100$ and is obtained by adding 1 to the 1's-complement value.

Since 10^n is a number represented by a 1 followed by n 0's, $10^n - N$, which is the 10's complement of N, can be formed also by leaving all least significant 0's unchanged, subtracting the first nonzero least significant digit from 10, and subtracting all higher significant digits from 9.

The 10's complement of 012398 is 987602.

The 10's complement of 246700 is 753300.

The 10's complement of the first number is obtained by subtracting 8 from 10 in the least significant position and subtracting all other digits from 9. The 10's complement of the second number is obtained by leaving the two least significant 0's unchanged, subtracting 7 from 10, and subtracting the other three digits from 9.

Similarly, the 2's complement can be formed by leaving all least significant 0's and the first 1 unchanged, and replacing 1's with 0's and 0's with 1's in all other higher significant digits.

The 2's complement of 1101100 is 0010100.

The 2's complement of 0110111 is 1001001.

The 2's complement of the first number is obtained by leaving the two least significant 0's and the first 1 unchanged, and then replacing 1's with 0's and 0's with 1's in the other four most-significant digits. The 2's complement of the second number is obtained by leaving the least significant 1 unchanged and complementing all other digits.

In the previous definitions, it was assumed that the numbers do not have a radix point. If the original number N contains a radix point, the point should be removed temporarily in order to form the r's or $(r-1)$'s complement. The radix point is then restored to the complemented number in the same relative position. It is also worth mentioning that the complement of the complement restores the number to its original value. The r's complement of N is $r^n - N$. The complement of the complement is $r^n - (r^n - N) = N$, giving back the original number.

Subtraction with Complements

The direct method of subtraction taught in elementary schools uses the borrow concept. In this method, we borrow a 1 from a higher significant position when the minuend digit is smaller than the subtrahend digit. This seems to be easiest when people perform subtraction with paper and pencil. When subtraction is implemented with digital hardware, this method is found to be less efficient than the method that uses complements.

The subtraction of two n-digit unsigned numbers $M - N$ in base r can be done as follows:

1. Add the minuend M to the r's complement of the subtrahend N. This performs $M + (r^n - N) = M - N + r^n$.
2. If $M \geq N$, the sum will produce an end carry, r^n, which is discarded; what is left is the result $M - N$.
3. If $M < N$, the sum does not produce an end carry and is equal to $r^n - (N - M)$, which is the r's complement of $(N - M)$. To obtain the answer in a familiar form, take the r's complement of the sum and place a negative sign in front.

The following examples illustrate the procedure.

**Example
1-5**   Using 10's complement, subtract $72532 - 3250$.

$$M = \qquad 72532$$
$$\text{10's complement of } N = \quad +\ \underline{96750}$$
$$\text{Sum} = \qquad 169282$$
$$\text{Discard end carry } 10^5 = \quad -\underline{100000}$$
$$\text{Answer} = \qquad 69282 \qquad \blacksquare$$

Note that M has 5 digits and N has only 4 digits. Both numbers must have the same number of digits; so we can write N as 03250. Taking the 10's complement of N produces a 9 in the most significant position. The occurrence of the end carry signifies that $M \geq N$ and the result is positive.

Example 1-6

Using 10's complement, subtract $3250 - 72532$.

$$M = \qquad 03250$$

$$\text{10's complement of } N = \qquad +\ \underline{27468}$$

$$\text{Sum} = \qquad 30718$$

There is no end carry.

Answer: $-(\text{10's complement of } 30718) = -69282$ ∎

Note that since $3250 < 72532$, the result is negative. Since we are dealing with unsigned numbers, there is really no way to get an unsigned result for this case. When subtracting with complements, the negative answer is recognized from the absence of the end carry and the complemented result. When working with paper and pencil, we can change the answer to a signed negative number in order to put it in a familiar form.

Subtraction with complements is done with binary numbers in a similar manner using the same procedure outlined before.

Example 1-7

Given the two binary numbers $X = 1010100$ and $Y = 1000011$, perform the subtraction (a) $X - Y$ and (b) $Y - X$ using 2's complements.

(a)

$$X = \qquad 1010100$$

$$\text{2's complement of } Y = \qquad +\ \underline{0111101}$$

$$\text{Sum} = \qquad 10010001$$

$$\text{Discard end carry } 2^7 = \qquad -\underline{10000000}$$

$$\text{Answer: } X - Y = \qquad 0010001$$

(b)

$$Y = \qquad 1000011$$

$$\text{2's complement of } X = \qquad +\ \underline{0101100}$$

$$\text{Sum} = \qquad 1101111$$

There is no end carry.

Answer: $Y - X = -(\text{2's complement of } 1101111) = -0010001$ ∎

Subtraction of unsigned numbers can be done also by means of the $(r - 1)$'s complement. Remember that the $(r - 1)$'s complement is one less than the r's complement. Because of this, the result of adding the minuend to the complement of the subtrahend produces a sum that is 1 less than the correct difference when an end carry occurs. Removing the end carry and adding 1 to the sum is referred to as an *end-around carry*.

Example 1-8

Repeat Example 1-7 using 1's complement.

(a) $X - Y = 1010100 - 1000011$

$$
\begin{array}{rr}
X = & 1010100 \\
\text{1's complement of } Y = & +\ 0111100 \\
\hline
\text{Sum} = & 10010000 \\
\text{End-around carry} & +\ 1 \\
\hline
\text{Answer: } X - Y = & 0010001
\end{array}
$$

(b) $Y - X = 1000011 - 1010100$

$$
\begin{array}{rr}
Y = & 1000011 \\
\text{1's complement of } X = & +\ 0101011 \\
\hline
\text{Sum} = & 1101110
\end{array}
$$

There is no end carry.

Answer: $Y - X = -(1\text{'s complement of } 1101110) = -0010001$

Note that the negative result is obtained by taking the 1's complement of the sum since this is the type of complement used. The procedure with end-around carry is also applicable for subtracting unsigned decimal numbers with 9's complement.

1-6 SIGNED BINARY NUMBERS

Positive integers including zero can be represented as unsigned numbers. However, to represent negative integers, we need a notation for negative values. In ordinary arithmetic, a negative number is indicated by a minus sign and a positive number by a plus sign. Because of hardware limitations, computers must represent everything with binary digits, commonly referred to as *bits*. It is customary to represent the sign with a bit placed in the leftmost position of the number. The convention is to make the sign bit 0 for positive and 1 for negative.

It is important to realize that both signed and unsigned binary numbers consist of a string of bits when represented in a computer. The user determines whether the number is signed or unsigned. If the binary number is signed, then the leftmost bit represents the sign and the rest of the bits represent the number. If the binary number is assumed to be unsigned, then the leftmost bit is the most significant bit of the number. For example, the string of bits 01001 can be considered as 9 (unsigned binary) or a +9 (signed binary) because the leftmost bit is 0. The string of bits 11001 represent the binary equivalent of 25 when considered as an unsigned number or as − 9 when considered as a signed number because of the 1 in the leftmost position, which designates negative, and the other four bits, which represent binary 9. Usually, there is no confusion in identifying the bits if the type of representation for the number is known in advance.

The representation of the signed numbers in the last example is referred to as the *signed-magnitude* convention. In this notation, the number consists of a magnitude and a symbol (+ or −) or a bit (0 or 1) indicating the sign. This is the representation of signed numbers used in ordinary arithmetic. When arithmetic operations are implemented in a computer, it is more convenient to use a different system for representing negative numbers, referred to as the *signed-complement* system. In this system, a negative number is indicated by its complement. Whereas the signed-magnitude system negates a number by changing its sign, the signed-complement system negates a number by taking its complement. Since positive numbers always start with 0 (plus) in the leftmost position, the complement will always start with a 1, indicating a negative number. The signed-complement system can use either the 1's or the 2's complement, but the 2's complement is the most common.

As an example, consider the number 9 represented in binary with eight bits. +9 is represented with a sign bit of 0 in the leftmost position followed by the binary equivalent of 9 to give 00001001. Note that all eight bits must have a value and, therefore, 0's are inserted following the sign bit up to the first 1. Although there is only one way to represent +9, there are three different ways to represent − 9 with eight bits:

In signed-magnitude representation:	10001001
In signed-1's-complement representation:	11110110
In signed-2's-complement representation:	11110111

In signed-magnitude, −9 is obtained from +9 by changing the sign bit in the leftmost position from 0 to 1. In signed-1's complement, −9 is obtained by complementing all the bits of +9, including the sign bit. The signed-2's-complement representation of −9 is obtained by taking the 2's complement of the positive number, including the sign bit.

The signed-magnitude system is used in ordinary arithmetic, but is awkward when employed in computer arithmetic. Therefore, the signed-complement is normally used. The 1's complement imposes some difficulties and is seldom used for arithmetic operations except in some older computers. The 1's complement is useful as a logical operation since the change of 1 to 0 or 0 to 1 is equivalent to a logical complement operation, as will be shown in the next chapter. The following discussion of signed binary arithmetic deals exclusively with the signed-2's-complement representation of negative numbers. The same procedures can be applied to the signed-1's-complement system by including the end-around carry as done with unsigned numbers.

Arithmetic Addition

The addition of two numbers in the signed-magnitude system follows the rules of ordinary arithmetic. If the signs are the same, we add the two magnitudes and give the sum the common sign. If the signs are different, we subtract the smaller magnitude from the larger and give the result the sign of the larger magnitude. For example, $(+25) + (-37) = -(37 - 25) = -12$ and is done by subtracting the smaller magnitude 25 from the larger magnitude 37 and using the sign of 37 for the sign of the result. This is a process that requires the comparison of the signs and the magnitudes and then performing either addition or subtraction. The same procedure applies to binary numbers in signed-magnitude representation. In contrast, the rule for adding numbers in the signed-complement system does not require a comparison or subtraction, but only addition. The procedure is very simple and can be stated as follows for binary numbers.

The addition of two signed binary numbers with negative numbers represented in signed-2's-complement form is obtained from the addition of the two numbers, including their sign bits. A carry out of the sign-bit position is discarded.

Numerical examples for addition follow. Note that negative numbers must be initially in 2's complement and that the sum obtained after the addition if negative is in 2's-complement form.

$+\ 6$	00000110	$-\ 6$	11111010
$+13$	00001101	$+13$	00001101
$+19$	00010011	$+\ 7$	00000111
$+\ 6$	00000110	$-\ 6$	11111010
-13	11110011	-13	11110011
$-\ 7$	11111001	-19	11101101

In each of the four cases, the operation performed is addition with the sign bit included. Any carry out of the sign-bit position is discarded, and negative results are automatically in 2's-complement form.

In order to obtain a correct answer, we must ensure that the result has a sufficient number of bits to accommodate the sum. If we start with two n-bit numbers and the sum occupies $n + 1$ bits, we say that an overflow occurs. When one performs the addition with paper and pencil, an overflow is not a problem since we are not limited by the width of the page. We just add another 0 to a positive number and another 1 to a negative number in the most-significant position to extend them to $n + 1$ bits and then perform the addition. Overflow is a problem in computers because the number of bits that hold a number is finite, and a result that exceeds the finite value by 1 cannot be accommodated.

The complement form of representing negative numbers is unfamiliar to those used to the signed-magnitude system. To determine the value of a negative number when in signed-2's complement, it is necessary to convert it to a positive number to place it in a more familiar form. For example, the signed binary number 11111001 is negative because the leftmost bit is 1. Its 2's complement is 00000111, which is the binary equivalent of $+7$. We therefore recognize the original negative number to be equal to -7.

Arithmetic Subtraction

Subtraction of two signed binary numbers when negative numbers are in 2's-complement form is very simple and can be stated as follows:

Take the 2's complement of the subtrahend (including the sign bit) and add it to the minuend (including the sign bit). A carry out of the sign-bit position is discarded.

This procedure occurs because a subtraction operation can be changed to an addition operation if the sign of the subtrahend is changed. This is demonstrated by the following relationship:

$$(\pm A) \; - \; (+B) = (\pm A) + (-B)$$

$$(\pm A) \; - \; (-B) = (\pm A) + (+B)$$

But changing a positive number to a negative number is easily done by taking its 2's complement. The reverse is also true because the complement of a negative number in complement form produces the equivalent positive number. Consider the subtraction of $(-6) - (-13) = +7$. In binary with eight bits, this is written as (11111010 − 11110011). The subtraction is changed to addition by taking the 2's complement of the subtrahend (-13) to give $(+13)$. In binary, this is $11111010 + 00001101 = 100000111$. Removing the end carry, we obtain the correct answer 00000111 $(+7)$.

It is worth noting that binary numbers in the signed-complement system are added and subtracted by the same basic addition and subtraction rules as unsigned numbers. Therefore, computers need only one common hardware circuit to handle both types of arithmetic. The user or programmer must interpret the results of such addition or subtraction differently, depending on whether it is assumed that the numbers are signed or unsigned.

1-7 BINARY CODES

Electronic digital systems use signals that have two distinct values and circuit elements that have two stable states. There is a direct analogy among binary signals, binary circuit elements, and binary digits. A binary number of n digits, for example, may be represented by n binary circuit elements, each having an output signal equivalent to a 0 or a 1. Digital systems represent and manipulate not only binary numbers, but also many other discrete elements of information. Any discrete element of information distinct among a group of quantities can be represented by a binary code. Binary codes play an important role in digital computers. The codes must be in binary because computers can only hold 1's and 0's. It must be realized that binary codes merely change the symbols, not the meaning of the elements of information that they represent. If we inspect the bits of a computer at random, we will find that most of the time they represent some type of coded information rather than binary numbers.

A *bit*, by definition, is a binary digit. When used in conjunction with a binary code, it is better to think of it as denoting a binary quantity equal to 0 or 1. To represent a group of 2^n distinct elements in a binary code requires a minimum of n bits. This is because it is possible to arrange n bits in 2^n distinct ways. For example, a group of four distinct quantities can be represented by a two-bit code, with each quantity assigned one of the following bit combinations: 00, 01, 10, 11. A group of eight elements requires a three-bit code, with each element assigned to one and only one of the following: 000, 001, 010, 011, 100, 101, 110, 111. The examples show that the distinct bit combinations of an n-bit code can be found by counting in binary from 0 to $(2^n - 1)$. Some bit combinations are unassigned when the number of elements of the group to be coded is not a multiple of the power of 2. The ten decimal digits 0, 1, 2, . . . , 9 are an example of such a group. A binary code that distinguishes among ten elements must contain at least four bits; three bits can distinguish a maximum of eight elements. Four bits can form 16 distinct combinations, but since only ten digits are coded, the remaining six combinations are unassigned and not used.

Although the *minimum* number of bits required to code 2^n distinct quantities is n, there is no *maximum* number of bits that may be used for a binary code. For example, the ten decimal digits can be coded with ten bits, and each decimal digit assigned a bit combination of nine 0's and a 1. In this particular binary code, the digit 6 is assigned the bit combination 0001000000.

Decimal Codes

Binary codes for decimal digits require a minimum of four bits. Numerous different codes can be obtained by arranging four or more bits in ten distinct possible combinations. A few possibilities are shown in Table 1-2.

TABLE 1-2
Binary codes for the decimal digits

Decimal digit	(BCD) 8421	Excess-3	84-2-1	2421	(Biquinary) 5043210
0	0000	0011	0000	0000	0100001
1	0001	0100	0111	0001	0100010
2	0010	0101	0110	0010	0100100
3	0011	0110	0101	0011	0101000
4	0100	0111	0100	0100	0110000
5	0101	1000	1011	1011	1000001
6	0110	1001	1010	1100	1000010
7	0111	1010	1001	1101	1000100
8	1000	1011	1000	1110	1001000
9	1001	1100	1111	1111	1010000

The BCD (binary-code decimal) is a straight assignment of the binary equivalent. It is possible to assign weights to the binary bits according to their positions. The weights in the BCD code are 8, 4, 2, 1. The bit assignment 0110, for example, can be interpreted by the weights to represent the decimal digit 6 because $0 \times 8 + 1 \times 4 +$

$1 \times 2 + 0 \times 1 = 6$. It is also possible to assign negative weights to a decimal code, as shown by the 8, 4, -2, -1 code. In this case, the bit combination 0110 is interpreted as the decimal digit 2, as obtained from $0 \times 8 + 1 \times 4 + 1 \times (-2) + 0 \times (-1) = 2$. Two other weighted codes shown in the table are the 2421 and the 5043210. A decimal code that has been used in some old computers is the excess-3 code. This is an unweighted code; its code assignment is obtained from the corresponding value of BCD after the addition of 3.

Numbers are represented in digital computers either in binary or in decimal through a binary code. When specifying data, the user likes to give the data in decimal form. The input decimal numbers are stored internally in the computer by means of a decimal code. Each decimal digit requires at least four binary storage elements. The decimal numbers are converted to binary when arithmetic operations are done internally with numbers represented in binary. It is also possible to perform the arithmetic operations directly in decimal with all numbers left in a coded form throughout. For example, the decimal number 395, when converted to binary, is equal to 110001011 and consists of nine binary digits. The same number, when represented internally in the BCD code, occupies four bits for each decimal digit, for a total of 12 bits: 001110010101. The first four bits represent a 3, the next four a 9, and the last four a 5.

It is very important to understand the difference between *conversion* of a decimal number to binary and the binary *coding* of a decimal number. In each case, the final result is a series of bits. The bits obtained from conversion are binary digits. Bits obtained from coding are combinations of 1's and 0's arranged according to the rules of the code used. Therefore, it is extremely important to realize that a series of 1's and 0's in a digital system may sometimes represent a binary number and at other times represent some other discrete quantity of information as specified by a given binary code. The BCD code, for example, has been chosen to be both a code and a direct binary conversion, as long as the decimal numbers are integers from 0 to 9. For numbers greater than 9, the conversion and the coding are completely different. This concept is so important that it is worth repeating with another example. The binary conversion of decimal 13 is 1101; the coding of decimal 13 with BCD is 00010011.

From the five binary codes listed in Table 1-2, the BCD seems the most natural to use and is indeed the one most commonly encountered. The other four-bit codes listed have one characteristic in common that is not found in BCD. The excess-3, the 2, 4, 2, 1, and the 8, 4, -2, -1 are self-complementing codes, that is, the 9's complement of the decimal number is easily obtained by changing 1's to 0's and 0's to 1's. For example, the decimal 395 is represented in the 2, 4, 2, 1 code by 001111111011. Its 9's complement 604 is represented by 110000000100, which is easily obtained from the replacement of 1's by 0's and 0's by 1's. This property is useful when arithmetic operations are internally done with decimal numbers (in a binary code) and subtraction is calculated by means of 9's complement.

The biquinary code shown in Table 1-2 is an example of a seven-bit code with error-detection properties. Each decimal digit consists of five 0's and two 1's placed in the corresponding weighted columns. The error-detection property of this code may be understood if one realizes that digital systems represent binary 1 by one distinct signal and binary 0 by a second distinct signal. During transmission of signals from one location to another, an error may occur. One or more bits may change value. A circuit in the receiving side can detect the presence of more (or less) than two 1's and if the received combination of bits does not agree with the allowable combination, an error is detected.

Error-Detection Code

Binary information can be transmitted from one location to another by electric wires or other communication medium. Any external noise introduced into the physical communication medium may change some of the bits from 0 to 1 or vice versa. The purpose of an error-detection code is to detect such bit-reversal errors. One of the most common ways to achieve error detection is by means of a *parity bit*. A parity bit is an extra bit included with a message to make the total number of 1's transmitted either odd or even. A message of four bits and a parity bit P are shown in Table 1-3. If an odd parity is adopted, the P bit is chosen such that the total number of 1's is odd in the five bits that constitute the message and P. If an even parity is adopted, the P bit is chosen so that the total number of 1's in the five bits is even. In a particular situation, one or the other parity is adopted, with even parity being more common.

The parity bit is helpful in detecting errors during the transmission of information from one location to another. This is done in the following manner. An even parity bit is generated in the sending end for each message transmission. The message, together with the parity bit, is transmitted to its destination. The parity of the received data is

TABLE 1-3
Parity bit

Odd parity		Even parity	
Message	P	Message	P
0000	1	0000	0
0001	0	0001	1
0010	0	0010	1
0011	1	0011	0
0100	0	0100	1
0101	1	0101	0
0110	1	0110	0
0111	0	0111	1
1000	0	1000	1
1001	1	1001	0
1010	1	1010	0
1011	0	1011	1
1100	1	1100	0
1101	0	1101	1
1110	0	1110	1
1111	1	1111	0

checked in the receiving end. If the parity of the received information is not even, it means that at least one bit has changed value during the transmission. This method detects one, three, or any odd combination of errors in each message that is transmitted. An even combination of errors is undetected. Additional error-detection schemes may be needed to take care of an even combination of errors.

What is done after an error is detected depends on the particular application. One possibility is to request retransmission of the message on the assumption that the error was random and will not occur again. Thus, if the receiver detects a parity error, it sends back a negative acknowledge message. If no error is detected, the receiver sends back an acknowledge message. The sending end will respond to a previous error by transmitting the message again until the correct parity is received. If, after a number of attempts, the transmission is still in error, a message can be sent to the human operator to check for malfunctions in the transmission path.

Gray Code

Digital systems can be designed to process data in discrete form only. Many physical systems supply continuous output data. These data must be converted into digital form before they are applied to a digital system. Continuous or analog information is converted into digital form by means of an analog-to-digital converter. It is sometimes convenient to use the Gray code shown in Table 1-4 to represent the digital data when it is converted from analog data. The advantage of the Gray code over binary numbers is that only one bit in the code group changes when going from one number to the next. For example, in going from 7 to 8, the Gray code changes from 0100 to 1100. Only the

TABLE 1-4
Four-bit Gray code

Gray code	Decimal equivalent
0000	0
0001	1
0011	2
0010	3
0110	4
0111	5
0101	6
0100	7
1100	8
1101	9
1111	10
1110	11
1010	12
1011	13
1001	14
1000	15

first bit from the left changes from 0 to 1; the other three bits remain the same. When comparing this with binary numbers, the change from 7 to 8 will be from 0111 to 1000, which causes all four bits to change values.

The Gray code is used in applications where the normal sequence of binary numbers may produce an error or ambiguity during the transition from one number to the next. If binary numbers are used, a change from 0111 to 1000 may produce an intermediate erroneous number 1001 if the rightmost bit takes more time to change than the other three bits. The Gray code eliminates this problem since only one bit changes in value during any transition between two numbers.

A typical application of the Gray code occurs when analog data are represented by continuous change of a shaft position. The shaft is partitioned into segments, and each segment is assigned a number. If adjacent segments are made to correspond with the Gray-code sequence, ambiguity is eliminated when detection is sensed in the line that separates any two segments.

ASCII Character Code

Many applications of digital computers require the handling of data not only of numbers, but also of letters. For instance, an insurance company with thousands of policy holders will use a computer to process its files. To represent the names and other pertinent information, it is necessary to formulate a binary code for the letters of the alphabet. In addition, the same binary code must represent numerals and special characters such as $. An alphanumeric character set is a set of elements that includes the 10 decimal digits, the 26 letters of the alphabet, and a number of special characters. Such a set contains between 36 and 64 elements if only capital letters are included, or between 64 and 128 elements if both uppercase and lowercase letters are included. In the first case, we need a binary code of six bits, and in the second we need a binary code of seven bits.

The standard binary code for the alphanumeric characters is ASCII (American Standard Code for Information Interchange). It uses seven bits to code 128 characters, as shown in Table 1-5. The seven bits of the code are designated by b_1 through b_7, with b_7 being the most-significant bit. The letter A, for example, is represented in ASCII as 1000001 (column 100, row 0001). The ASCII code contains 94 graphic characters that can be printed and 34 nonprinting characters used for various control functions. The graphic characters consist of the 26 uppercase letters (A through Z), the 26 lowercase letters (a through z), the 10 numerals (0 through 9), and 32 special printable characters such as %, *, and $.

The 34 control characters are designated in the ASCII table with abbreviated names. They are listed in the table with their full functional names. The control characters are used for routing data and arranging the printed text into a prescribed format. There are three types of control characters: format effectors, information separators, and communication-control characters. Format effectors are characters that control the layout of printing. They include the familiar typewriter controls such as backspace (BS), horizontal tabulation (HT), and carriage return (CR). Information separators are used to

TABLE 1-5
American Standard Code for Information Interchange (ASCII)

$b_4b_3b_2b_1$	$b_7b_6b_5$							
	000	001	010	011	100	101	110	111
0000	NUL	DLE	SP	0	@	P	`	p
0001	SOH	DC1	!	1	A	Q	a	q
0010	STX	DC2	"	2	B	R	b	r
0011	ETX	DC3	#	3	C	S	c	s
0100	EOT	DC4	$	4	D	T	d	t
0101	ENQ	NAK	%	5	E	U	e	u
0110	ACK	SYN	&	6	F	V	f	v
0111	BEL	ETB	'	7	G	W	g	w
1000	BS	CAN	(	8	H	X	h	x
1001	HT	EM	)	9	I	Y	i	y
1010	LF	SUB	*	:	J	Z	j	z
1011	VT	ESC	+	;	K	[	k	{
1100	FF	FS	,	<	L	\	l	¦
1101	CR	GS	–	=	M	]	m	}
1110	SO	RS	.	>	N	∧	n	~
1111	SI	US	/	?	O	–	o	DEL

Control characters

NUL	Null	DLE	Data-link escape
SOH	Start of heading	DC1	Device control 1
STX	Start of text	DC2	Device control 2
ETX	End of text	DC3	Device control 3
EOT	End of transmission	DC4	Device control 4
ENQ	Enquiry	NAK	Negative acknowledge
ACK	Acknowledge	SYN	Synchronous idle
BEL	Bell	ETB	End-of-transmission block
BS	Backspace	CAN	Cancel
HT	Horizontal tab	EM	End of medium
LF	Line feed	SUB	Substitute
VT	Vertical tab	ESC	Escape
FF	Form feed	FS	File separator
CR	Carriage return	GS	Group separator
SO	Shift out	RS	Record separator
SI	Shift in	US	Unit separator
SP	Space	DEL	Delete

separate the data into divisions such as paragraphs and pages. They include characters such as record separator (RS) and file separator (FS). The communication-control characters are useful during the transmission of text between remote terminals. Examples of communication-control characters are STX (start of text) and ETX (end of text), which are used to frame a text message when transmitted through telephone wires.

ASCII is a 7-bit code, but most computers manipulate an 8-bit quantity as a single unit called a *byte*. Therefore, ASCII characters most often are stored one per byte. The extra bit is sometimes used for other purposes, depending on the application. For example, some printers recognize 8-bit ASCII characters with the most-significant bit set to 0. Additional 128 8-bit characters with the most-significant bit set to 1 are used for other symbols such as the Greek alphabet or italic type font. When used in data communication, the eighth bit may be employed to indicate the parity of the character.

Other Alphanumeric Codes

Another alphanumeric code used in IBM equipment is the EBCDIC (Extended Binary-Coded Decimal Interchange Code). It uses eight bits for each character. EBCDIC has the same character symbols as ASCII, but the bit assignment for characters is different. As the name implies, the binary code for the letters and numerals is an extension of the binary-coded decimal (BCD) code. This means that the last four bits of the code range from 0000 though 1001 as in BCD.

When characters are used internally in a computer for data processing (not for transmission purposes), it is sometimes convenient to use a 6-bit code to represent 64 characters. A 6-bit code can specify 64 characters consisting of the 26 capital letters, the 10 numerals, and up to 28 special characters. This set of characters is usually sufficient for data-processing purposes. Using fewer bits to code characters has the advantage of reducing the space needed to store large quantities of alphanumeric data.

A code developed in the early stages of teletype transmission is the 5-bit Baudot code. Although five bits can specify only 32 characters, the Baudot code represents 58 characters by using two modes of operation. In the mode called *letters,* the five bits encode the 26 letters of the alphabet. In the mode called *figures,* the five bits encode the numerals and other characters. There are two special characters that are recognized by both modes and used to shift from one mode to the other. The *letter-shift* character places the reception station in the letters mode, after which all subsequent character codes are interpreted as letters. The *figure-shift* character places the system in the figures mode. The shift operation is analogous to the shifting operation on a typewriter with a shift lock key.

When alphanumeric information is transferred to the computer using punched cards, the alphanumeric characters are coded with 12 bits. Programs and data in the past were prepared on punched cards using the Hollerith code. A punched card consists of 80 columns and 12 rows. Each column represents an alphanumeric character of 12 bits with holes punched in the appropriate rows. A hole is sensed as a 1 and the absence of a hole is sensed as a 0. The 12 rows are marked, starting from the top, as 12, 11, 0, 1,

2, . . . , 9. The first three are called the zone punch and the last nine are called the numeric punch. Decimal digits are represented by a single hole in a numeric punch. The letters of the alphabet are represented by two holes in a column, one in the zone punch and the other the numeric punch. Special characters are represented by one, two, or three holes in a column. The 12-bit card code is ineffecient in its use of bits. Consequently, computers that receive input from a card reader convert the input 12-bit card code into an internal six-bit code to conserve bits of storage.

1-8   BINARY STORAGE AND REGISTERS

The discrete elements of information in a digital computer must have a physical existence in some information-storage medium. Furthermore, when discrete elements of information are represented in binary form, the information-storage medium must contain binary storage elements for storing individual bits. A *binary cell* is a device that possesses two stable states and is capable of storing one bit of information. The input to the cell receives excitation signals that set it to one of the two states. The output of the cell is a physical quantity that distinguishes between the two states. The information stored in a cell is a 1 when it is in one stable state and a 0 when in the other stable state. Examples of binary cells are electronic flip-flop circuits, ferrite cores used in memories, and positions punched with a hole or not punched in a card.

Registers

A *register* is a group of binary cells. Since a cell stores one bit of information, it follows that a register with n cells can store any discrete quantity of information that contains n bits. The *state* of a register is an n-tuple number of 1's and 0's, with each bit designating the state of one cell in the register. The *content* of a register is a function of the interpretation given to the information stored in it. Consider, for example, the following 16-cell register:

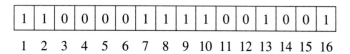

Physically, one may think of the register as composed of 16 binary cells, with each cell storing either a 1 or a 0. Suppose that the bit configuration stored in the register is as shown. The state of the register is the 16-tuple number 1100001111001001. Clearly, a register with n cells can be in one of 2^n possible states. Now, if one assumes that the content of the register represents a binary integer, then obviously the register can store any binary number from 0 to $2^{16} - 1$. For the particular example shown, the content of the register is the binary equivalent of the decimal number 50121. If it is assumed that the register stores alphanumeric characters of an eight-bit code, the content of the register is any two meaningful characters. For the ASCII code with an even parity placed in the eighth most-significant bit position the previous example represents the two characters C (left eight bits) and I (right eight bits). On the other hand, if one interprets the content of the register to be four decimal digits represented by a four-bit code, the content of the register is a four-digit decimal number. In the excess-3 code, the previous example is the decimal number 9096. The content of the register is meaningless in BCD since the bit combination 1100 is not assigned to any decimal digit. From this example, it is clear that a register can store one or more discrete elements of information and that the same bit configuration may be interpreted differently for different types of elements of information. It is important that the user store meaningful information in registers and that the computer be programmed to process this information according to the *type* of information stored.

Register Transfer

A digital computer is characterized by its registers. The memory unit (Fig. 1-1) is merely a collection of thousands of registers for storing digital information. The processor unit is composed of various registers that store operands upon which operations are performed. The control unit uses registers to keep track of various computer sequences, and every input or output device must have at least one register to store the information transferred to or from the device. An *interregister transfer* operation, a basic operation in digital systems, consists of a transfer of the information stored in one register into another. Figure 1-2 illustrates the transfer of information among registers and demonstrates pictorially the transfer of binary information from a keyboard into a register in the memory unit. The input unit is assumed to have a keyboard, a control circuit, and an input register. Each time a key is struck, the control enters into the input register an equivalent eight-bit alphanumeric character code. We shall assume that the code used is the ASCII code with an odd-parity eighth bit. The information from the input register is transferred into the eight least significant cells of a processor register. After every transfer, the input register is cleared to enable the control to insert a new eight-bit code when the keyboard is struck again. Each eight-bit character transferred to the processor register is preceded by a shift of the previous character to the next eight cells on its left. When a transfer of four characters is completed, the processor register is full, and its contents are transferred into a memory register. The content stored in the memory register shown in Fig. 1-2 came from the transfer of the characters JOHN after the four appropriate keys were struck.

To process discrete quantities of information in binary form, a computer must be provided with (1) devices that hold the data to be processed and (2) circuit elements that manipulate individual bits of information. The device most commonly used for holding data is a register. Manipulation of binary variables is done by means of digital logic circuits. Figure 1-3 illustrates the process of adding two 10-bit binary numbers. The memory unit, which normally consists of thousands of registers, is shown in the

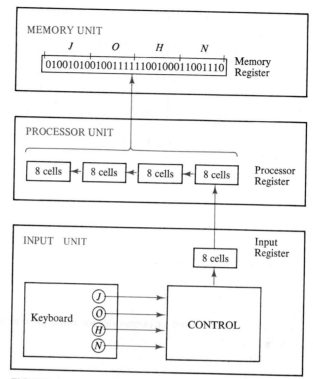

FIGURE 1-2

Transfer of information with registers

diagram with only three of its registers. The part of the processor unit shown consists of three registers, R1, R2, and R3, together with digital logic circuits that manipulate the bits of R1 and R2 and transfer into R3 a binary number equal to their arithmetic sum. Memory registers store information and are incapable of processing the two operands. However, the information stored in memory can be transferred to processor registers. Results obtained in processor registers can be transferred back into a memory register for storage until needed again. The diagram shows the contents of two operands transferred from two memory registers into R1 and R2. The digital logic circuits produce the sum, which is transferred to register R3. The contents of R3 can now be transferred back to one of the memory registers.

The last two examples demonstrated the information-flow capabilities of a digital system in a very simple manner. The registers of the system are the basic elements for storing and holding the binary information. The digital logic circuits process the information. Digital logic circuits and their manipulative capabilities are introduced in the next section. Registers and memory are presented in Chapter 7.

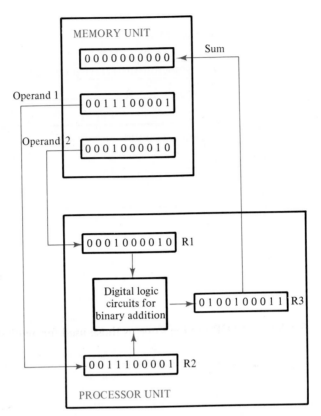

FIGURE 1-3
Example of binary information processing

1-9   BINARY LOGIC

Binary logic deals with variables that take on two discrete values and with operations that assume logical meaning. The two values the variables take may be called by different names (e.g., *true* and *false, yes* and *no,* etc.), but for our purpose, it is convenient to think in terms of bits and assign the values of 1 and 0. Binary logic is used to describe, in a mathematical way, the manipulation and processing of binary information. It is particularly suited for the analysis and design of digital systems. For example, the digital logic circuits of Fig. 1-3 that perform the binary arithmetic are circuits whose behavior is most conveniently expressed by means of binary variables and logical operations. The binary logic to be introduced in this section is equivalent to an algebra called Boolean algebra. The formal presentation of a two-valued Boolean algebra is covered in more detail in Chapter 2. The purpose of this section is to introduce Boolean algebra in a heuristic manner and relate it to digital logic circuits and binary signals.

Definition of Binary Logic

Binary logic consists of binary variables and logical operations. The variables are designated by letters of the alphabet such as A, B, C, x, y, z, etc., with each variable having two and only two distinct possible values: 1 and 0. There are three basic logical operations: AND, OR, and NOT.

1. AND: This operation is represented by a dot or by the absence of an operator. For example, $x \cdot y = z$ or $xy = z$ is read "x AND y is equal to z." The logical operation AND is interpreted to mean that $z = 1$ if and only if $x = 1$ *and* $y = 1$; otherwise $z = 0$. (Remember that x, y, and z are binary variables and can be equal either to 1 or 0, and nothing else.)

2. OR: This operation is represented by a plus sign. For example, $x + y = z$ is read "x OR y is equal to z," meaning that $z = 1$ if $x = 1$ *or* if $y = 1$ *or* if both $x = 1$ and $y = 1$. If both $x = 0$ and $y = 0$, then $z = 0$.

3. NOT: This operation is represented by a prime (sometimes by a bar). For example, $x' = z$ (or $\overline{x} = z$) is read "not x is equal to z," meaning that z is what x is not. In other words, if $x = 1$, then $z = 0$; but if $x = 0$, then $z = 1$.

Binary logic resembles binary arithmetic, and the operations AND and OR have some similarities to multiplication and addition, respectively. In fact, the symbols used for AND and OR are the same as those used for multiplication and addition. However, binary logic should not be confused with binary arithmetic. One should realize that an arithmetic variable designates a number that may consist of many digits. A logic variable is always either a 1 or a 0. For example, in binary arithmetic, we have $1 + 1 = 10$ (read: "one plus one is equal to 2"), whereas in binary logic, we have $1 + 1 = 1$ (read: "one OR one is equal to one").

For each combination of the values of x and y, there is a value of z specified by the definition of the logical operation. These definitions may be listed in a compact form using *truth tables*. A truth table is a table of all possible combinations of the variables showing the relation between the values that the variables may take and the result of the operation. For example, the truth tables for the operations AND and OR with variables x and y are obtained by listing all possible values that the variables may have when combined in pairs. The result of the operation for each combination is then listed in a separate row. The truth tables for AND, OR, and NOT are listed in Table 1-6. These tables clearly demonstrate the definition of the operations.

Switching Circuits and Binary Signals

The use of binary variables and the application of binary logic are demonstrated by the simple switching circuits of Fig. 1-4. Let the manual switches A and B represent two binary variables with values equal to 0 when the switch is open and 1 when the switch is closed. Similarly, let the lamp L represent a third binary variable equal to 1 when the light is on and 0 when off. For the switches in series, the light turns on if A *and* B are

TABLE 1-6
Truth Tables of Logical Operations

AND			OR			NOT	
x y		$x \cdot y$	x y		$x + y$	x	x'
0 0		0	0 0		0	0	1
0 1		0	0 1		1	1	0
1 0		0	1 0		1		
1 1		1	1 1		1		

closed. For the switches in parallel, the light turns on if *A or B* is closed. It is obvious that the two circuits can be expressed by means of binary logic with the AND and OR operations, respectively:

$$L = A \cdot B \qquad \text{for the circuit of Fig. 1-4(a)}$$

$$L = A + B \qquad \text{for the circuit of Fig. 1-4(b)}$$

Electronic digital circuits are sometimes called *switching circuits* because they behave like a switch, with the active element such as a transistor either conducting (switch closed) or not conducting (switch open). Instead of changing the switch manually, an electronic switching circuit uses binary signals to control the conduction or nonconduction state of the active element. Electrical signals such as voltages or currents exist throughout a digital system in either one of two recognizable values (except during transition). Voltage-operated circuits, for example, respond to two separate voltage levels, which represent a binary variable equal to logic-1 or logic-0. For example, a particular digital system may define logic-1 as a signal with a nominal value of 3 volts and logic-0 as a signal with a nominal value of 0 volt. As shown in Fig. 1-5, each voltage level has an acceptable deviation from the nominal. The intermediate region between the allowed regions is crossed only during state transitions. The input terminals of digital circuits accept binary signals within the allowable tolerances and respond at the output terminal with binary signals that fall within the specified tolerances.

Logic Gates

Electronic digital circuits are also called *logic circuits* because, with the proper input, they establish logical manipulation paths. Any desired information for computing or

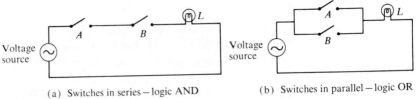

(a) Switches in series – logic AND   (b) Switches in parallel – logic OR

FIGURE 1-4
Switching circuits that demonstrate binary logic

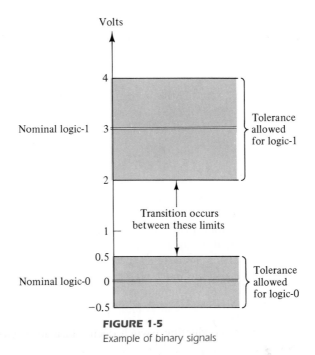

Volts

Nominal logic-1   3

Nominal logic-0   0

FIGURE 1-5
Example of binary signals

control can be operated upon by passing binary signals through various combinations of logic circuits, each signal representing a variable and carrying one bit of information. Logic circuits that perform the logical operations of AND, OR, and NOT are shown with their symbols in Fig. 1-6. These circuits, called *gates,* are blocks of hardware that produce a logic-1 or logic-0 output signal if input logic requirements are satisfied. Note that four different names have been used for the same type of circuits: digital circuits, switching circuits, logic circuits, and gates. All four names are widely used, but we shall refer to the circuits as AND, OR, and NOT gates. The NOT gate is sometimes called an *inverter circuit* since it inverts a binary signal.

The input signals x and y in the two-input gates of Fig. 1-6 may exist in one of four possible states: 00, 10, 11, or 01. These input signals are shown in Fig. 1-7, together with the output signals for the AND and OR gates. The timing diagrams in Fig. 1-7 il-

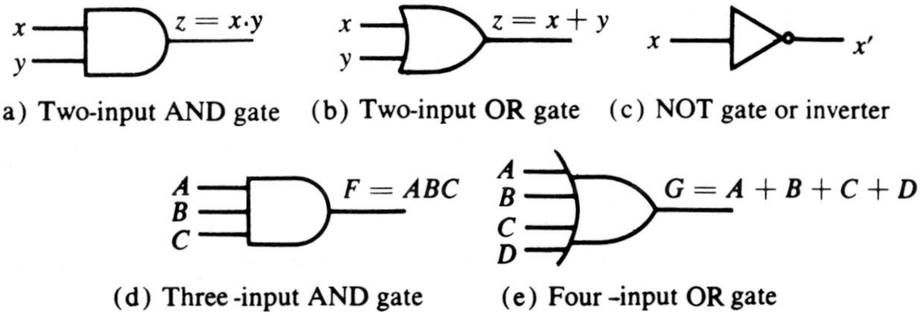

x ⟶ $z = x \cdot y$ x ⟶ $z = x + y$ x ⟶ x'
y y

(a) Two-input AND gate   (b) Two-input OR gate   (c) NOT gate or inverter

A A
B ⟶ $F = ABC$ B ⟶ $G = A + B + C + D$
C C
                          D

(d) Three-input AND gate   (e) Four-input OR gate

FIGURE 1-6
Symbols for digital logic circuits

Chapter 1 Binary Systems

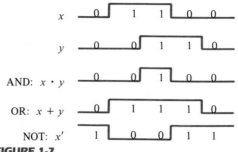

FIGURE 1-7
Input–output signals for gates (a), (b), and (c) of Fig. 1-6

lustrate the response of each circuit to each of the four possible input binary combinations. The reason for the name "inverter" for the NOT gate is apparent from a comparison of the signal x (input of inverter) and that of x' (output of inverter).

AND and OR gates may have more than two inputs. An AND gate with three inputs and an OR gate with four inputs are shown in Fig. 1-6. The three-input AND gate responds with a logic-1 output if all three input signals are logic-1. The output produces a logic-0 signal if any input is logic-0. The four-input OR gate responds with a logic-1 when any input is a logic-1. Its output becomes logic-0 if all input signals are logic-0.

The mathematical system of binary logic is better known as Boolean, or switching, algebra. This algebra is conveniently used to describe the operation of complex networks of digital circuits. Designers of digital systems use Boolean algebra to transform circuit diagrams to algebraic expressions and vice versa. Chapters 2 and 3 are devoted to the study of Boolean algebra, its properties, and manipulative capabilities. Chapter 4 shows how Boolean algebra may be used to express mathematically the interconnections among networks of gates.

REFERENCES

1. CAVANAGH, J. J., *Digital Computer Arithmetic,* New York: McGraw-Hill, 1984.
2. HWANG, K., *Computer Arithmetic,* New York: John Wiley, 1979.
3. SCHMID, H., *Decimal Computation,* New York: John Wiley, 1974.
4. KNUTH, D. E., *The Art of Computer Programming: Seminumerical Algorithms.* Reading, MA: Addison-Wesley, 1969.
5. FLORES, I., *The Logic of Computer Arithmetic.* Englewood Cliffs, NJ: Prentice-Hall, 1963.
6. RICHARD, R. K., *Arithmetic Operations in Digital Computers.* New York: Van Nostrand, 1955.
7. MANO, M. M., *Computer Engineering: Hardware Design.* Englewood Cliffs, NJ: Prentice-Hall, 1988.
8. CHU, Y., *Computer Organization and Programming.* Englewood Cliffs, NJ: Prentice-Hall, 1972.

PROBLEMS

1-1 List the first 16 numbers in base 12. Use the letters A and B to represent the last two digits.

1-2 What is the largest binary number that can be obtained with 16 bits? What is its decimal equivalent?

1-3 Convert the following binary numbers to decimal: 101110; 1110101.11; and 110110100.

1-4 Convert the following numbers with the indicated bases to decimal: $(12121)_3$; $(4310)_5$; $(50)_7$; and $(198)_{12}$.

1-5 Convert the following decimal numbers to binary: 1231; 673.23; 10^4; and 1998.

1-6 Convert the following decimal numbers to the indicated bases:
(a) 7562.45 to octal.
(b) 1938.257 to hexadecimal.
(c) 175.175 to binary.

1-7 Convert the hexadecimal number F3A7C2 to binary and octal.

1-8 Convert the following numbers from the given base to the other three bases indicated.
(a) Decimal 225 to binary, octal, and hexadecimal.
(b) Binary 11010111 to decimal, octal, and hexadecimal.
(c) Octal 623 to decimal, binary, and hexadecimal.
(d) Hexadecimal 2AC5 to decimal, octal, and binary.

1-9 Add and multiply the following numbers without converting to decimal.
(a) $(367)_8$ and $(715)_8$.
(b) $(15F)_{16}$ and $(A7)_{16}$.
(c) $(110110)_2$ and $(110101)_2$.

1-10 Perform the following division in binary: $11111111/101$.

1-11 Determine the value of base x if $(211)_x = (152)_8$.

1-12 Noting that $3^2 = 9$, formulate a simple procedure for converting base-3 numbers directly to base-9. Use the procedure to convert $(2110201102220112)_3$ to base 9.

1-13 Find the 9's complement of the following 8-digit decimal numbers: 12349876; 00980100; 90009951; and 00000000.

1-14 Find the 10's complement of the following 6-digit decimal numbers: 123900; 090657; 100000; and 000000.

1-15 Find the 1's and 2's complements of the following 8-digit binary numbers: 10101110; 10000001; 10000000; 00000001; and 00000000.

1-16 Perform subtraction with the following unsigned decimal numbers by taking the 10's complement of the subtrahend.
(a) $5250 - 1321$
(b) $1753 - 8640$
(c) $20 - 100$
(d) $1200 - 250$

1-17 Perform the subtraction with the following unsigned binary numbers by taking the 2's complement of the subtrahend.

(a) $11010 - 10000$
(b) $11010 - 1101$
(c) $100 - 110000$
(d) $1010100 - 1010100$

1-18 Perform the arithmetic operations $(+42) + (-13)$ and $(-42) - (-13)$ in binary using the signed-2's-complement representation for negative numbers.

1-19 The binary numbers listed have a sign in the leftmost position and, if negative, are in 2's-complement form. Perform the arithmetic operations indicated and verify the answers.
(a) $101011 + 111000$
(b) $001110 + 110010$
(c) $111001 - 001010$
(d) $101011 - 100110$

1-20 Represent the following decimal numbers in BCD: 13597; 93286; and 99880.

1-21 Determine the binary code for each of the ten decimal digits using a weighted code with weights 7, 4, 2, and 1.

1-22 The $(r - 1)$'s complement of base-6 numbers is called the 5's complement.
(a) Determine a procedure for obtaining the 5's complement of base-6 numbers.
(b) Obtain the 5's complement of $(543210)_6$.
(c) Design a 3-bit code to represent each of the six digits of the base-6 number system. Make the binary code self-complementing so that the 5's complement is obtained by changing 1's to 0's and 0's to 1's in all the bits of the coded number.

1-23 Represent decimal number 8620 in (a) BCD, (b) excess-3 code, (c)2421 code, and (d) as a binary number.

1-24 Represent decimal 3864 in the 2421 code of Table 1-2. Show that the code is self-complementing by taking the 9's complement of 3864.

1-25 Assign a binary code in some orderly manner to the 52 playing cards. Use the minimum number of bits.

1-26 List the ten BCD digits with an even parity in the leftmost position. (Total of five bits per digit.) Repeat with an odd-parity bit.

1-27 Write your full name in ASCII using an eight-bit code with the leftmost bit always 0. Include a space between names and a period after a middle initial.

1-28 Decode the following ASCII code: 1001010 1101111 1101000 1101110 0100000 1000100 1101111 1100101.

1-29 Show the bit configuration that represents the decimal number 295 (a) in binary, (b) in BCD, and (c) in ASCII.

1-30 How many printing characters are there in ASCII? How many of them are not letters or numerals?

1-31 The state of a 12-bit register is 010110010111. What is its content if it represents:
(a) three decimal digits in BCD;
(b) three decimal digits in the excess-3 code;
(c) three decimal digits in the 2421 code?

1-32 Show the contents of all registers in Fig. 1-3 if the two binary numbers added have the decimal equivalent of 257 and 514.

1-33 Show the signals (by means of diagram similar to Fig. 1-7) of the outputs F and G in the two gates of Figs. 1-6(d) and (e). Use all 16 possible combinations of the input signals A, B, C, and D.

1-34 Express the switching circuit shown in the figure in binary logic notation.

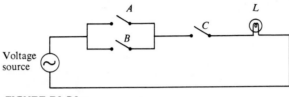

FIGURE P1-34

Boolean Algebra and Logic Gates

2-1 BASIC DEFINITIONS

Boolean algebra, like any other deductive mathematical system, may be defined with a set of elements, a set of operators, and a number of unproved axioms or postulates. A *set* of elements is any collection of objects having a common property. If S is a set, and x and y are certain objects, then $x \in S$ denotes that x is a member of the set S, and $y \notin S$ denotes that y is not an element of S. A set with a denumerable number of elements is specified by braces: A = {1, 2, 3, 4}, i.e., the elements of set A are the numbers 1, 2, 3, and 4. A *binary operator* defined on a set S of elements is a rule that assigns to each pair of elements from S a unique element from S. As an example, consider the relation $a * b = c$. We say that $*$ is a binary operator if it specifies a rule for finding c from the pair (a, b) and also if $a, b, c \in S$. However, $*$ is not a binary operator if $a, b \in S$, whereas the rule finds $c \notin S$.

The postulates of a mathematical system form the basic assumptions from which it is possible to deduce the rules, theorems, and properties of the system. The most common postulates used to formulate various algebraic structures are:

1. *Closure*. A set S is closed with respect to a binary operator if, for every pair of elements of S, the binary opertor specifies a rule for obtaining a unique element of S. For example, the set of natural numbers N = {1, 2, 3, 4, . . . } is closed with respect to the binary operator plus (+) by the rules of arithmetic addition, since for any $a, b \in N$ we obtain a unique $c \in N$ by the operation $a + b = c$. The set of natural numbers is not closed with respect to the binary operator minus (−) by the rules of arithmetic subtraction because $2 − 3 = −1$ and $2, 3 \in N$, while $(−1) \notin N$.

2. *Associative law.* A binary operator $*$ on a set S is said to be associative whenever

$$(x * y) * z = x * (y * z) \qquad \text{for all } x, y, z, \in S$$

3. *Commutative law.* A binary operator $*$ on a set S is said to be commutative whenever

$$x * y = y * x \qquad \text{for all } x, y \in S$$

4. *Identity element.* A set S is said to have an identity element with respect to a binary operation $*$ on S if there exists an element $e \in S$ with the property

$$e * x = x * e = x \qquad \text{for every } x \in S$$

Example: The element 0 is an identity element with respect to operation $+$ on the set of integers $I = \{ \ldots, -3, -2, -1, 0, 1, 2, 3, \ldots \}$ since

$$x + 0 = 0 + x = x \qquad \text{for any } x \in I$$

The set of natural numbers N has no identity element since 0 is excluded from the set.

5. *Inverse.* A set S having the identity element e with respect to a binary operator $*$ is said to have an inverse whenever, for every $x \in S$, there exists an element $y \in S$ such that

$$x * y = e$$

Example: In the set of integers I with $e = 0$, the inverse of an element a is $(-a)$ since $a + (-a) = 0$.

6. *Distributive law.* If $*$ and $\cdot$ are two binary operators on a set S, $*$ is said to be distributive over $\cdot$ whenever

$$x * (y \cdot z) = (x * y) \cdot (x * z)$$

An example of an algebraic structure is a *field*. A field is a set of elements, together with two binary operators, each having properties 1 to 5 and both operators combined to give property 6. The set of real numbers together with the binary operators $+$ and $\cdot$ form the field of real numbers. The field of real numbers is the basis for arithmetic and ordinary algebra. The operators and postulates have the following meanings:

The binary operator $+$ defines addition.

The additive identity is 0.

The additive inverse defines subtraction.

The binary operator $\cdot$ defines multiplication.

The multiplicative identity is 1.

The multiplicative inverse of $a = 1/a$ defines division, i.e., $a \cdot 1/a = 1$.

The only distributive law applicable is that of $\cdot$ over $+$:

$$a \cdot (b + c) = (a \cdot b) + (a \cdot c)$$

2-2   AXIOMATIC DEFINITION OF BOOLEAN ALGEBRA

In 1854 George Boole introduced a systematic treatment of logic and developed for this purpose an algebraic system now called *Boolean algebra*. In 1938 C. E. Shannon introduced a two-valued Boolean algebra called *switching algebra,* in which he demonstrated that the properties of bistable electrical switching circuits can be represented by this algebra. For the formal definition of Boolean algebra, we shall employ the postulates formulated by E. V. Huntington in 1904.

Boolean algebra is an algebraic structure defined on a set of elements B together with two binary operators $+$ and $\cdot$ provided the following (Huntington) postulates are satisfied:

1. (a) Closure with respect to the operator $+$.
  (b) Closure with respect to the operator $\cdot$.

2. (a) An identity element with respect to $+$, designated by 0: $x + 0 = 0 + x = x$.
  (b) An identity element with respect to $\cdot$, designated by 1: $x \cdot 1 = 1 \cdot x = x$.

3. (a) Commutative with respect to $+$: $x + y = y + x$.
  (b) Commutative with respect to $\cdot$: $x \cdot y = y \cdot x$.

4. (a) $\cdot$ is distributive over $+$: $x \cdot (y + z) = (x \cdot y) + (x \cdot z)$.
  (b) $+$ is distributive over $\cdot$: $x + (y \cdot z) = (x + y) \cdot (x + z)$.

5. For every element $x \in B$, there exists an element $x' \in B$ (called the complement of x) such that (a) $x + x' = 1$ and (b) $x \cdot x' = 0$.

6. There exists at least two elements $x, y \in B$ such that $x \neq y$.

Comparing Boolean algebra with arithmetic and ordinary algebra (the field of real numbers), we note the following differences:

1. Huntington postulates do not include the associative law. However, this law holds for Boolean algebra and can be derived (for both operators) from the other postulates.

2. The distributive law of $+$ over $\cdot$, i.e., $x + (y \cdot z) = (x + y) \cdot (x + z)$, is valid for Boolean algebra, but not for ordinary algebra.

3. Boolean algebra does not have additive or multiplicative inverses; therefore, there are no subtraction or division operations.

4. Postulate 5 defines an operator called *complement* that is not available in ordinary algebra.

5. Ordinary algebra deals with the real numbers, which consitute an infinite set of elements. Boolean algebra deals with the as yet undefined set of elements B, but in the two-valued Boolean algebra defined below (and of interest in our subsequent use of this algebra), B is defined as a set with only two elements, 0 and 1.

Boolean algebra resembles ordinary algebra in some respects. The choice of symbols $+$ and $\cdot$ is intentional to facilitate Boolean algebraic manipulations by persons already familiar with ordinary algebra. Although one can use some knowledge from ordinary algebra to deal with Boolean algebra, the beginner must be careful not to substitute the rules of ordinary algebra where they are not applicable.

It is important to distinguish between the elements of the set of an algebraic structure and the variables of an algebraic system. For example, the elements of the field of real numbers are numbers, whereas variables such as a, b, c, etc., used in ordinary algebra, are symbols that stand for real numbers. Similarly in Boolean algebra, one defines the elements of the set B, and variables such as x, y, z are merely symbols that represent the elements. At this point, it is important to realize that in order to have a Boolean algebra, one must show:

1. the elements of the set B,
2. the rules of operation for the two binary operators, and
3. that the set of elements B, together with the two operators, satisfies the six Huntington postulates.

One can formulate many Boolean algebras, depending on the choice of elements of B and the rules of operation. In our subsequent work, we deal only with a two-valued Boolean algebra, i.e., one with only two elements. Two-valued Boolean algebra has applications in set theory (the algebra of classes) and in propositional logic. Our interest here is with the application of Boolean algebra to gate-type circuits.

Two-Valued Boolean Algebra

A two-valued Boolean algebra is defined on a set of two elements, $B = \{0, 1\}$, with rules for the two binary operators $+$ and $\cdot$ as shown in the following operator tables (the rule for the complement operator is for verification of postulate 5):

x	y	$x \cdot y$
0	0	0
0	1	0
1	0	0
1	1	1

x	y	$x + y$
0	0	0
0	1	1
1	0	1
1	1	1

x	x'
0	1
1	0

These rules are exactly the same as the AND, OR, and NOT operations, respectively, defined in Table 1-6. We must now show that the Huntington postulates are valid for the set $B = \{0, 1\}$ and the two binary operators defined before.

1. *Closure* is obvious from the tables since the result of each operation is either 1 or 0 and 1, 0 $\in B$.
2. From the tables we see that
   (a) $0 + 0 = 0$ $0 + 1 = 1 + 0 = 1$
   (b) $1 \cdot 1 = 1$ $1 \cdot 0 = 0 \cdot 1 = 0$
   which establishes the two *identity elements* 0 for $+$ and 1 for $\cdot$ as defined by postulate 2.

3. The *commutative* laws are obvious from the symmetry of the binary operator tables.

4. **(a)** The *distributive* law $x \cdot (y + z) = (x \cdot y) + (x \cdot z)$ can be shown to hold true from the operator tables by forming a truth table of all possible values of x, y, and z. For each combination, we derive $x \cdot (y + z)$ and show that the value is the same as $(x \cdot y) + (x \cdot z)$.

x	y	z	$y + z$	$x \cdot (y + z)$	$x \cdot y$	$x \cdot z$	$(x \cdot y) + (x \cdot z)$
0	0	0	0	0	0	0	0
0	0	1	1	0	0	0	0
0	1	0	1	0	0	0	0
0	1	1	1	0	0	0	0
1	0	0	0	0	0	0	0
1	0	1	1	1	0	1	1
1	1	0	1	1	1	0	1
1	1	1	1	1	1	1	1

(b) The *distributive* law of + over · can be shown to hold true by means of a truth table similar to the one above.

5. From the complement table it is easily shown that
   (a) $x + x' = 1$, since $0 + 0' = 0 + 1 = 1$ and $1 + 1' = 1 + 0 = 1$.
   (b) $x \cdot x' = 0$, since $0 \cdot 0' = 0 \cdot 1 = 0$ and $1 \cdot 1' = 1 \cdot 0 = 0$, which verifies postulate 5.

6. Postulate 6 is satisfied because the two-valued Boolean algebra has two distinct elements, 1 and 0, with $1 \neq 0$.

We have just established a two-valued Boolean algebra having a set of two elements, 1 and 0, two binary operators with operation rules equivalent to the AND and OR operations, and a complement operator equivalent to the NOT operator. Thus, Boolean algebra has been defined in a formal mathematical manner and has been shown to be equivalent to the binary logic presented heuristically in Section 1-9. The heuristic presentation is helpful in understanding the application of Boolean algebra to gate-type circuits. The formal presentation is necessary for developing the theorems and properties of the algebraic system. The two-valued Boolean algebra defined in this section is also called "switching algebra" by engineers. To emphasize the similarities between two-valued Boolean algebra and other binary systems, this algebra was called "binary logic" in Section 1-9. From here on, we shall drop the adjective "two-valued" from Boolean algebra in subsequent discussions.

2-3 BASIC THEOREMS AND PROPERTIES OF BOOLEAN ALGEBRA

Duality

The Huntington postulates have been listed in pairs and designated by part (a) and part (b). One part may be obtained from the other if the binary operators and the identity elements are interchanged. This important property of Boolean algebra is called the *duality principle*. It states that every algebraic expression deducible from the postulates of Boolean algebra remains valid if the operators and identity elements are interchanged. In a two-valued Boolean algebra, the identity elements and the elements of the set B are the same: 1 and 0. The duality principle has many applications. If the *dual* of an algebraic expression is desired, we simply interchange OR and AND operators and replace 1's by 0's and 0's by 1's.

Basic Theorems

Table 2-1 lists six theorems of Boolean algebra and four of its postulates. The notation is simplified by omitting the · whenever this does not lead to confusion. The theorems and postulates listed are the most basic relationships in Boolean algebra. The reader is advised to become familiar with them as soon as possible. The theorems, like the postulates, are listed in pairs; each relation is the dual of the one paired with it. The postulates are basic axioms of the algebraic structure and need no proof. The theorems must be proven from the postulates. The proofs of the theorems with one variable are presented below. At the right is listed the number of the postulate that justifies each step of the proof.

TABLE 2-1
Postulates and Theorems of Boolean Algebra

Postulate 2	(a) $x + 0 = x$	(b) $x \cdot 1 = x$
Postulate 5	(a) $x + x' = 1$	(b) $x \cdot x' = 0$
Theorem 1	(a) $x + x = x$	(b) $x \cdot x = x$
Theorem 2	(a) $x + 1 = 1$	(b) $x \cdot 0 = 0$
Theorem 3, involution	$(x')' = x$	
Postulate 3, commutative	(a) $x + y = y + x$	(b) $xy = yx$
Theorem 4, associative	(a) $x + (y + z) = (x + y) + z$	(b) $x(yz) = (xy)z$
Postulate 4, distributive	(a) $x(y + z) = xy + xz$	(b) $x + yz = (x + y)(x + z)$
Theorem 5, DeMorgan	(a) $(x + y)' = x'y'$	(b) $(xy)' = x' + y'$
Theorem 6, absorption	(a) $x + xy = x$	(b) $x(x + y) = x$

THEOREM 1(a): $x + x = x$.

$$
\begin{aligned}
x + x &= (x + x) \cdot 1 &&\text{by postulate:} && 2(b) \\
&= (x + x)(x + x') && && 5(a) \\
&= x + xx' && && 4(b) \\
&= x + 0 && && 5(b) \\
&= x && && 2(a)
\end{aligned}
$$

THEOREM 1(b): $x \cdot x = x$.

$$
\begin{aligned}
x \cdot x &= xx + 0 &&\text{by postulate:} && 2(a) \\
&= xx + xx' && && 5(b) \\
&= x(x + x') && && 4(a) \\
&= x \cdot 1 && && 5(a) \\
&= x && && 2(b)
\end{aligned}
$$

Note that theorem 1(b) is the dual of theorem 1(a) and that each step of the proof in part (b) is the dual of part (a). Any dual theorem can be similarly derived from the proof of its corresponding pair.

THEOREM 2(a): $x + 1 = 1$.

$$
\begin{aligned}
x + 1 &= 1 \cdot (x + 1) &&\text{by postulate:} && 2(b) \\
&= (x + x')(x + 1) && && 5(a) \\
&= x + x' \cdot 1 && && 4(b) \\
&= x + x' && && 2(b) \\
&= 1 && && 5(a)
\end{aligned}
$$

THEOREM 2(b): $x \cdot 0 = 0$ by duality.

THEOREM 3: $(x')' = x$. From postulate 5, we have $x + x' = 1$ and $x \cdot x' = 0$, which defines the complement of x. The complement of x' is x and is also $(x')'$. Therefore, since the complement is unique, we have that $(x')' = x$.

The theorems involving two or three variables may be proven algebraically from the postulates and the theorems that have already been proven. Take, for example, the absorption theorem.

THEOREM 6(a): $x + xy = x$.

$$
\begin{aligned}
x + xy &= x \cdot 1 + xy &&\text{by postulate:} &&\text{2(b)} \\
&= x(1 + y) &&&&\text{4(a)} \\
&= x(y + 1) &&&&\text{3(a)} \\
&= x \cdot 1 &&&&\text{2(a)} \\
&= x &&&&\text{2(b)}
\end{aligned}
$$

THEOREM 6(b): $x(x + y) = x$ by duality.

The theorems of Boolean algebra can be shown to hold true by means of truth tables. In truth tables, both sides of the relation are checked to yield identical results for all possible combinations of variables involved. The following truth table verifies the first absorption theorem.

x	y	xy	$x + xy$
0	0	0	0
0	1	0	0
1	0	0	1
1	1	1	1

The algebraic proofs of the associative law and DeMorgan's theorem are long and will not be shown here. However, their validity is easily shown with truth tables. For example, the truth table for the first DeMorgan's theorem $(x + y)' = x'y'$ is shown below.

x	y	$x + y$	$(x + y)'$	x'	y'	$x'y'$
0	0	0	1	1	1	1
0	1	1	0	1	0	0
1	0	1	0	0	1	0
1	1	1	0	0	0	0

Operator Precedence

The operator precedence for evaluating Boolean expressions is (1) parentheses, (2) NOT, (3) AND, and (4) OR. In other words, the expression inside the parentheses must be evaluated before all other operations. The next operation that holds precedence is the complement, then follows the AND, and finally the OR. As an example, consider the truth table for DeMorgan's theorem. The left side of the expression is $(x + y)'$. Therefore, the expression inside the parentheses is evaluated first and the result then complemented. The right side of the expression is $x'y'$. Therefore, the complement of x and the complement of y are both evaluated first and the result is then ANDed. Note that in ordinary arithmetic, the same precedence holds (except for the complement) when multiplication and addition are replaced by AND and OR, respectively.

Venn Diagram

A helpful illustration that may be used to visualize the relationships among the variables of a Boolean expression is the *Venn diagram*. This diagram consists of a rectangle such as shown in Fig. 2-1, inside of which are drawn overlapping circles, one for each variable. Each circle is labeled by a variable. We designate all points inside a circle as belonging to the named variable and all points outside a circle as not belonging to the variable. Take, for example, the circle labeled x. If we are inside the circle, we say that $x = 1$; when outside, we say $x = 0$. Now, with two overlapping circles, there are four distinct areas inside the rectangle: the area not belonging to either x or y ($x'y'$), the area inside circle y but outside x ($x'y$), the area inside circle x but outside y (xy'), and the area inside both circles (xy).

Venn diagrams may be used to illustrate the postulates of Boolean algebra or to show the validity of theorems. Figure 2-2, for example, illustrates that the area belonging to xy is inside the circle x and therefore $x + xy = x$. Figure 2-3 illustrates the distributive law $x(y + z) = xy + xz$. In this diagram, we have three overlapping circles, one for each of the variables x, y, and z. It is possible to distinguish eight distinct areas in a three-variable Venn diagram. For this particular example, the distributive law is demonstrated by noting that the area intersecting the circle x with the area enclosing y or z is the same area belonging to xy or xz.

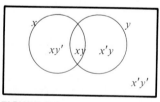

FIGURE 2-1
Venn diagram for two variables

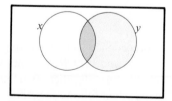

FIGURE 2-2
Venn diagram illustration $x = xy + x$

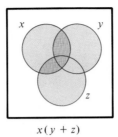

$$x(y + z)$$

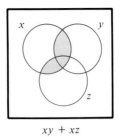

$$xy + xz$$

FIGURE 2-3
Venn diagram illustration of the distributive law

2-4 BOOLEAN FUNCTIONS

A binary variable can take the value of 0 or 1. A Boolean function is an expression formed with binary variables, the two binary operators OR and AND, and unary operator NOT, parentheses, and an equal sign. For a given value of the variables, the function can be either 0 or 1. Consider, for example, the Boolean function

$$F_1 = xyz'$$

The function F_1 is equal to 1 if $x = 1$ *and* $y = 1$ *and* $z' = 1$; otherwise $F_1 = 0$. The above is an example of a Boolean function represented as an algebraic expression. A Boolean function may also be represented in a truth table. To represent a function in a truth table, we need a list of the 2^n combinations of 1's and 0's of the n binary variables, and a column showing the combinations for which the function is equal to 1 or 0. As shown in Table 2-2, there are eight possible distinct combinations for assigning bits to three variables. The column labeled F_1 contains either a 0 or a 1 for each of these combinations. The table shows that the function F_1 is equal to 1 only when $x = 1$, $y = 1$, and $z = 0$. It is equal to 0 otherwise. (Note that the statement $z' = 1$ is equivalent to saying that $z = 0$.) Consider now the function

TABLE 2-2
Truth Tables for $F_1 = xyz'$, $F_2 = x + y'z$,
$F_3 = x'y'z + x'yz + xy'$, and $F_4 = xy' + x'z$

x	y	z	F_1	F_2	F_3	F_4
0	0	0	0	0	0	0
0	0	1	0	1	1	1
0	1	0	0	0	0	0
0	1	1	0	0	1	1
1	0	0	0	1	1	1
1	0	1	0	1	1	1
1	1	0	1	1	0	0
1	1	1	0	1	0	0

$$F_2 = x + y'z$$

$F_2 = 1$ if $x = 1$ or if $y = 0$, while $z = 1$. In Table 2-2, $x = 1$ in the last four rows and $yz = 01$ in rows 001 and 101. The latter combination applies also for $x = 1$. Therefore, there are five combinations that make $F_2 = 1$. As a third example, consider the function

$$F_3 = x'y'z + x'yz + xy'$$

This is shown in Table 2-2 with four 1's and four 0's. F_4 is the same as F_3 and is considered below.

Any Boolean function can be represented in a truth table. The number of rows in the table is 2^n, where n is the number of binary variables in the function. The 1's and 0's combinations for each row is easily obtained from the binary numbers by counting from 0 to $2^n - 1$. For each row of the table, there is a value for the function equal to either 1 or 0. The question now arises, is it possible to find two algebraic expressions that specify the same function? The answer to this question is yes. As a matter of fact, the manipulation of Boolean algebra is applied mostly to the problem of finding simpler expressions for the same function. Consider, for example, the function:

$$F_4 = xy' + x'z$$

From Table 2-2, we find that F_4 is the same as F_3, since both have identical 1's and 0's for each combination of values of the three binary variables. In general, two functions of n binary variables are said to be equal if they have the same value for all possible 2^n combinations of the n variables.

A Boolean function may be transformed from an algebraic expression into a logic diagram composed of AND, OR, and NOT gates. The implementation of the four functions introduced in the previous discussion is shown in Fig. 2-4. The logic diagram includes an inverter circuit for every variable present in its complement form. (The inverter is unnecessary if the complement of the variable is available.) There is an AND gate for each term in the expression, and an OR gate is used to combine two or more terms. From the diagrams, it is obvious that the implementation of F_4 requires fewer gates and fewer inputs than F_3. Since F_4 and F_3 are equal Boolean functions, it is more economical to implement the F_4 form than the F_3 form. To find simpler circuits, one must know how to manipulate Boolean functions to obtain equal and simpler expressions. What constitutes the best form of a Boolean function depends on the particular application. In this section, consideration is given to the criterion of equipment minimization.

Algebraic Manipulation

A *literal* is a primed or unprimed variable. When a Boolean function is implemented with logic gates, each literal in the function designates an input to a gate, and each term is implemented with a gate. The minimization of the number of literals and the number of terms results in a circuit with less equipment. It is not always possible to minimize both simultaneously; usually, further criteria must be available. At the moment, we

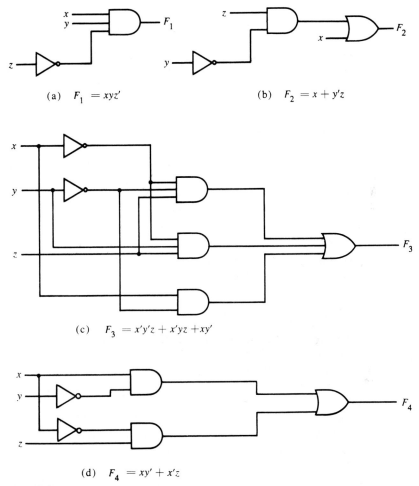

(a) $F_1 = xyz'$

(b) $F_2 = x + y'z$

(c) $F_3 = x'y'z + x'yz + xy'$

(d) $F_4 = xy' + x'z$

FIGURE 2-4
Implementation of Boolean functions with gates

shall narrow the minimization criterion to literal minimization. We shall discuss other criteria in Chapter 5. The number of literals in a Boolean function can be minimized by algebraic manipulations. Unfortunately, there are no specific rules to follow that will guarantee the final answer. The only method available is a cut-and-try procedure employing the postulates, the basic theorems, and any other manipulation method that becomes familiar with use. The following examples illustrate this procedure.

Example 2-1

Simplify the following Boolean functions to a minimum number of literals.

1. $x + x'y = (x + x')(x + y) = 1 \cdot (x + y) = x + y$
2. $x(x' + y) = xx' + xy = 0 + xy = xy$

3. $x'y'z + x'yz + xy' = x'z(y' + y) + xy' = x'z + xy'$

4. $xy + x'z + yz = xy + x'z + yz(x + x')$

$$= xy + x'z + xyz + x'yz$$
$$= xy(1 + z) + x'z(1 + y)$$
$$= xy + x'z$$

5. $(x + y)(x' + z)(y + z) = (x + y)(x' + z)$ by duality from function 4.   ■

Functions 1 and 2 are the duals of each other and use dual expressions in corresponding steps. Function 3 shows the equality of the functions F_3 and F_4 discussed previously. The fourth illustrates the fact that an increase in the number of literals sometimes leads to a final simpler expression. Function 5 is not minimized directly but can be derived from the dual of the steps used to derive function 4.

Complement of a Function

The complement of a function F is F' and is obtained from an interchange of 0's for 1's and 1's for 0's in the value of F. The complement of a function may be derived algebraically through DeMorgan's theorem. This pair of theorems is listed in Table 2-1 for two variables. DeMorgan's theorems can be extended to three or more variables. The three-variable form of the first DeMorgan's theorem is derived below. The postulates and theorems are those listed in Table 2-1.

$$(A + B + C)' = (A + X)' \qquad \text{let } B + C = X$$

$$= A'X' \qquad \text{by theorem 5(a) (DeMorgan)}$$

$$= A' \cdot (B + C)' \qquad \text{substitute } B + C = X$$

$$= A' \cdot (B'C') \qquad \text{by theorem 5(a) (DeMorgan)}$$

$$= A'B'C' \qquad \text{by theorem 4(b) (associative)}$$

DeMorgan's theorems for any number of variables resemble in form the two variable case and can be derived by successive substitutions similar to the method used in the above derivation. These theorems can be generalized as follows:

$$(A + B + C + D + \cdots + F)' = A'B'C'D' \cdots F'$$

$$(ABCD \cdots F)' = A' + B' + C' + D' + \cdots + F'$$

The generalized form of DeMorgan's theorem states that the complement of a function is obtained by interchanging AND and OR operators and complementing each literal.

Example
2-2

Find the complement of the functions $F_1 = x'yz' + x'y'z$ and $F_2 = x(y'z' + yz)$. By applying DeMorgan's theorem as many times as necessary, the complements are obtained as follows:

$$F_1' = (x'yz' + x'y'z)' = (x'yz')'(x'y'z)' = (x + y' + z)(x + y + z')$$

$$F_2' = [x(y'z' + yz)]' = x' + (y'z' + yz)' = x' + (y'z')' \cdot (yz)'$$

$$= x' + (y + z)(y' + z')$$ ∎

A simpler procedure for deriving the complement of a function is to take the dual of the function and complement each literal. This method follows from the generalized DeMorgan's theorem. Remember that the dual of a function is obtained from the interchange of AND and OR operators and 1's and 0's.

Example 2-3   Find the complement of the functions F_1 and F_2 of Example 2-2 by taking their duals and complementing each literal.

1. $F_1 = x'yz' + x'y'z$.
    The dual of F_1 is $(x' + y + z')(x' + y' + z)$.
    Complement each literal: $(x + y' + z)(x + y + z') = F_1'$.
2. $F_2 = x(y'z' + yz)$.
    The dual of F_2 is $x + (y' + z')(y + z)$.
    Complement each literal: $x' + (y + z)(y' + z') = F_2'$. ∎

2-5 CANONICAL AND STANDARD FORMS

Minterms and Maxterms

A binary variable may appear either in its normal form (x) or in its complement form (x'). Now consider two binary variabes x and y combined with an AND operation. Since each variable may appear in either form, there are four possible combinations: $x'y'$, $x'y$, xy', and xy. Each of these four AND terms represents one of the distinct areas in the Venn diagram of Fig. 2-1 and is called a *minterm,* or a *standard product*. In a similar manner, n variables can be combined to form 2^n minterms. The 2^n different minterms may be determined by a method similar to the one shown in Table 2-3 for three variables. The binary numbers from 0 to $2^n - 1$ are listed under the n variables. Each minterm is obtained from an AND term of the n variables, with each variable being primed if the corresponding bit of the binary number is a 0 and unprimed if a 1. A symbol for each minterm is also shown in the table and is of the form m_j, where j denotes the decimal equivalent of the binary number of the minterm designated.

In a similar fashion, n variables forming an OR term, with each variable being primed or unprimed, provide 2^n possible combinations, called *maxterms,* or *standard sums*. The eight maxterms for three variables, together with their symbolic designation, are listed in Table 2-3. Any 2^n maxterms for n variables may be determined similarly. Each maxterm is obtained from an OR term of the n variables, with each variable being unprimed if the corresponding bit is a 0 and primed if a 1. Note that each maxterm is the complement of its corresponding minterm, and vice versa.

TABLE 2-3
Minterms and Maxterms for Three Binary Variables

x	y	z	Minterms		Maxterms	
			Term	Designation	Term	Designation
0	0	0	$x'y'z'$	m_0	$x + y + z$	M_0
0	0	1	$x'y'z$	m_1	$x + y + z'$	M_1
0	1	0	$x'yz'$	m_2	$x + y' + z$	M_2
0	1	1	$x'yz$	m_3	$x + y' + z'$	M_3
1	0	0	$xy'z'$	m_4	$x' + y + z$	M_4
1	0	1	$xy'z$	m_5	$x' + y + z'$	M_5
1	1	0	xyz'	m_6	$x' + y' + z$	M_6
1	1	1	xyz	m_7	$x' + y' + z'$	M_7

A Boolean function may be expressed algebraically from a given truth table by forming a minterm for each combination of the variables that produces a 1 in the function, and then taking the OR of all those terms. For example, the function f_1 in Table 2-4 is determined by expressing the combinations 001, 100, and 111 as $x'y'z$, $xy'z'$, and xyz, respectively. Since each one of these minterms results in $f_1 = 1$, we should have

$$f_1 = x'y'z + xy'z' + xyz = m_1 + m_4 + m_7$$

Similarly, it may be easily verified that

$$f_2 = x'yz + xy'z + xyz' + xyz = m_3 + m_5 + m_6 + m_7$$

These examples demonstrate an important property of Boolean algebra: Any Boolean function can be expressed as a sum of minterms (by "sum" is meant the ORing of terms).

TABLE 2-4
Functions of Three Variables

x	y	z	Function f_1	Function f_2
0	0	0	0	0
0	0	1	1	0
0	1	0	0	0
0	1	1	0	1
1	0	0	1	0
1	0	1	0	1
1	1	0	0	1
1	1	1	1	1

Now consider the complement of a Boolean function. It may be read from the truth table by forming a minterm for each combination that produces a 0 in the function and then ORing those terms. The complement of f_1 is read as

$$f_1' = x'y'z' + x'yz' + x'yz + xy'z + xyz'$$

If we take the complement of f_1', we obtain the function f_1:

$$f_1 = (x + y + z)(x + y' + z)(x + y' + z')(x' + y + z')(x' + y' + z)$$

$$= M_0 \cdot M_2 \cdot M_3 \cdot M_5 \cdot M_6$$

Similarly, it is possible to read the expression for f_2 from the table:

$$f_2 = (x + y + z)(x + y + z')(x + y' + z)(x' + y + z)$$

$$= M_0 M_1 M_2 M_4$$

These examples demonstrate a second important property of Boolean algebra: Any Boolean function can be expressed as a product of maxterms (by "product" is meant the ANDing of terms). The procedure for obtaining the product of maxterms directly from the truth table is as follows. Form a maxterm for each combination of the variables that produces a 0 in the function, and then form the AND of all those maxterms. Boolean functions expressed as a sum of minterms or product of maxterms are said to be in *canonical form*.

Sum of Minterms

It was previously stated that for n binary variables, one can obtain 2^n distinct minterms, and that any Boolean function can be expressed as a sum of minterms. The minterms whose sum defines the Boolean function are those that give the 1's of the function in a truth table. Since the function can be either 1 or 0 for each minterm, and since there are 2^n minterms, one can calculate the possible functions that can be formed with n variables to be 2^{2^n}. It is sometimes convenient to express the Boolean function in its sum of minterms form. If not in this form, it can be made so by first expanding the expression into a sum of AND terms. Each term is then inspected to see if it contains all the variables. If it misses one or more variables, it is ANDed with an expression such as $x + x'$, where x is one of the missing variables. The following examples clarifies this procedure.

Example 2-4   Express the Boolean function $F = A + B'C$ in a sum of minterms. The function has three variables, A, B, and C. The first term A is missing two variables; therefore:

$$A = A(B + B') = AB + AB'$$

This is still missing one variable:

$$A = AB(C + C') + AB'(C + C')$$
$$= ABC + ABC' + AB'C + AB'C'$$

The second term $B'C$ is missing one variable:

$$B'C = B'C(A + A') = AB'C + A'B'C$$

Combining all terms, we have

$$F = A + B'C$$
$$= ABC + ABC' + AB'C + AB'C' + AB'C + A'B'C$$

But $AB'C$ appears twice, and according to theorem 1 ($x + x = x$), it is possible to re-move one of them. Rearranging the minterms in ascending order, we finally obtain

$$F = A'B'C + AB'C' + AB'C + ABC' + ABC$$
$$= m_1 + m_4 + m_5 + m_6 + m_7 \qquad \blacksquare$$

It is sometimes convenient to express the Boolean function, when in its sum of minterms, in the following short notation:

$$F(A, B, C) = \Sigma(1, 4, 5, 6, 7)$$

The summation symbol Σ stands for the ORing of terms; the numbers following it are the minterms of the function. The letters in parentheses following F form a list of the variables in the order taken when the minterm is converted to an AND term.

An alternate procedure for deriving the minterms of a Boolean function is to obtain the truth table of the function directly from the algebraic expression and then read the minterms from the truth table. Consider the Boolean function given in Example 2-4:

$$F = A + B'C$$

The truth table shown in Table 2-5 can be derived directly from the algebraic expression by listing the eight binary combinations under variables A, B, and C and inserting

TABLE 2-5
Truth Table for $F = A + B'C$

A	B	C	F
0	0	0	0
0	0	1	1
0	1	0	0
0	1	1	0
1	0	0	1
1	0	1	1
1	1	0	1
1	1	1	1

1's under F for those combinations where $A = 1$, and $BC = 01$. From the truth table, we can then read the five minterms of the function to be 1, 4, 5, 6, and 7.

Product of Maxterms

Each of the 2^{2^n} functions of n binary variables can be also expressed as a product of maxterms. To express the Boolean function as a product of maxterms, it must first be brought into a form of OR terms. This may be done by using the distributive law, $x + yz = (x + y)(x + z)$. Then any missing variable x in each OR term is ORed with xx'. This procedure is clarified by the following example.

**Example
2-5**

Express the Boolean function $F = xy + x'z$ in a product of maxterm form. First, convert the function into OR terms using the distributive law:

$$F = xy + x'z = (xy + x')(xy + z)$$
$$= (x + x')(y + x')(x + z)(y + z)$$
$$= (x' + y)(x + z)(y + z)$$

The function has three variables: x, y, and z. Each OR term is missing one variable; therefore:

$$x' + y = x' + y + zz' = (x' + y + z)(x' + y + z')$$
$$x + z = x + z + yy' = (x + y + z)(x + y' + z)$$
$$y + z = y + z + xx' = (x + y + z)(x' + y + z)$$

Combining all the terms and removing those that appear more than once, we finally obtain:

$$F = (x + y + z)(x + y' + z)(x' + y + z)(x' + y + z')$$
$$= M_0 M_2 M_4 M_5$$ ∎

A convenient way to express this function is as follows:

$$F(x, y, z) = \Pi(0, 2, 4, 5)$$

The product symbol, Π, denotes the ANDing of maxterms; the numbers are the maxterms of the function.

Conversion between Canonical Forms

The complement of a function expressed as the sum of minterms equals the sum of minterms missing from the original function. This is because the original function is expressed by those minterms that make the function equal to 1, whereas its complement is a 1 for those minterms that the function is a 0. As an example, consider the function

$$F(A, B, C) = \Sigma(1, 4, 5, 6, 7)$$

This has a complement that can be expressed as

$$F'(A, B, C) = \Sigma(0, 2, 3) = m_0 + m_2 + m_3$$

Now, if we take the complement of F' by DeMorgan's theorem, we obtain F in a different form:

$$F = (m_0 + m_2 + m_3)' = m_0' \cdot m_2' \cdot m_3' = M_0 M_2 M_3 = \Pi(0, 2, 3)$$

The last conversion follows from the definition of minterms and maxterms as shown in Table 2-3. From the table, it is clear that the following relation holds true:

$$m_j' = M_j$$

That is, the maxterm with subscript j is a complement of the minterm with the same subscript j, and vice versa.

The last example demonstrates the conversion between a function expressed in sum of minterms and its equivalent in product of maxterms. A similar argument will show that the conversion between the product of maxterms and the sum of minterms is similar. We now state a general conversion procedure. To convert from one canonical form to another, interchange the symbols Σ and Π and list those numbers missing from the original form. In order to find the missing terms, one must realize that the total number of minterms or maxterms is 2^n, where n is the number of binary variables in the function.

A Boolean function can be converted from an algebraic expression to a product of maxterms by using a truth table and the canonical conversion procedure. Consider, for example, the Boolean expression

$$F = xy + x'z$$

First, we derive the truth table of the function, as shown in Table 2-6. The 1's under F in the table are determined from the combination of the variable where $xy = 11$ and

TABLE 2-6
Truth Table for $F = xy + x'z$

x	y	z	F
0	0	0	0
0	0	1	1
0	1	0	0
0	1	1	1
1	0	0	0
1	0	1	0
1	1	0	1
1	1	1	1

$xz = 01$. The minterms of the function are read from the truth table to be 1, 3, 6, and 7. The function expressed in sum of minterms is

$$F(x, y, z) = \Sigma(1, 3, 6, 7)$$

Since there are a total of eight minterms or maxterms in a function of three variable, we determine the missing terms to be 0, 2, 4, and 5. The function expressed in product of maxterm is

$$F(x, y, z) = \Pi(0, 2, 4, 5)$$

This is the same answer obtained in Example 2-5.

Standard Forms

The two canonical forms of Boolean algebra are basic forms that one obtains from reading a function from the truth table. These forms are very seldom the ones with the least number of literals, because each minterm or maxterm must contain, by definition, *all* the variables either complemented or uncomplemented.

Another way to express Boolean functions is in *standard* form. In this configuration, the terms that form the function may contain one, two, or any number of literals. There are two types of standard forms: the sum of products and product of sums.

The *sum of products* is a Boolean expression containing AND terms, called *product terms,* of one or more literals each. The *sum* denotes the ORing of these terms. An example of a function expressed in sum of products is

$$F_1 = y' + xy + x'yz'$$

The expression has three product terms of one, two, and three literals each, respectively. Their sum is in effect an OR operation.

A *product of sums* is a Boolean expression containing OR terms, called *sum terms.* Each term may have any number of literals. The *product* denotes the ANDing of these terms. An example of a function expressed in product of sums is

$$F_2 = x(y' + z)(x' + y + z' + w)$$

This expression has three sum terms of one, two, and four literals each, respectively. The product is an AND operation. The use of the words *product* and *sum* stems from the smiilarity of the AND operation to the arithmetic product (multiplication) and the similarity of the OR operation to the arithmetic sum (addition).

A Boolean function may be expressed in a nonstandard form. For example, the function

$$F_3 = (AB + CD)(A'B' + C'D')$$

is neither in sum of products nor in product of sums. It can be changed to a standard form by using the distribuive law to remove the parentheses:

$$F_3 = A'B'CD + ABC'D'$$

2-6 OTHER LOGIC OPERATIONS

When the binary operators AND and OR are placed between two variables, x and y, they form two Boolean functions, $x \cdot y$ and $x + y$, respectively. It was stated previously that there are 2^{2^n} functions for n binary variables. For two variables, $n = 2$, and the number of possible Boolean functions is 16. Therefore, the AND and OR functions are only two of a total of 16 possible functions formed with two binary variables. It would be instructive to find the other 14 functions and investigate their properties.

The truth tables for the 16 functions formed with two binary variables, x and y, are listed in Table 2-7. In this table, each of the 16 columns, F_0 to F_{15}, represents a truth table of one possible function for the two given variables, x and y. Note that the functions are determined from the 16 binary combinations that can be assigned to F. Some of the functions are shown with an operator symbol. For example, F_1 represents the truth table for AND and F_7 represents the truth table for OR. The operator symbols for these functions are $\cdot$ and $+$, respectively.

TABLE 2-7
Truth Tables for the 16 Functions of Two Binary Variables

x	y	F_0	F_1	F_2	F_3	F_4	F_5	F_6	F_7	F_8	F_9	F_{10}	F_{11}	F_{12}	F_{13}	F_{14}	F_{15}
0	0	0	0	0	0	0	0	0	0	1	1	1	1	1	1	1	1
0	1	0	0	0	0	1	1	1	1	0	0	0	0	1	1	1	1
1	0	0	0	1	1	0	0	1	1	0	0	1	1	0	0	1	1
1	1	0	1	0	1	0	1	0	1	0	1	0	1	0	1	0	1
Operator symbol			$\cdot$	$/$		$/$		$\oplus$	$+$	$\downarrow$	$\odot$	$'$	$\subset$	$'$	$\supset$	$\uparrow$	

The 16 functions listed in truth table form can be expressed algebraically by means of Boolean expressions. This is shown in the first column of Table 2-8. The Boolean expressions listed are simplified to their minimum number of literals.

Although each function can be expressed in terms of the Boolean operators AND, OR, and NOT, there is no reason one cannot assign special operator symbols for expressing the other functions. Such operator symbols are listed in the second column of Table 2-8. However, all the new symbols shown, except for the exclusive-OR symbol, $\oplus$, are not in common use by digital designers.

Each of the functions in Table 2-8 is listed with an accompanying name and a comment that explains the function in some way. The 16 functions listed can be subdivided into three categories:

1. Two functions that produce a constant 0 or 1.
2. Four functions with unary operations: complement and transfer.
3. Ten functions with binary operators that define eight different operations: AND, OR, NAND, NOR, exclusive-OR, equivalence, inhibition, and implication.

TABLE 2-8
Boolean Expressions for the 16 Functions of Two Variables

Boolean functions	Operator symbol	Name	Comments
$F_0 = 0$		Null	Binary constant 0
$F_1 = xy$	$x \cdot y$	AND	x and y
$F_2 = xy'$	x/y	Inhibition	x but not y
$F_3 = x$		Transfer	x
$F_4 = x'y$	y/x	Inhibition	y but not x
$F_5 = y$		Transfer	y
$F_6 = xy' + x'y$	$x \oplus y$	Exclusive-OR	x or y but not both
$F_7 = x + y$	$x + y$	OR	x or y
$F_8 = (x + y)'$	$x \downarrow y$	NOR	Not-OR
$F_9 = xy + x'y'$	$x \odot y$	Equivalence	x equals y
$F_{10} = y'$	y'	Complement	Not y
$F_{11} = x + y'$	$x \subset y$	Implication	If y then x
$F_{12} = x'$	x'	Complement	Not x
$F_{13} = x' + y$	$x \supset y$	Implication	If x then y
$F_{14} = (xy)'$	$x \uparrow y$	NAND	Not-AND
$F_{15} = 1$		Identity	Binary constant 1

Any function can be equal to a constant, but a binary function can be equal to only 1 or 0. The complement function produces the complement of each of the binary variables. A function that is equal to an input variable has been given the name *transfer*, because the variable x or y is transferred through the gate that forms the function without changing its value. Of the eight binary operators, two (inhibition and implication) are used by logicians but are seldom used in computer logic. The AND and OR operators have been mentioned in conjunction with Boolean algebra. The other four functions are extensively used in the design of digital systems.

The NOR function is the complement of the OR function and its name is an abbreviation of *not-OR*. Similarly, NAND is the complement of AND and is an abbreviation of *not-AND*. The exclusive-OR, abbreviated XOR or EOR, is similar to OR but excludes the combination of *both* x and y being equal to 1. The equivalence is a function that is 1 when the two binary variables are equal, i.e., when both are 0 or both are 1. The exclusive-OR and equivalence functions are the complements of each other. This can be easily verified by inspecting Table 2-7. The truth table for the exclusive-OR is F_6 and for the equivalence is F_9, and these two functions are the complements of each other. For this reason, the equivalence function is often called exclusive-NOR, i.e., exclusive-OR-NOT.

Boolean algebra, as defined in Section 2-2, has two binary operators, which we have called AND and OR, and a unary operator, NOT (complement). From the definitions, we have deduced a number of properties of these operators and now have defined other binary operators in terms of them. There is nothing unique about this procedure. We could have just as well started with the operator NOR ($\downarrow$), for example, and later defined AND, OR, and NOT in terms of it. There are, nevertheless, good reasons for introducing Boolean algebra in the way it has been introduced. The concepts of "and," "or," and "not" are familiar and are used by people to express everyday logical ideas. Moreover, the Huntington postulates reflect the dual nature of the algebra, emphasizing the symmetry of $+$ and $\cdot$ with respect to each other.

2-7 DIGITAL LOGIC GATES

Since Boolean functions are expressed in terms of AND, OR, and NOT operations, it is easier to implement a Boolean function with these types of gates. The possibility of constructing gates for the other logic operations is of practial interest. Factors to be weighed when considering the construction of other types of logic gates are (1) the feasibility and economy of producing the gate with physical components, (2) the possibility of extending the gate to more than two inputs, (3) the basic properties of the binary operator such as commutativity and associativity, and (4) the ability of the gate to implement Boolean functions alone or in conjunction with other gates.

Of the 16 functions defined in Table 2-8, two are equal to a constant and four others are repeated twice. There are only ten functions left to be considered as candidates for logic gates. Two, inhibition and implication, are not commutative or associative and thus are impractical to use as standard logic gates. The other eight: complement, transfer, AND, OR, NAND, NOR, exclusive-OR, and equivalence, are used as standard gates in digital design.

The graphic symbols and truth tables of the eight gates are shown in Fig. 2-5. Each gate has one or two binary input variables designated by x and y and one binary output variable designated by F. The AND, OR, and inverter circuits were defined in Fig. 1-6. The inverter circuit inverts the logic sense of a binary variable. It produces the NOT, or complement, function. The small circle in the output of the graphic symbol of an inverter designates the logic complement. The triangle symbol by itself designates a buffer circuit. A buffer produces the *transfer* function but does not produce any particular logic operation, since the binary value of the output is equal to the binary value of the input. This circuit is used merely for power amplification of the signal and is equivalent to two inverters connected in cascade.

The NAND function is the complement of the AND function, as indicated by a graphic symbol that consists of an AND graphic symbol followed by a small circle. The NOR function is the complement of the OR function and uses an OR graphic symbol followed by a small circle. The NAND and NOR gates are extensively used as standard logic gates and are in fact far more popular than the AND and OR gates. This is because NAND and NOR gates are easily constructed with transistor circuits and because Boolean functions can be easily implemented with them.

Name	Graphic symbol	Algebraic function	Truth table

AND	x, y → F	$F = xy$	x y | F 0 0 | 0 0 1 | 0 1 0 | 0 1 1 | 1
OR	x, y → F	$F = x + y$	x y | F 0 0 | 0 0 1 | 1 1 0 | 1 1 1 | 1
Inverter	x → F	$F = x'$	x | F 0 | 1 1 | 0
Buffer	x → F	$F = x$	x | F 0 | 0 1 | 1
NAND	x, y → F	$F = (xy)'$	x y | F 0 0 | 1 0 1 | 1 1 0 | 1 1 1 | 0
NOR	x, y → F	$F = (x + y)'$	x y | F 0 0 | 1 0 1 | 0 1 0 | 0 1 1 | 0
Exclusive-OR (XOR)	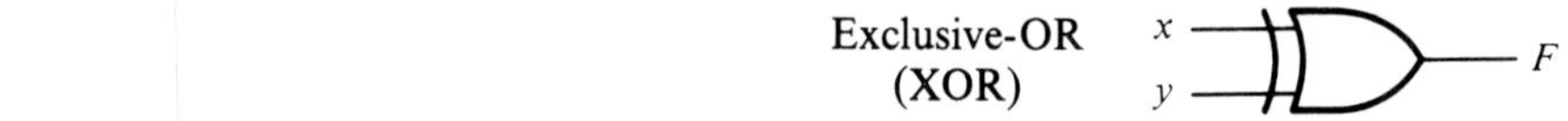 x, y → F	$F = xy' + x'y$ $= x \oplus y$	x y | F 0 0 | 0 0 1 | 1 1 0 | 1 1 1 | 0
Exclusive-NOR or equivalence	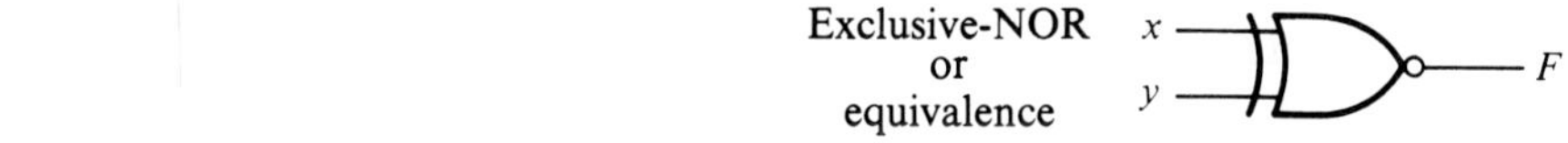 x, y → F	$F = xy + x'y'$ $= x \odot y$	x y | F 0 0 | 1 0 1 | 0 1 0 | 0 1 1 | 1

FIGURE 2-5
Digital logic gates

Exter

x	y	z
L	L	L
L	H	L
H	L	L
H	H	H

(a) Truth table
    with H and L

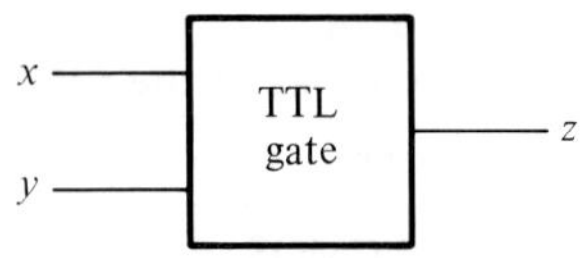

(b) Gate block diagram

x	y	z
0	0	0
0	1	0
1	0	0
1	1	1

(c) Truth table for
    positive logic

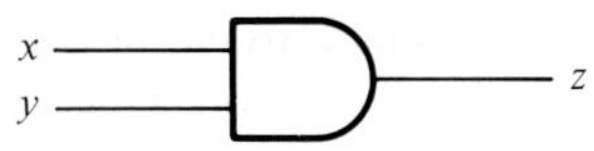

(d) Positive logic AND gate

x	y	z
1	1	1
1	0	1
0	1	1
0	0	0

(e) Truth table for
    negative logic

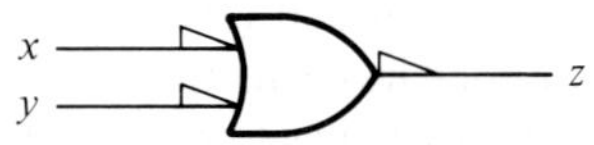

(f) Negative logic OR gate

FIGURE 2-11

Demonstration of positive and negative logic

designate a *polarity indicator*. The presence of this polarity indicator along a terminal signifies that negative logic is assumed for the signal. Thus, the same physical gate can operate either as a positive logic AND gate or as a negative logic OR gate.

The conversion from positive logic to negative logic, and vice versa, is essentially an operation that changes 1's to 0's and 0's to 1's in both the inputs and the output of a gate. Since this operation produces the dual of a function, the change of all terminals from one polarity to the other results in taking the dual of the function. The result of this conversion is that all AND operations are converted to OR operations (or graphic symbols) and vice versa. In addition, one must not forget to include the polarity-indicator triangle in the graphic symbols when negative logic is assumed. In this book, we will not use negative logic gates and assume that all gates operate with a positive logic assignment.

REFERENCES

1. BOOLE, G., *An Investigation of the Laws of Thought*. New York: Dover, 1954.
2. SHANNON, C. E., "A Symbolic Analysis of Relay and Switching Circuits." *Trans. AIEE,* **57** (1938), 713–723.
3. HUNTINGTON, E. V., "Sets of Independent Postulates for the Algebra of Logic." *Trans. Am. Math. Soc.,* **5** (1904). 288–309.
4. BIRKHOFF, G., and T. C. Bartee, *Modern Applied Algebra*. New York: McGraw-Hill, 1970.
5. HOHN, F. E., *Applied Boolean Algebra,* 2nd Ed. New York: Macmillan, 1966.
6. WHITESITT, J. E., *Boolean Algebra and Its Application*. Reading, MA: Addison-Wesley, 1961.
7. FRIEDMAN, A. D., and P. R. MENON, *Theory and Design of Switching Circuits*. Rockville, MD: Computer Science Press, 1975.
8. *The TTL Data Book*. Dallas: Texas Instruments, 1988.
9. TOCCI, R. J., *Digital Systems Principles and Applications,* 4th Ed. Englewood Cliffs, NJ: Prentice-Hall, 1988.

PROBLEMS

2-1 Demonstrate by means of truth tables the validity of the following identities:
(a) DeMorgan's theorem for three variables: $(xyz)' = x' + y' + z'$.
(b) The second distributive law: $x + yz = (x + y)(x + z)$.
(c) The consensus theorem: $xy + x'z + yz = xy + x'z$. (This is done algebraically in Example 2-1, part 4.)

2-2 Simplify the following Boolean expressions to a minimum number of literals.
(a) $x'y' + xy + x'y$
(b) $(x + y)(x + y')$
(c) $x'y + xy' + xy + x'y'$
(d) $x' + xy + xz' + xy'z'$
(e) $xy' + y'z' + x'z'$ [use the consensus theorem, Problem 2-1(c)].

2-3 Simplify the following Boolean expressions to a minimum number of literals:
(a) $ABC + A'B + ABC'$
(b) $x'yz + xz$
(c) $(x + y)'(x' + y')$
(d) $xy + x(wz + wz')$
(e) $(BC' + A'D)(AB' + CD')$

2-4 Reduce the following Boolean expressions to the indicated number of literals:
(a) $A'C' + ABC + AC'$ to three literals
(b) $(x'y' + z)' + z + xy + wz$ to three literals
(c) $A'B(D' + C'D) + B(A + A'CD)$ to one literal
(d) $(A' + C)(A' + C')(A + B + C'D)$ to four literals

2-5 Find the complement of $F = x + yz$; then show that $F \cdot F' = 0$ and $F + F' = 1$.

2-6   Find the complement of the following expressions:
(a) $xy' + x'y$
(b) $(AB' + C)D' + E$
(c) $AB(C'D + CD') + A'B'(C' + D)(C + D')$
(d) $(x + y' + z)(x' + z')(x + y)$

2-7   Using DeMorgan's theorem, convert the following Boolean expressions to equivalent expressions that have only OR and complement operations. Show that the functions can be implemented with logic diagrams that have only OR gates and inverters.
(a) $F = x'y' + x'z + y'z$
(b) $F = (y + z')(x + y)(y' + z)$

2-8   Using DeMorgan's theorem, convert the two Boolean expressions listed in Problem 2-7 to equivalent expressions that have only AND and complement operations. Show that the functions can be implemented with only AND gates and inverters.

2-9   Obtain the truth table of the following functions and express each function in sum of minterms and product of maxterms:
(a) $(xy + z)(y + xz)$
(b) $(A' + B)(B' + C)$
(c) $y'z + wxy' + wxz' + w'x'z$

2-10 For the Boolean function F given in the truth table, find the following:
(a) List the minterms of the function.
(b) List the minterms of F'.
(c) Express F in sum of minterms in algebraic form.
(d) Simplify the function to an expression with a minimum number of literals.

x	y	z	F
0	0	0	0
0	0	1	0
0	1	0	1
0	1	1	1
1	0	0	0
1	0	1	0
1	1	0	1
1	1	1	1

2-11 Given the following Boolean function:

$$F = xy'z + x'y'z + w'xy + wx'y + wxy$$

(a) Obtain the truth table of the function.
(b) Draw the logic diagram using the original Boolean expression.
(c) Simplify the function to a minimum number of literals using Boolean algebra.
(d) Obtain the truth table of the function from the simplified expression and show that it is the same as the one in part (a).

(e) Draw the logic diagram from the simplified expression and compare the total number of gates with the diagram of part (b).

2-12 Express the following functions in sum of minterms and product of maxterms:
(a) $F(A, B, C, D) = B'D + A'D + BD$
(b) $F(x, y, z) = (xy + z)(xz + y)$

2-13 Express the complement of the following functions in sum of minterms:
(a) $F(A, B, C, D) = \Sigma(0, 2, 6, 11, 13, 14)$
(b) $F(x, y, z) = \Pi(0, 3, 6, 7)$

2-14 Convert the following to the other canonical form:
(a) $F(x, y, z) = \Sigma(1, 3, 7)$
(b) $F(A, B, C, D) = \Pi(0, 1, 2, 3, 4, 6, 12)$

2-15 The sum of all the minterms of a Boolean function of n variables is equal to 1.
(a) Prove the above statement for $n = 3$.
(b) Suggest a procedure for a general proof.

2-16 Convert the following expressions into sum of products and product of sums:
(a) $(AB + C)(B + C'D)$
(b) $x' + x(x + y')(y + z')$

2-17 Draw the logic diagram corresponding to the following Boolean expressions without simplifying them:
(a) $BC' + AB + ACD$
(b) $(A + B)(C + D)(A' + B + D)$
(c) $(AB + A'B')(CD' + C'D)$

2-18 Show that the dual of the exclusive-OR is equal to its complement.

2-19 By substituting the Boolean expression equivalent of the binary operations as defined in Table 2-8, show the following:
(a) The inhibition operation is neither commutative nor associative.
(b) The exclusive-OR operation is commutative and associative.

2-20 Verify the truth table for the three-variable exclusive-OR function listed in Fig. 2-8(c). Do that by listing all eight combinations of x, y, and z; then evaluate $A = x \oplus y$; and then evaluate $F = A \oplus z = x \oplus y \oplus z$.

2-21 TTL SSI come mostly in 14-pin packages. Two pins are reserved for power and the other 12 pins are used for input and output terminals. Determine the number of gates that can be enclosed in one package if it contains the folowing type of gates:
(a) Two-input exclusive-OR gates
(b) Three-input AND gates
(c) Four-input NAND gates
(d) Five-input NOR gates
(e) Eight-input NAND gates

2-22 Show that a positive logic NAND gate is a negative logic NOR gate and vice versa.

2-23 An integrated-circuit logic family has NAND gates with fan-out of 5 and buffer gates with fan-out of 10. Show how the output signal of a single NAND gate can be applied to 50 other NAND-gate inputs without overloading the output gate. Use buffers to satisfy the fan-out requirements.

Simplification of Boolean Functions

3-1 THE MAP METHOD

The complexity of the digital logic gates that implement a Boolean function is directly related to the complexity of the algebraic expression from which the function is implemented. Although the truth table representation of a function is unique, expressed algebraically, it can appear in many different forms. Boolean functions may be simplified by algebraic means as discussed in Section 2-4. However, this procedure of minimization is awkward because it lacks specific rules to predict each succeeding step in the manipulative process. The map method provides a simple straightforward procedure for minimizing Boolean functions. This method may be regarded either as a pictorial form of a truth table or as an extension of the Venn diagram. The map method, first proposed by Veitch and modified by Karnaugh, is also known as the "Veitch diagram" or the "Karnaugh map."

The map is a diagram made up of squares. Each square represents one minterm. Since any Boolean function can be expressed as a sum of minterms, it follows that a Boolean function is recognized graphically in the map from the area enclosed by those squares whose minterms are included in the function. In fact, the map presents a visual diagram of all possible ways a function may be expressed in a standard form. By recognizing various patterns, the user can derive alternative algebraic expressions for the same function, from which he can select the simplest one. We shall assume that the simplest algebraic expression is any one in a sum of products or product of sums that has a minimum number of literals. (This expression is not necessarily unique.)

3-2   TWO- AND THREE-VARIABLE MAPS

A two-variable map is shown in Fig. 3-1(a). There are four minterms for two variables; hence, the map consists of four squares, one for each minterm. The map is redrawn in (b) to show the relationship between the squares and the two variables. The 0's and 1's marked for each row and each column designate the values of variables x and y, respectively. Notice that x appears primed in row 0 and unprimed in row 1. Similarly, y appears primed in column 0 and unprimed in column 1.

If we mark the squares whose minterms belong to a given function, the two-variable map becomes another useful way to represent any one of the 16 Boolean functions of two variables. As an example, the function xy is shown in Fig. 3-2(a). Since xy is equal to m_3, a 1 is placed inside the square that belongs to m_3. Similarly, the function $x + y$ is represented in the map of Fig. 3-2(b) by three squares marked with 1's. These squares are found from the minterms of the function:

$$x + y = x'y + xy' + xy = m_1 + m_2 + m_3$$

The three squares could have also been determined from the intersection of variable x in the second row and variable y in the second column, which encloses the area belonging to x or y.

A three-variable map is shown in Fig. 3-3. There are eight minterms for three binary variables. Therefore, a map consists of eight squares. Note that the minterms are not arranged in a binary sequence, but in a sequence similar to the Gray code listed in Table 1-4. The characteristic of this sequence is that only one bit changes from 1 to 0

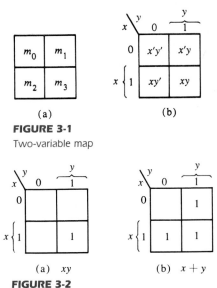

(a)

FIGURE 3-1

Two-variable map

(a) xy (b) $x + y$

FIGURE 3-2

Representation of functions in the map

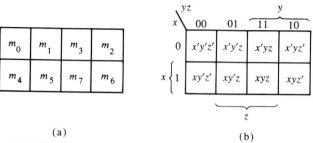

(a)   (b)

FIGURE 3-3

Three-variable map

or from 0 to 1 in the listing sequence. The map drawn in part (b) is marked with numbers in each row and each column to show the relationship between the squares and the three variables. For example, the square assigned to m_5 corresponds to row 1 and column 01. When these two numbers are concatenated, they give the binary number 101, whose decimal equivalent is 5. Another way of looking at square $m_5 = xy'z$ is to consider it to be in the row marked x and the column belonging to $y'z$ (column 01). Note that there are four squares where each variable is equal to 1 and four where each is equal to 0. The variable appears unprimed in those four squares where it is equal to 1 and primed in those squares where it is equal to 0. For convenience, we write the variable with its letter symbol under the four squares where it is unprimed.

To understand the usefulness of the map for simplifying Boolean functions, we must recognize the basic property possessed by adjacent squares. Any two adjacent squares in the map differ by only one variable, which is primed in one square and unprimed in the other. For example, m_5 and m_7 lie in two adjacent squares. Variable y is primed in m_5 and unprimed in m_7, whereas the other two variables are the same in both squares. From the postulates of Boolean algebra, it follows that the sum of two minterms in adjacent squares can be simplified to a single AND term consisting of only two literals. To clarify this, consider the sum of two adjacent squares such as m_5 and m_7:

$$m_5 + m_7 = xy'z + xyz = xz(y' + y) = xz$$

Here the two squares differ by the variable y, which can be removed when the sum of the two minterms is formed. Thus, any two minterms in adjacent squares that are ORed together will cause a removal of the different variable. The following example explains the procedure for minimizing a Boolean function with a map.

Example 3-1

Simplify the Boolean function

$$F(x, y, z) = \Sigma(2, 3, 4, 5)$$

First, a 1 is marked in each minterm that represents the function. This is shown in Fig. 3-4, where the squares for minterms 010, 011, 100, and 101 are marked with 1's. The next step is to find possible adjacent squares. These are indicated in the map by two rectangles, each enclosing two 1's. The upper right rectangle represents the area en-

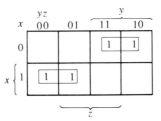

FIGURE 3-4
Map for Example 3-1; $F(x, y, z) =$
$\Sigma\,(2, 3, 4, 5) = x'y + xy'$

closed by $x'y$. This is determined by observing that the two-square area is in row 0, corresponding to x', and the last two columns, corresponding to y. Similarly, the lower left rectangle represents the product term xy'. (The second row represents x and the two left columns represent y'.) The logical sum of these two product terms gives the simplified expression

$$F = x'y + xy'$$ ■

There are cases where two squares in the map are considered to be adjacent even though they do not touch each other. In Fig. 3-3, m_0 is adjacent to m_2 and m_4 is adjacent to m_6 because the minterms differ by one variable. This can be readily verified algebraically.

$$m_0 + m_2 = x'y'z' + x'yz' = x'z'(y' + y) = x'z'$$

$$m_4 + m_6 = xy'z' + xyz' = xz' + (y' + y) = xz'$$

Consequently, we must modify the definition of adjacent squares to include this and other similar cases. This is done by considering the map as being drawn on a surface where the right and left edges touch each other to form adjacent squares.

Example 3-2

Simplify the Boolean function

$$F(x, y, z) = \Sigma\,(3, 4, 6, 7)$$

The map for this function is shown in Fig. 3-5. There are four squares marked with 1's, one for each minterm of the function. Two adjacent squares are combined in the third column to give a two-literal term yz. The remaining two squares with 1's are also adjacent by the new definition and are shown in the diagram with their values enclosed in half rectangles. These two squares when combined, give the two-literal term xz'. The simplified function becomes

$$F = yz + xz'$$ ■

Consider now any combination of four adjacent squares in the three-variable map. Any such combination represents the logical sum of four minterms and results in an ex-

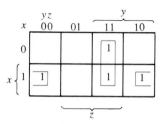

FIGURE 3-5
Map for Example 3-2; $F(x, y, z) =$
$\Sigma\,(3, 4, 6, 7) = yz + xz'$

pression of only one literal. As an example, the logical sum of the four adjacent minterms 0, 2, 4, and 6 reduces to a single literal term z'.

$$m_0 + m_2 + m_4 + m_6 = x'y'z' + x'yz' + xy'z' + xyz'$$
$$= x'z'(y' + y) + xz'(y' + y)$$
$$= x'z' + xz' = z'(x' + x) = z'$$

The number of adjacent squares that may be combined must always represent a number that is a power of two such as 1, 2, 4, and 8. As a larger number of adjacent squares are combined, we obtain a product term with fewer literals.

One square represents one minterm, giving a term of three literals.

Two adjacent squares represent a term of two literals.

Four adjacent squares represent a term of one literal.

Eight adjacent squares encompass the entire map and produce a function that is always equal to 1.

**Example
3-3**

Simplify the Boolean function

$$F(x, y, z) = \Sigma\,(0, 2, 4, 5, 6)$$

The map for F is shown in Fig. 3-6. First, we combine the four adjacent squares in the first and last columns to give the single literal term z'. The remaining single square representing minterm 5 is combined with an adjacent square that has already been used once. This is not only permissible, but rather desirable since the two adjacent squares give the two-literal term xy' and the single square represents the three-literal minterm $xy'z$. The simplified function is

$$F = z' + xy'$$ ∎

If a function is not expressed in sum of minterms, it is possible to use the map to obtain the minterms of the function and then simplify the function to an expression with a minimum number of terms. It is necessary to make sure that the algebraic ex-

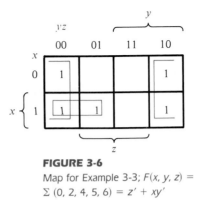

FIGURE 3-6
Map for Example 3-3; $F(x, y, z) =$
$\Sigma (0, 2, 4, 5, 6) = z' + xy'$

pression is in sum of products form. Each product term can be plotted in the map in one, two, or more squares. The minterms of the function are then read directly from the map.

Example 3-4

Given the following Boolean function:

$$F = A'C + A'B + AB'C + BC$$

(a) Express it in sum of minterms.
(b) Find the minimal sum of products expression.

Three product terms in the expression have two literals and are represented in a three-variable map by two squares each. The two squares corresponding to the first term $A'C$ are found in Fig. 3-7 from the coincidence of A' (first row) and C (two middle columns) to give squares 001 and 011. Note that when marking 1's in the squares, it is possible to find a 1 already placed there from a preceding term. This happens with the second term $A'B$, which has 1's in squares 011 and 010, but square 011 is common with the first term $A'C$, so only one 1 is marked in it. Continuing in this fashion, we determine that the term $AB'C$ belongs in square 101, corresponding to minterm 5, and the term BC has two 1's in squares 011 and 111. The function has a total of five minterms, as indicated by the five 1's in the map of Fig. 3-7. The minterms are read

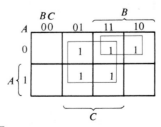

FIGURE 3-7
Map for Example 3-4; $A'C + A'B + AB'C + BC = C + A'B$

directly from the map to be 1, 2, 3, 5, and 7. The function can be expressed in sum of minterms form:

$$F(A, B, C) = \Sigma(1, 2, 3, 5, 7)$$

The sum of products expression as originally given has too many terms. It can be simplified, as shown in the map, to an expression with only two terms:

$$F = C + A'B$$ ■

3-3 FOUR-VARIABLE MAP

The map for Boolean functions of four binary variables is shown in Fig. 3-8. In (a) are listed the 16 minterms and the squares assigned to each. In (b) the map is redrawn to show the relationship with the four variables. The rows and columns are numbered in a reflected-code sequence, with only one digit changing value between two adjacent rows or columns. The minterm corresponding to each square can be obtained from the concatenation of the row number with the column number. For example, the numbers of the third row (11) and the second column (01), when concatenated, give the binary number 1101, the binary equivalent of decimal 13. Thus, the square in the third row and second column represents minterm m_{13}.

The map minimization of four-variable Boolean functions is similar to the method used to minimize three-variable functions. Adjacent squares are defined to be squares next to each other. In addition, the map is considered to lie on a surface with the top and bottom edges, as well as the right and left edges, touching each other to form adjacent squares. For example, m_0 and m_2 form adjacent squares, as do m_3 and m_{11}. The combination of adjacent squares that is useful during the simplification process is easily determined from inspection of the four-variable map:

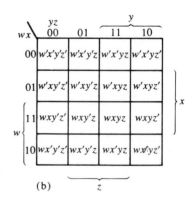

m_0	m_1	m_3	m_2
m_4	m_5	m_7	m_6
m_{12}	m_{13}	m_{15}	m_{14}
m_8	m_9	m_{11}	m_{10}

(a)

(b)

FIGURE 3-8
Four-variable map

One square represents one minterm, giving a term of four literals.

Two adjacent squares represent a term of three literals.

Four adjacent squares represent a term of two literals.

Eight adjacent squares represent a term of one literal.

Sixteen adjacent squares represent the function equal to 1.

No other combination of squares can simplify the function. The following two examples show the procedure used to simplify four-variable Boolean functions.

Example 3-5

Simplify the Boolean function

$$F(w, x, y, z) = \Sigma(0, 1, 2, 4, 5, 6, 8, 9, 12, 13, 14)$$

Since the function has four variables, a four-variable map must be used. The minterms listed in the sum are marked by 1's in the map of Fig. 3-9. Eight adjacent squares marked with 1's can be combined to form the one literal term y'. The remaining three 1's on the right cannot be combined to give a simplified term. They must be combined as two or four adjacent squares. The larger the number of squares combined, the smaller the number of literals in the term. In this example, the top two 1's on the right are combined with the top two 1's on the left to give the term $w'z'$. Note that it is permissible to use the same square more than once. We are now left with a square marked by 1 in the third row and fourth column (square 1110). Instead of taking this square alone (which will give a term of four literals), we combine it with squares already used to form an area of four adjacent squares. These squares comprise the two middle rows and the two end columns, giving the term xz'. The simplified function is

$$F = y' + w'z' + xz'$$

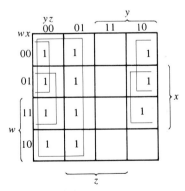

FIGURE 3-9

Map for Example 3-5; $F(w, x, y, z) =$
$\Sigma(0, 1, 2, 4, 5, 6, 8, 9, 12, 13, 14) =$
$y' + w'z' + xz'$

Example 3-6

Simplify the Boolean function

$$F = A'B'C' + B'CD' + A'BCD' + AB'C'$$

The area in the map covered by this function consists of the squares marked with 1's in Fig. 3-10. This function has four variables and, as expressed, consists of three terms, each with three literals, and one term of four literals. Each term of three literals is represented in the map by two squares. For example, $A'B'C'$ is represented in squares 0000 and 0001. The function can be simplified in the map by taking the 1's in the four corners to give the term $B'D'$. This is possible because these four squares are adjacent when the map is drawn in a surface with top and bottom or left and right edges touching one another. The two left-hand 1's in the top row are combined with the two 1's in the bottom row to give the term $B'C'$. The remaining 1 may be combined in a two-square area to give the term $A'CD'$. The simplified function is

$$F = B'D' + B'C' + A'CD'$$ ∎

Prime Implicants

When choosing adjacent squares in a map, we must ensure that all the minterms of the function are covered when combining the squares. At the same time, it is necessary to minimize the number of terms in the expression and avoid any redundant terms whose minterms are already covered by other terms. Sometimes there may be two or more expressions that satisfy the simplification criteria. The procedure for combining squares in the map may be made more systematic if we understand the meaning of the terms referred to as prime implicant and essential prime implicant. A *prime implicant* is a product term obtained by combining the maximum possible number of adjacent squares in the map. If a minterm in a square is covered by only one prime implicant, that prime implicant is said to be *essential*. A more satisfactory definition of prime implicant is given in Section 3-10. Here we will use it to help us find all possible simplified expressions of a Boolean function by means of a map.

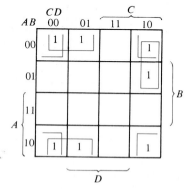

FIGURE 3-10

Map for Example 3-6; $A'B'C' + B'CD' + A'BCD' + AB'C' = B'D' + B'C' + A'CD'$

The prime implicants of a function can be obtained from the map by combining all possible maximum numbers of squares. This means that a single 1 on a map represents a prime implicant if it is not adjacent to any other 1's. Two adjacent 1's form a prime implicant provided they are not within a group of four adjacent squares. Four adjacent 1's form a prime implicant if they are not within a group of eight adjacent squares, and so on. The essential prime implicants are found by looking at each square marked with a 1 and checking the number of prime implicants that cover it. The prime implicant is essential if it is the only prime implicant that covers the minterm.

Consider the following four-variable Boolean function:

$$F(A, B, C, D) = \Sigma(0, 2, 3, 5, 7, 8, 9, 10, 11, 13, 15)$$

The minterms of the function are marked with 1's in the maps of Fig. 3-11. Part (a) of the figure shows two essential prime implicants. One term is essential because there is only one way to include minterms m_0 within four adjacent squares. These four squares define the term $B'D'$. Similarly, there is only one way that minterm m_5 can be combined with four adjacent squares and this gives the second term BD. The two essential prime implicants cover eight minterms. The remaining three minterms, m_3, m_9, and m_{11}, must be considered next.

Figure 3-11(b) shows all possible ways that the three minterms can be covered with prime implicants. Minterm m_3 can be covered with either prime implicant CD or $B'C$. Minterm m_9 can be covered with either AD or AB'. Minterm m_{11} is covered with any one of the four prime implicants. The simplified expression is obtained from the logical sum of the two essential prime implicants and any two prime implicants that cover minterms m_3, m_9, and m_{11}. There are four possible ways that the function can be expressed with four product terms of two literals each:

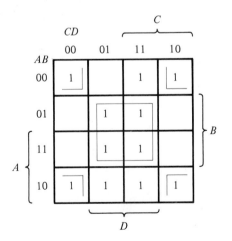

(a) Essential prime implicants
    BD and $B'D'$

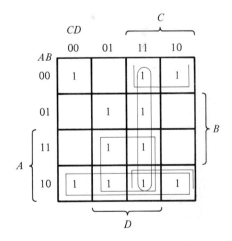

(b) Prime implicants CD, $B'C$,
    AD, and AB'

FIGURE 3-11

Simplification using prime implicants

$$F = BD + B'D' + CD + AD$$
$$= BD + B'D' + CD + AB'$$
$$= BD + B'D' + B'C + AD$$
$$= BD + B'D' + B'C + AB'$$

The above example has demonstrated that the identification of the prime implicants in the map helps in determining the alternatives that are available for obtaining a simplified expression.

The procedure for finding the simplified expression from the map requires that we first determine all the essential prime implicants. The simplified expression is obtained from the logical sum of all the essential prime implicants plus other prime implicants that may be needed to cover any remaining minterms not covered by the essential prime implicants. Occasionally, there may be more than one way of combining squares and each combination may produce an equally simplified expression.

3-4 FIVE-VARIABLE MAP

Maps for more than four variables are not as simple to use. A five-variable map needs 32 squares and a six-variable map needs 64 squares. When the number of variables becomes large, the number of squares becomes excessively large and the geometry for combining adjacent squares becomes more involved.

The five-variable map is shown in Fig. 3-12. It consists of 2 four-variable maps with variables A, B, C, D, and E. Variable A distinguishes between the two maps, as indicated on the top of the diagram. The left-hand four-variable map represents the 16

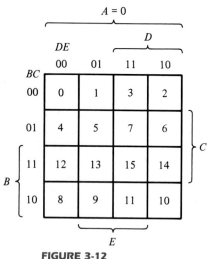
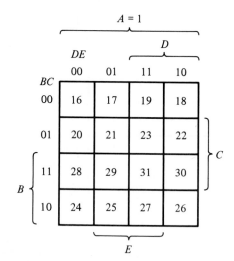

FIGURE 3-12

Five-variable map

squares where $A = 0$, and the other four-variable map represents the squares where $A = 1$. Minterms 0 through 15 belong with $A = 0$ and minterms 16 through 31 with $A = 1$. Each four-variable map retains the previously defined adjacency when taken separately. In addition, each square in the $A = 0$ map is adjacent to the corresponding square in the $A = 1$ map. For example, minterm 4 is adjacent to minterm 20 and minterm 15 to 31. The best way to visualize this new rule for adjacent squares is to consider the two half maps as being one on top of the other. Any two squares that fall one over the other are considered adjacent.

By following the procedure used for the five-variable map, it is possible to construct a six-variable map with 4 four-variable maps to obtain the required 64 squares. Maps with six or more variables need too many squares and are impractical to use. The alternative is to employ computer programs specifically written to facilitate the simplification of Boolean functions with a large number of variables.

From inspection, and taking into account the new definition of adjacent squares, it is possible to show that any 2^k adjacent squares, for $k = 0, 1, 2, \ldots, n$, in an n-variable map, will represent an area that gives a term of $n - k$ literals. For the above statement to have any meaning, n must be larger than k. When $n = k$, the entire area of the map is combined to give the identity function. Table 3-1 shows the relationship between the number of adjacent squares and the number of literals in the term. For example, eight adjacent squares combine an area in the five-variable map to give a term of two literals.

TABLE 3-1
The Relationship Between the Number of Adjacent Squares and the Number of Literals in the Term

k	Number of adjacent squares 2^k	Number of literals in a term in an n-variable map					
		$n = 2$	$n = 3$	$n = 4$	$n = 5$	$n = 6$	$n = 7$
0	1	2	3	4	5	6	7
1	2	1	2	3	4	5	6
2	4	0	1	2	3	4	5
3	8		0	1	2	3	4
4	16			0	1	2	3
5	32				0	1	2
6	64					0	1

Example 3-7   Simplify the Boolean function

$$F(A, B, C, D, E) = (0, 2, 4, 6, 9, 13, 21, 23, 25, 29, 31)$$

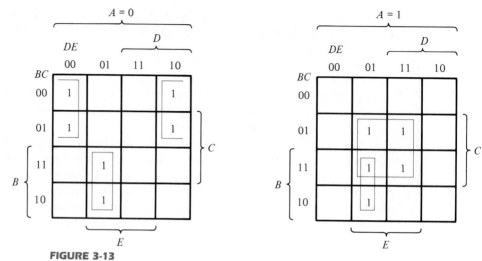

FIGURE 3-13
Map for Example 3-7; $F = A'B'E' + BD'E + ACE$

The five-variable map for this function is shown in Fig. 3-13. There are six minterms from 0 to 15 that belong to the part of the map with $A = 0$. The other five minterms belong with $A = 1$. Four adjacent squares in the $A = 0$ map are combined to give the three-literal term $A'B'E'$. Note that it is necessary to include A' with the term because all the squares are associated with $A = 0$. The two squares in column 01 and the last two rows are common to both parts of the map. Therefore, they constitute four adjacent squares and give the three-literal term $BD'E$. Variable A is not included here because the adjacent squares belong to both $A = 0$ and $A = 1$. The term ACE is obtained from the four adjacent squares that are entirely within the $A = 1$ map. The simplified function is the logical sum of the three terms:

$$F = A'B'E' + BD'E + ACE$$ ◼

3-5 PRODUCT OF SUMS SIMPLIFICATION

The minimized Boolean functions derived from the map in all previous examples were expressed in the sum of products form. With a minor modification, the product of sums form can be obtained.

The procedure for obtaining a minimized function in product of sums follows from the basic properties of Boolean functions. The 1's placed in the squares of the map represent the minterms of the function. The minterms not included in the function denote the complement of the function. From this we see that the complement of a function is represented in the map by the squares not marked by 1's. If we mark the empty squares by 0's and combine them into valid adjacent squares, we obtain a simplified expression of the complement of the function, i.e., of F'. The complement of F' gives us back the function F. Because of the generalized DeMorgan's theorem, the function so obtained is automatically in the product of sums form. The best way to show this is by example.

Example
3-8 Simplify the following Boolean function in (a) sum of products and (b) product of sums.

$$F(A, B, C, D) = \Sigma(0, 1, 2, 5, 8, 9, 10)$$

The 1's marked in the map of Fig. 3-14 represent all the minterms of the function. The squares marked with 0's represent the minterms not included in F and, therefore, denote the complement of F. Combining the squares with 1's gives the simplified function in sum of products:

(a) $$F = B'D' + B'C' + A'C'D$$

If the squares marked with 0's are combined, as shown in the diagram, we obtain the simplified complemented function:

$$F' = AB + CD + BD'$$

Applying DeMorgan's theorem (by taking the dual and complementing each literal as described in Section 2-4), we obtain the simplified function in product of sums:

(b) $$F = (A' + B')(C' + D')(B' + D) \qquad \blacksquare$$

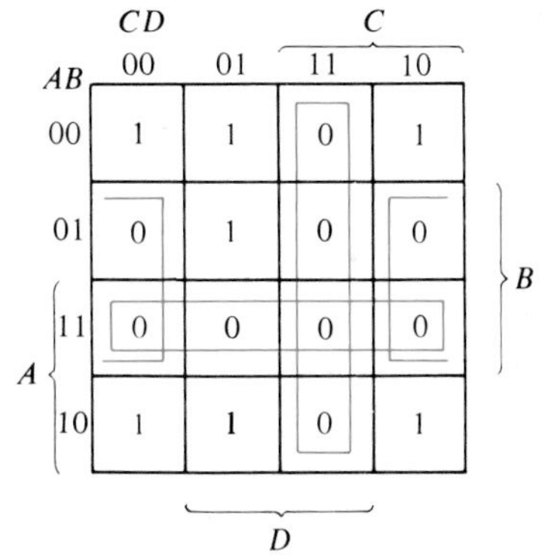

FIGURE 3-14
Map for Example 3-8; $F(A, B, C, D) = \Sigma(0, 1, 2, 5, 8, 9, 10) = B'D' + B'C' + A'C'D = (A' + B')(C' + D')(B' + D)$

The implementation of the simplified expressions obtained in Example 3-8 is shown in Fig. 3-15. The sum of products expression is implemented in (a) with a group of AND gates, one for each AND term. The outputs of the AND gates are connected to the inputs of a single OR gate. The same function is implemented in (b) in its product of sums form with a group of OR gates, one for each OR term. The outputs of the OR

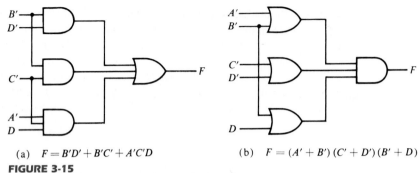

(a) $F = B'D' + B'C' + A'C'D$ (b) $F = (A' + B')(C' + D')(B' + D)$

FIGURE 3-15

Gate implementation of the function of Example 3-8

gates are connected to the inputs of a single AND gate. In each case, it is assumed that the input variables are directly available in their complement, so inverters are not needed. The configuration pattern established in Fig. 3-15 is the general form by which any Boolean function is implemented when expressed in one of the standard forms. AND gates are connected to a single OR gate when in sum of products; OR gates are connected to a single AND gate when in product of sums. Either configuration forms two levels of gates. Thus, the implementation of a function in a standard form is said to be a two-level implementation.

Example 3-8 showed the procedure for obtaining the product of sums simplification when the function is originally expressed in the sum of minterms canonical form. The procedure is also valid when the function is originally expressed in the product of maxterms canonical form. Consider, for example, the truth table that defines the function F in Table 3-2. In sum of minterms, this function is expressed as

$$F(x, y, z) = \Sigma(1, 3, 4, 6)$$

In product of maxterms, it is expressed as

$$F(x, y, z) = \Pi(0, 2, 5, 7)$$

TABLE 3-2
Truth Table of Function F

x	y	z	F
0	0	0	0
0	0	1	1
0	1	0	0
0	1	1	1
1	0	0	1
1	0	1	0
1	1	0	1
1	1	1	0

In other words, the 1's of the function represent the minterms, and the 0's represent the maxterms. The map for this function is shown in Fig. 3-16. One can start simplifying this function by first marking the 1's for each minterm that the function is a 1. The remaining squares are marked by 0's. If, on the other hand, the product of maxterms is initially given, one can start marking 0's in those squares listed in the function; the remaining squares are then marked by 1's. Once the 1's and 0's are marked, the function can be simplified in either one of the standard forms. For the sum of products, we combine the 1's to obtain

$$F = x'z + xz'$$

For the product of sums, we combine the 0's to obtain the simplified complemented function:

$$F' = xz + x'z'$$

which shows that the exclusive-OR function is the complement of the equivalence function (Section 2-6). Taking the complement of F', we obtain the simplified function in product of sums:

$$F = (x' + z')(x + z)$$

To enter a function expressed in product of sums in the map, take the complement of the function and from it find the squares to be marked by 0's. For example, the function

$$F = (A' + B' + C')(B + D)$$

can be entered in the map by first taking its complement:

$$F' = ABC + B'D'$$

and then marking 0's in the squares representing the minterms of F'. The remaining squares are marked with 1's.

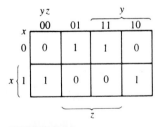

FIGURE 3-16
Map for the function of Table 3-2

3-6   NAND AND NOR IMPLEMENTATION

Digital circuits are more frequently constructed with NAND or NOR gates than with AND and OR gates. NAND and NOR gates are easier to fabricate with electronic components and are the basic gates used in all IC digital logic families. Because of the prominence of NAND and NOR gates in the design of digital circuits, rules and procedures have been developed for the conversion from Boolean functions given in terms of AND, OR, and NOT into equivalent NAND and NOR logic diagrams. The procedure for two-level implementation is presented in this section. Multilevel implementation is discussed in Section 4-7.

To facilitate the conversion to NAND and NOR logic, it is convenient to define two other graphic symbols for these gates. Two equivalent symbols for the NAND gate are shown in Fig. 3-17(a). The AND-invert symbol has been defined previously and consists of an AND graphic symbol followed by a small circle. Instead, it is possible to represent a NAND gate by an OR graphic symbol preceded by small circles in all the inputs. The invert-OR symbol for the NAND gate follows from DeMorgan's theorem and from the convention that small circles denote complementation.

Similarly, there are two graphic symbols for the NOR gate, as shown in Fig. 3-17(b). The OR-invert is the conventional symbol. The invert-AND is a convenient alternative that utilizes DeMorgan's theorem and the convention that small circles in the inputs denote complementation.

A one-input NAND or NOR gate behaves like an inverter. As a consequence, an inverter gate can be drawn in three different ways, as shown in Fig. 3-17(c). The small circles in all inverter symbols can be transferred to the input terminal without changing the logic of the gate.

$$x \quad y \quad z \qquad F = (xyz)' \qquad\qquad x \quad y \quad z \qquad F = x' + y' + z' = (xyz)'$$

AND-invert   Invert-OR

(a) Two graphic symbols for NAND gate.

$$x \quad y \quad z \qquad F = (x + y + z)' \qquad\qquad x \quad y \quad z \qquad F = x'y'z' = (x + y + z)'$$

OR-invert   Invert-AND

(b) Two graphic symbols for NOR gate.

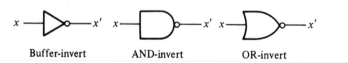

$$x \qquad x' \qquad x \qquad x' \qquad x \qquad x'$$

Buffer-invert   AND-invert   OR-invert

(c) Three graphic symbols for inverter.

FIGURE 3-17
Graphic symbols for NAND and NOR gates

It should be pointed out that the alternate symbols for the NAND and NOR gates could be drawn with small triangles in all input terminals instead of the circles. A small triangle is a negative-logic polarity indicator (see Section 2-8 and Fig. 2-11). With small triangles in the input terminals, the graphic symbol denotes a negative-logic polarity for the inputs, but the output of the gate (not having a triangle) would have a positive-logic assignment. In this book, we prefer to stay with positive logic throughout and employ small circles when necessary to denote complementation.

NAND Implementation

The implementation of a Boolean function with NAND gates requires that the function be simplified in the sum of products form. To see the relationship between a sum of products expression and its equivalent NAND implementation, consider the logic diagrams of Fig. 3-18. All three diagrams are equivalent and implement the function:

$$F = AB + CD + E$$

The function is implemented in Fig. 3-18(a) in sum of products form with AND and OR gates. In (b) the AND gates are replaced by NAND gates and the OR gate is replaced by a NAND gate with an invert-OR symbol. The single variable E is complemented and applied to the second-level invert-OR gate. Remember that a small circle denotes complementation. Therefore, two circles on the same line represent double complementation and both can be removed. The complement of E goes through a small

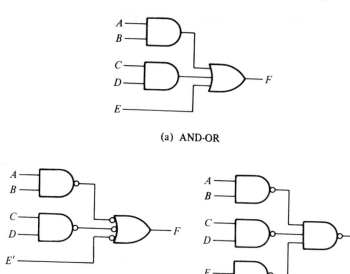

(a) AND-OR

(b) NAND-NAND   (c) NAND-NAND

FIGURE 3-18

Three ways to implement $F = AB + CD + E$

circle that complements the variable again to produce the normal value of E. Removing the small circles in the gates of Fig. 3-18(b) produces the circuit in (a). Therefore, the two diagrams implement the same function and are equivalent.

In Fig. 3-18(c), the output NAND gate is redrawn with the conventional symbol. The one-input NAND gate complements variable E. It is possible to remove this inverter and apply E' directly to the input of the second-level NAND gate. The diagram in (c) is equivalent to the one in (b), which in turn is equivalent to the diagram in (a). Note the similarity between the diagrams in (a) and (c). The AND and OR gates have been changed to NAND gates, but an additional NAND gate has been included with the single variable E. When drawing NAND logic diagrams, the circuit shown in either (b) or (c) is acceptable. The one in (b), however, represents a more direct relationship to the Boolean expression it implements.

The NAND implementation in Fig. 3-18(c) can be verified algebraically. The NAND function it implements can be easily converted to a sum of products form by using DeMorgan's theorem:

$$F = [(AB)' \cdot (CD)' \cdot E']' = AB + CD + E$$

From the transformation shown in Fig. 3-18, we conclude that a Boolean function can be implemented with two levels of NAND gates. The rule for obtaining the NAND logic diagram from a Boolean function is as follows:

1. Simplify the function and express it in sum of products.
2. Draw a NAND gate for each product term of the function that has at least two literals. The inputs to each NAND gate are the literals of the term. This constitutes a group of first-level gates.
3. Draw a single NAND gate (using the AND-invert or invert-OR graphic symbol) in the second level, with inputs coming from outputs of first-level gates.
4. A term with a single literal requires an inverter in the first level or may be complemented and applied as an input to the second-level NAND gate.

Before applying these rules to a specific example, it should be mentioned that there is a second way to implement a Boolean function with NAND gates. Remember that if we combine the 0's in a map, we obtain the simplified expression of the *complement* of the function in sum of products. The complement of the function can then be implemented with two levels of NAND gates using the rules stated above. If the normal output is desired, it would be necessary to insert a one-input NAND or inverter gate to generate the true value of the output variable. There are occasions where the designer may want to generate the complement of the function; so this second method may be preferable.

Example 3-9   Implement the following function with NAND gates:

$$F(x, y, z) = \Sigma(0, 6)$$

The first step is to simplify the function in sum of products form. This is attempted with the map shown in Fig. 3-19(a). There are only two 1's in the map, and they can-

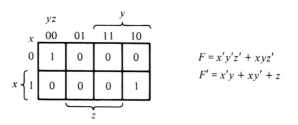

$$F = x'y'z' + xyz'$$
$$F' = x'y + xy' + z$$

(a) Map simplification in sum of products.

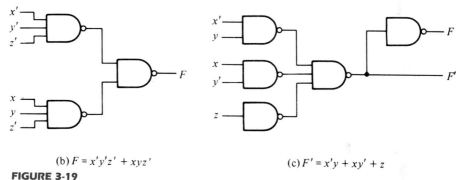

(b) $F = x'y'z' + xyz'$ (c) $F' = x'y + xy' + z$

FIGURE 3-19

Implementation of the function of Example 3-9 with NAND gates

not be combined. The simplified function in sum of products for this example is

$$F = x'y'z' + xyz'$$

The two-level NAND implementation is shown in Fig. 3-19(b). Next we try to simplify the complement of the function in sum of products. This is done by combining the 0's in the map:

$$F' = x'y + xy' + z$$

The two-level NAND gate for generating F' is shown in Fig. 3-19(c). If output F is required, it is necessary to add a one-input NAND gate to invert the function. This gives a three-level implementation. In each case, it is assumed that the input variables are available in both the normal and complement forms. If they were available in only one form, it would be necessary to insert inverters in the inputs, which would add another level to the circuits. The one-input NAND gate associated with the single variable z can be removed provided the input is changed to z'. ∎

NOR Implementation

The NOR function is the dual of the NAND function. For this reason, all procedures and rules for NOR logic are the duals of the corresponding procedures and rules developed for NAND logic.

The implementation of a Boolean function with NOR gates requires that the function be simplified in product of sums form. A product of sums expression specifies a group of OR gates for the sum terms, followed by an AND gate to produce the product. The transformation from the OR-AND to the NOR-NOR diagram is depicted in Fig. 3-20. It is similar to the NAND transformation discussed previously, except that now we use the product of sums expression

$$F = (A + B)(C + D)E$$

The rule for obtaining the NOR logic diagram from a Boolean function can be derived from this transformation. It is similar to the three-step NAND rule, except that the simplified expression must be in the product of sums and the terms for the first-level NOR gates are the sum terms. A term with a single literal requires a one-input NOR or inverter gate or may be complemented and applied directly to the second-level NOR gate.

A second way to implement a function with NOR gates would be to use the expression for the complement of the function in product of sums. This will give a two-level implementation for F' and a three-level implementation if the normal output F is required.

To obtain the simplified product of sums from a map, it is necessary to combine the 0's in the map and then complement the function. To obtain the simplified product of sums expression for the complement of the function, it is necessary to combine the 1's in the map and then complement the function. The following example demonstrates the procedure for NOR implementation.

Example 3-10

Implement the function of Example 3-9 with NOR gates.

The map of this function is drawn in Fig. 3-19(a). First, combine the 0's in the map to obtain

$$F' = x'y + xy' + z$$

This is the complement of the function in sum of products. Complement F' to obtain the simplified function in product of sums as required for NOR implementation:

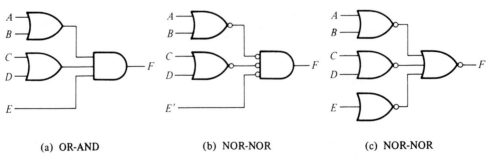

(a) OR-AND   (b) NOR-NOR   (c) NOR-NOR

FIGURE 3-20
Three ways to implement $F = (A + B)(C + D)E$

$$F = (x + y')(x' + y)z'$$

The two-level implementation with NOR gates is shown in Fig. 3-21(a). The term with a single literal z' requires a one-input NOR or inverter gate. This gate can be removed and input z applied directly to the input of the second-level NOR gate.

A second implementation is possible from the complement of the function in product of sums. For this case, first combine the 1's in the map to obtain

$$F = x'y'z' + xyz'$$

This is the simplified expression in sum of products. Complement this function to obtain the complement of the function in product of sums as required for NOR implementation:

$$F' = (x + y + z)(x' + y' + z)$$

The two-level implementation for $F'$is shown in Fig. 3-21(b). If output F is desired, it can be generated with an inverter in the third level. ■

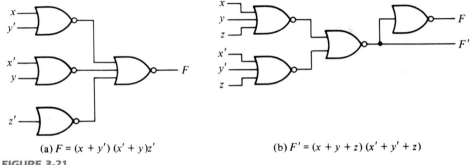

(a) $F = (x + y')(x' + y)z'$ (b) $F' = (x + y + z)(x' + y' + z)$

FIGURE 3-21
Implementation with NOR gates

Table 3-3 summarizes the procedures for NAND or NOR implementation. One should not forget to always simplify the function in order to reduce the number of gates in the implementation. The standard forms obtained from the map-simplification procedures apply directly and are very useful when dealing with NAND or NOR logic.

TABLE 3-3
Rules for NAND and NOR Implementation

Case	Function to simplify	Standard form to use	How to derive	Implement with	Number of levels to F
(a)	F	Sum of products	Combine 1's in map	NAND	2
(b)	F'	Sum of products	Combine 0's in map	NAND	3
(c)	F	Product of sums	Complement F' in (b)	NOR	2
(d)	F'	Product of sums	Complement F in (a)	NOR	3

3-7 OTHER TWO-LEVEL IMPLEMENTATIONS

The types of gates most often found in integrated circuits are NAND and NOR. For this reason, NAND and NOR logic implementations are the most important from a practical point of view. Some NAND or NOR gates (but not all) allow the possibility of a wire connection between the outputs of two gates to provide a specific logic function. This type of logic is called *wired logic*. For example, open-collector TTL NAND gates, when tied together, perform the wired-AND logic. (The open-collector TTL gate is shown in Chapter 10, Fig. 10-11.) The wired-AND logic performed with two NAND gates is depicted in Fig. 3-22(a). The AND gate is drawn with the lines going through the center of the gate to distinguish it from a conventional gate. The wired-AND gate is not a physical gate, but only a symbol to designate the function obtained from the indicated wired connection. The logic function implemented by the circuit of Fig. 3-22(a) is

$$F = (AB)' \cdot (CD)' = (AB + CD)'$$

and is called an AND-OR-INVERT function.

Similarly, the NOR output of ECL gates can be tied together to perform a wired-OR function. The logic function implemented by the circuit of Fig. 3-22(b) is

$$F = (A + B)' + (C + D)' = [(A + B)(C + D)]'$$

and is called an OR-AND-INVERT function.

A wired-logic gates does not produce a physical second-level gate since it is just a wire connection. Nevertheless, for discussion purposes, we will consider the circuits of Fig. 3-22 as two-level implementations. The first level consists of NAND (or NOR) gates and the second level has a single AND (or OR) gate. The wired connection in the graphic symbol will be omitted in subsequent discussions.

Nondegenerate Forms

It will be instructive from a theoretical point of view to find out how many two-level combinations of gates are possible. We consider four types of gates: AND, OR, NAND, and NOR. If we assign one type of gate for the first level and one type for the second

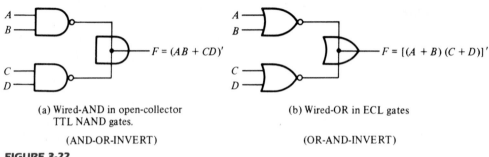

(a) Wired-AND in open-collector   (b) Wired-OR in ECL gates
    TTL NAND gates.

(AND-OR-INVERT)   (OR-AND-INVERT)

FIGURE 3-22
Wired logic

level, we find that there are 16 possible combinations of two-level forms. (The same type of gate can be in the first and second levels, as in NAND-NAND implementation.) Eight of these combinations are said to be *degenerate* forms because they degenerate to a single operation. This can be seen from a circuit with AND gates in the first level and an AND gate in the second level. The output of the circuit is merely the AND function of all input variables. The other eight *nondegenerate* forms produce an implementation in sum of products or product of sums. The eight nondegenerate forms are

AND-OR	OR-AND
NAND-NAND	NOR-NOR
NOR-OR	NAND-AND
OR-NAND	AND-NOR

The first gate listed in each of the forms constitutes a first level in the implementation. The second gate listed is a single gate placed in the second level. Note that any two forms listed in the same line are the duals of each other.

The AND-OR and OR-AND forms are the basic two-level forms discussed in Section 3-5. The NAND-NAND and NOR-NOR were introduced in Section 3-6. The remaining four forms are investigated in this section.

AND-OR-INVERT Implementation

The two forms NAND-AND and AND-NOR are equivalent forms and can be treated together. Both perform the AND-OR-INVERT function, as shown in Fig. 3-23. The AND-NOR form resembles the AND-OR form with an inversion done by the small circle in the output of the NOR gate. It implements the function

$$F = (AB + CD + E)'$$

By using the alternate graphic symbol for the NOR gate, we obtain the diagram of Fig. 3-23(b). Note that the single variable E is *not* complemented because the only change made is in the graphic symbol of the NOR gate. Now we move the circles from

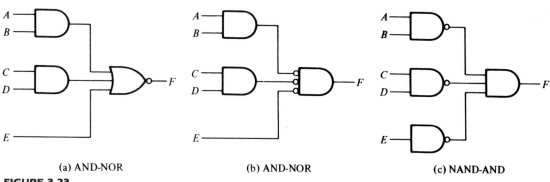

(a) AND-NOR   (b) AND-NOR   (c) NAND-AND

FIGURE 3-23
AND-OR-INVERT circuits; $F = (AB + CD + E)'$

the input terminal of the second-level gate to the output terminals of the first-level gates. An inverter is needed for the single variable to maintain the circle. Alternatively, the inverter can be removed provided input E is complemented. The circuit of Fig. 3-23(c) is a NAND-AND form and was shown in Fig. 3-22 to implement the AND-OR-INVERT function.

An AND-OR implementation requires an expression in sum of products. The AND-OR-INVERT implementation is similar except for the inversion. Therefore, if the *complement* of the function is simplified in sum of products (by combining the 0's in the map), it will be possible to implement F' with the AND-OR part of the function. When F' passes through the always present output inversion (the INVERT part), it will generate the ouput F of the function. An example for the AND-OR-INVERT implementation will be shown subsequently.

OR-AND-INVERT Implementation

The OR-NAND and NOR-OR forms perform the OR-AND-INVERT function. This is shown in Fig. 3-24. The OR-NAND form resembles the OR-AND form, except for the inversion done by the circle in the NAND gate. It implements the function

$$F = [(A + B)(C + D)E]'$$

By using the alternate graphic symbol for the NAND gate, we obtain the diagram of Fig. 3-24(b). The circuit in (c) is obtained by moving the small circles from the inputs of the second-level gate to the outputs of the first-level gates. The circuit of Fig. 3-24(c) is a NOR-OR form and was shown in Fig. 3-22 to implement the OR-AND-INVERT function.

The OR-AND-INVERT implementation requires an expression in product of sums. If the complement of the function is simplified in product of sums, we can implement F' with the OR-AND part of the function. When F' passes through the INVERT part, we obtain the complement of F', or F, in the output.

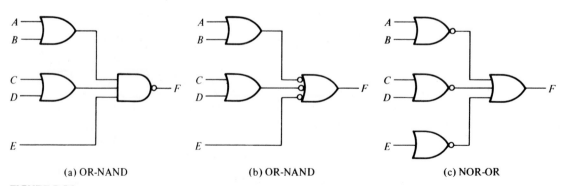

(a) OR-NAND   (b) OR-NAND   (c) NOR-OR

FIGURE 3-24
OR-AND-INVERT circuits; $F = [(A + B)(C + D)E]'$

Tabular Summary and Example

Table 3-4 summarizes the procedures for implementing a Boolean function in any one of the four two-level forms. Because of the INVERT part in each case, it is convenient to use the simplification of F' (the complement) of the function. When F' is implemented in one of these forms, we obtain the complement of the function in the AND-OR or OR-AND form. The four two-level forms invert this function, giving an output that is the complement of F'. This is the normal output F.

TABLE 3-4
Implementation with Other Two-Level Forms

Equivalent nondegenerate form		Implements the function	Simplify F' in	To get an output of
(a)	(b)*			
AND-NOR	NAND-AND	AND-OR-INVERT	Sum of products by combining 0's in the map	F
OR-NAND	NOR-OR	OR-AND-INVERT	Product of sums by combining 1's in the map and then complementing	F

*Form (b) requires a one-input NAND or NOR (inverter) gate for a single literal term.

Example 3-11

Implement the function of Fig. 3-19(a) with the four two-level forms listed in Table 3-4. The complement of the function is simplified in sum of products by combining the 0's in the map:

$$F' = x'y + xy' + z$$

The normal output for this function can be expressed as

$$F = (x'y + xy' + z)'$$

which is in the AND-OR-INVERT form. The AND-NOR and NAND-AND implementations are shown in Fig. 3-25(a). Note that a one-input NAND or inverter gate is needed in the NAND-AND implementation, but not in the AND-NOR case. The inverter can be removed if we apply the input variable z' instead of z.

The OR-AND-INVERT forms require a simplified expression of the complement of the function in product of sums. To obtain this expression, we must first combine the 1's in the map

$$F = x'y'z' + xyz'$$

Then we take the complement of the function

$$F' = (x + y + z)(x' + y' + z)$$

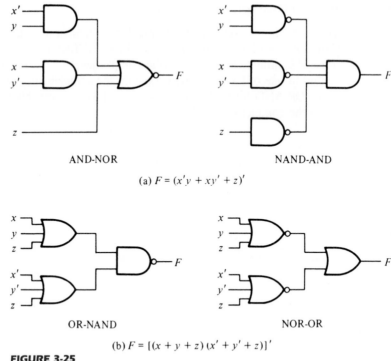

AND-NOR   NAND-AND

(a) $F = (x'y + xy' + z)'$

OR-NAND   NOR-OR

(b) $F = [(x + y + z)(x' + y' + z)]'$

FIGURE 3-25
Other two-level implementations

The normal output F can now be expressed in the form

$$F = [(x + y + z)(x' + y' + z)]'$$

which is in the OR-AND-INVERT form. From this expression, we can implement the function in the OR-NAND and NOR-OR forms, as shown in Fig. 3-25(b).   ■

3-8   DON'T-CARE CONDITIONS

The logical sum of the minterms associated with a Boolean function specifies the conditions under which the function is equal to 1. The function is equal to 0 for the rest of the minterms. This assumes that all the combinations of the values for the variables of the function are valid. In practice, there are some applications where the function is not specified for certain combinations of the variables. As an example, the four-bit binary code for the decimal digits has six combinations that are not used and consequently are considered as unspecified. Functions that have unspecified outputs for some input combinations are called incompletely specified functions. In most applications, we simply don't care what value is assumed by the function for the unspecified minterms. For this reason, it is customary to call the unspecified minterms of a function don't-care conditions. These don't-care conditions can be used on a map to provide further simplification of the Boolean expression.

It should be realized that a don't-care minterm is a combination of variables whose logical value is not specified. It cannot be marked with a 1 in the map because it would require that the function always be a 1 for such combination. Likewise, putting a 0 on the square requires the function to be 0. To distinguish the don't-care condition from 1's and 0's, an X is used. Thus, an X inside a square in the map indicates that we don't care whether the value of 0 or 1 is assigned to F for the particular minterm.

When choosing adjacent squares to simplify the function in a map, the don't-care minterms may be assumed to be either 0 or 1. When simplifying the function, we can choose to include each don't-care minterm with either the 1's or the 0's, depending on which combination gives the simplest expression.

Example 3-12

Simplify the Boolean function

$$F(w, x, y, z) = \Sigma(1, 3, 7, 11, 15)$$

that has the don't-care conditions

$$d(w, x, y, z) = \Sigma(0, 2, 5)$$

The minterms of F are the variable combinations that make the function equal to 1. The minterms of d are the don't-care minterms that may be assigned either 0 or 1. The map simplification is shown in Fig. 3-26. The minterms of F are marked by 1's, those of d are marked by X's, and the remaining squares are filled with 0's. To get the simplified expression in sum of products, we must include all the five 1's in the map, but

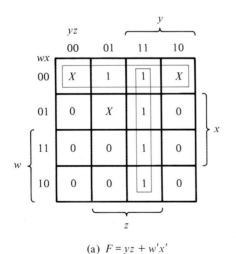

(a) $F = yz + w'x'$

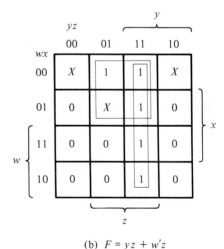

(b) $F = yz + w'z$

FIGURE 3-26

Example with don't-care conditions

we may or may not include any of the X's, depending on the way the function is simplified. The term yz covers the four minterms in the third column. The remaining minterm m_1 can be combined with minterm m_3 to give the three-literal term $w'x'z$. However, by including one or two adjacent X's we can combine four adjacent squares to give a two-literal term. In part (a) of the diagram, don't-care minterms 0 and 2 are included with the 1's, which results in the simplified function

$$F = yz + w'x'$$

In part (b), don't-care minterm 5 is included with the 1's and the simplified function now is

$$F = yz + w'z$$

Either one of the above expressions satisfies the conditions stated for this example. ■

The above example has shown that the don't-care minterms in the map are initially marked with X's and are considered as being either 0 or 1. The choice between 0 and 1 is made depending on the way the incompletely specified function is simplified. Once the choice is made, the simplified function so obtained will consist of a sum of minterms that includes those minterms that were initially unspecified and have been chosen to be included with the 1's. Consider the two simplified expressions obtained in Example 3-12:

$$F(w, x, y, z) = yz + w'x' = \Sigma(0, 1, 2, 3, 7, 11, 15)$$

$$F(w, x, y, z) = yz + w'z = \Sigma(1, 3, 5, 7, 11, 15)$$

Both expressions include minterms 1, 3, 7, 11, and 15 that make the function F equal to 1. The don't-care minterms 0, 2, and 5 are treated differently in each expression. The first expression includes minterms 0 and 2 with the 1's and leaves minterm 5 with the 0's. The second expression includes minterm 5 with the 1's and leaves minterms 0 and 2 with the 0's. The two expressions represent two functions that are algebraically unequal. Both cover the specified minterms of the function, but each covers different don't-care minterms. As far as the incompletely specified function is concerned, either expression is acceptable since the only difference is in the value of F for the don't-care minterms.

It is also possible to obtain a simplified product of sums expression for the function of Fig. 3-26. In this case, the only way to combine the 0's is to include don't-care minterms 0 and 2 with the 0's to give a simplified complemented function:

$$F' = z' + wy'$$

Taking the complement of F' gives the simplified expression in product of sums:

$$F(w, x, y, z) = z(w' + y) = \Sigma(1, 3, 5, 7, 11, 15)$$

For this case, we include minterms 0 and 2 with the 0's and minterm 5 with the 1's.

3-9 THE TABULATION METHOD

The map method of simplification is convenient as long as the number of variables does not exceed five or six. As the number of variables increases, the excessive number of squares prevents a reasonable selection of adjacent squares. The obvious disadvantage of the map is that it is essentially a trial-and-error procedure that relies on the ability of the human user to recognize certain patterns. For functions of six or more variables, it is difficult to be sure that the best selection has been made.

The tabulation method overcomes this difficulty. It is a specific step-by-step procedure that is guaranteed to produce a simplified standard-form expression for a function. It can be applied to problems with many variables and has the advantage of being suitable for machine computation. However, it is quite tedious for human use and is prone to mistakes because of its routine, monotonous process. The tabulation method was first formulated by Quine and later improved by McCluskey. It is also known as the Quine–McCluskey method.

The tabular method of simplification consists of two parts. The first is to find by an exhaustive search all the terms that are candidates for inclusion in the simplified function. These terms are called *prime implicants*. The second operation is to choose among the prime implicants those that give an expression with the least number of literals.

3-10 DETERMINATION OF PRIME IMPLICANTS

The starting point of the tabulation method is the list of minterms that specify the function. The first tabular operation is to find the prime implicants by using a matching process. This process compares each minterm with every other minterm. If two minterms differ in only one variable, that variable is removed and a term with one less literal is found. This process is repeated for every minterm until the exhaustive search is completed. The matching-process cycle is repeated for those new terms just found. Third and further cycles are continued until a single pass through a cycle yields no further elimination of literals. The remaining terms and all the terms that did not match during the process comprise the prime implicants. This tabulation method is illustrated by the following example.

Example 3-13

Simplify the following Boolean function by using the tabulation method:

$$F = \Sigma\,(0,\ 1,\ 2,\ 8,\ 10,\ 11,\ 14,\ 15)$$

Step 1: Group binary representation of the minterms according to the number of 1's contained, as shown in Table 3-5, column (a). This is done by grouping the minterms into five sections separated by horizontal lines. The first section contains the number with no 1's in it. The second section contains those numbers that have only one 1. The

TABLE 3-5
Determination of Prime Implicants for Example 3-13

(a)		(b)		(c)	
	$w\ x\ y\ z$		$w\ x\ y\ z$		$w\ x\ y\ z$
0	0 0 0 0 ✓	0, 1	0 0 0 −	0, 2, 8, 10	− 0 − 0
		0, 2	0 0 − 0 ✓	0, 8, 2, 10	− 0 − 0
1	0 0 0 1 ✓	0, 8	− 0 0 0 ✓	10, 11, 14, 15	1 − 1 −
2	0 0 1 0 ✓			10, 14, 11, 15	1 − 1 −
8	1 0 0 0 ✓	2, 10	− 0 1 0 ✓		
		8, 10	1 0 − 0 ✓		
10	1 0 1 0 ✓	10, 11	1 0 1 − ✓		
		10, 14	1 − 1 − ✓		
11	1 0 1 1 ✓				
14	1 1 1 0 ✓	11, 15	1 − 1 1 ✓		
15	1 1 1 1 ✓	14, 15	1 1 1 − ✓		

third, fourth, and fifth sections contain those binary numbers with two, three, and four 1's, respectively. The decimal equivalents of the minterms are also carried along for identification.

Step 2: Any two minterms that differ from each other by only one variable can be combined, and the unmatched variable removed. Two minterm numbers fit into this category if they both have the same bit value in all positions except one. The minterms in one section are compared with those of the next section down only, because two terms differing by more than one bit cannot match. The minterm in the first section is compared with each of the three minterms in the second section. If any two numbers are the same in every position but one, a check is placed to the right of both minterms to show that they have been used. The resulting term, together with the decimal equivalents, is listed in column (b) of the table. The variable eliminated during the matching is denoted by a dash in its original position. In this case, m_0 (0000) combines with m_1 (0001) to form (000−). This combination is equivalent to the algebraic operation

$$m_0 + m_1 = w'x'y'z' + w'x'y'z = w'x'y'$$

Minterm m_0 also combines with m_2 to form (00−0) and with m_8 to form (−000). The result of this comparison is entered into the first section of column (b). The minterms of sections two and three of column (a) are next compared to produce the terms listed in the second section of column (b). All other sections of (a) are similarly compared and subsequent sections formed in (b). This exhaustive comparing process results in the four sections of (b).

Step 3: The terms of column (b) have only three variables. A 1 under the variable means it is unprimed, a 0 means it is primed, and a dash means the variable is not included in the term. The searching and comparing process is repeated for the terms in column (b) to form the two-variable terms of column (c). Again, terms in each section need to be compared only if they have dashes in the same position. Note that the term $(000-)$ does not match with any other term. Therefore, it has no check mark at its right. The decimal equivalents are written on the left-hand side of each entry for identification purposes. The comparing process should be carried out again in column (c) and in subsequent columns as long as proper matching is encountered. In the present example, the operation stops at the third column.

Step 4: The unchecked terms in the table form the prime implicants. In this example, we have the term $w'x'y'$ $(000-)$ in column (b), and the terms $x'z'$ $(-0-0)$ and wy $(1-1-)$ in column (c). Note that each term in column (c) appears twice in the table, and as long as the term forms a prime implicant, it is unnecessary to use the same term twice. The sum of the prime implicants gives a simplified expression for the function. This is because each checked term in the table has been taken into account by an entry of a simpler term in a subsequent column. Therefore, the unchecked entries (prime implicants) are the terms left to formulate the function. For the present example, the sum of prime implicants gives the minimized function in sum of products:

$$F = w'x'y' + x'z' + wy \qquad \blacksquare$$

It is worth comparing this answer with that obtained by the map method. Figure 3-27 shows the map simplification of this function. The combinations of adjacent squares give the three prime implicants of the function. The sum of these three terms is the simplified expression in sum of products.

It is important to point out that Example 3-13 was purposely chosen to give the simplified function from the sum of prime implicants. In most other cases, the sum of prime implicants does not necessarily form the expression with the minimum number of terms. This is demonstrated in Example 3-14.

The tedious manipulation that one must undergo when using the tabulation method is reduced if the comparing is done with decimal numbers instead of binary. A method will now be shown that uses subtraction of decimal numbers instead of the comparing and matching of binary numbers. We note that each 1 in a binary number represents the

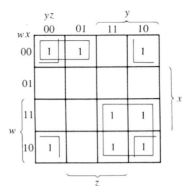

FIGURE 3-27
Map for the function of Example 3-13;
$F = w'x'y' + x'z' + wy$

coefficient multiplied by a power of 2. When two minterms are the same in every position except one, the minterm with the extra 1 must be larger than the number of the
other minterm by a power of 2. Therefore, two minterms can be combined if the number of the first minterm differs by a power of 2 from a second larger number in the next
section down the table. We shall illustrate this procedure by repeating Example
3-13.

As shown in Table 3-6, column (a), the minterms are arranged in sections as before,
except that now only the decimal equivalents of the minterms are listed. The process of
comparing minterms is as follows: Inspect every two decimal numbers in adjacent sections of the table. If the number in the section below is *greater* than the number in the
section above by a power of 2 (i.e., 1, 2, 4, 8, 16, etc.), check both numbers to show
that they have been used, and write them down in column (b). The pair of numbers
transferred to column (b) includes a third number in parentheses that designates the
power of 2 by which the numbers differ. The number in parentheses tells us the position of the dash in the binary notation. The results of all comparisons of column (a) are
shown in column (b).

The comparison between adjacent sections in column (b) is carried out in a similar
fashion, except that only those terms with the same number in parentheses are compared. The pair of numbers in one section must differ by a power of 2 from the pair of
numbers in the next section. And the numbers in the next section below must be
greater for the combination to take place. In column (c), write all four decimal numbers with the two numbers in parentheses designating the positions of the dashes. A
comparison of Tables 3-5 and 3-6 may be helpful in understanding the derivations in
Table 3-6.

TABLE 3-6
Determination of Prime Implicants of Example 3-13 with Decimal Notation

(a)	(b)	(c)
0 √	0, 1 (1)	0, 2, 8, 10 (2, 8)
	0, 2 (2) √	0, 2, 8, 10 (2, 8)
1 √	0, 8 (8) √	
2 √		10, 11, 14, 15 (1, 4)
8 √	2, 10 (8) √	10, 11, 14, 15 (1, 4)
	8, 10 (2) √	
10 √		
	10, 11 (1) √	
11 √	10, 14 (4) √	
14 √		
	11, 15 (4) √	
15 √	14, 15 (1) √	

The prime implicants are those terms not checked in the table. These are the same as before, except that they are given in decimal notation. To convert from decimal notation to binary, convert all decimal numbers in the term to binary and then insert a dash in those positions designated by the numbers in parentheses. Thus 0, 1 (1) is converted to binary as 0000, 0001; a dash in the first position of either number results in (000–). Similarly, 0, 2, 8, 10 (2, 8) is converted to the binary notation from 0000, 0010, 1000, and 1010, and a dash inserted in positions 2 and 8, to result in (–0–0).

Example 3-14

Determine the prime implicants of the function

$$F(w, x, y, z) = \Sigma(1, 4, 6, 7, 8, 9, 10, 11, 15)$$

The minterm numbers are grouped in sections, as shown in Table 3-7, column (a). The binary equivalent of the minterm is included for the purpose of counting the number of

TABLE 3-7
Determination of Prime Implicants for Example 3-14

(a)			(b)			(c)
0001	1	✓	1, 9	(8)		8, 9, 10, 11 (1, 2)
0100	4	✓	4, 6	(2)		8, 9, 10, 11 (1, 2)
1000	8	✓	8, 9	(1)	✓	
			8, 10	(2)	✓	
0110	6	✓				
1001	9	✓	6, 7	(1)		
1010	10	✓	9, 11	(2)	✓	
			10, 11	(1)	✓	
0111	7	✓				
1011	11	✓	7, 15	(8)		
			11, 15	(4)		
1111	15	✓				

	Prime implicants					
		Binary				
Decimal		w	x	y	z	Term
1, 9 (8)		–	0	0	1	$x'y'z$
4, 6 (2)		0	1	–	0	$w'xz'$
6, 7 (1)		0	1	1	–	$w'xy$
7, 15 (8)		–	1	1	1	xyz
11, 15 (4)		1	–	1	1	wyz
8, 9, 10, 11 (1, 2)		1	0	–	–	wx'

1's. The binary numbers in the first section have only one 1; in the second section, two 1's; etc. The minterm numbers are compared by the decimal method and a match is found if the number in the section below is greater than that in the section above. If the number in the section below is smaller than the one above, a match is not recorded even if the two numbers differ by a power of 2. The exhaustive search in column (a) results in the terms of column (b), with all minterms in column (a) being checked. There are only two matches of terms in column (b). Each gives the same two-literal term recorded in column (c). The prime implicants consist of all the unchecked terms in the table. The conversion from the decimal to the binary notation is shown at the bottom of the table. The prime implicants are found to be $x'y'z$, $w'xz'$, $w'xy$, xyz, wyz, and wx'.   ■

The sum of the prime implicants gives a valid algebraic expression for the function. However, this expression is not necessarily the one with the minimum number of terms. This can be demonstrated from inspection of the map for the function of Example 3-14. As shown in Fig. 3-28, the minimized function is recognized to be

$$F = x'y'z + w'xz' + xyz + wx'$$

which consists of the sum of four of the six prime implicants derived in Example 3-14. The tabular procedure for selecting the prime implicants that give the minimized function is the subject of the next section.

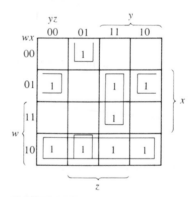

FIGURE 3-28
Map for the function of Example 3-14;
$F = x'y'z' + w'xz' + xyz + wx'$

3-11 SELECTION OF PRIME IMPLICANTS

The selection of prime implicants that form the minimized function is made from a prime implicant table. In this table, each prime implicant is represented in a row and each minterm in a column. X's are placed in each row to show the composition of minterms that make the prime implicants. A minimum set of prime implicants is then chosen that covers all the minterms in the function. This procedure is illustrated in Example 3-15.

**Example
3-15**

Minimize the function of Example 3-14. The prime-implicant table for this example is shown in Table 3-8. There are six rows, one for each prime implicant (derived in Example 3-14), and nine columns, each representing one minterm of the function. X's are placed in each row to indicate the minterms contained in the prime implicant of that row. For example, the two X's in the first row indicate that minterms 1 and 9 are contained in the prime implicant $x'y'z$. It is advisable to include the decimal equivalent of the prime implicant in each row, as it conveniently gives the minterms contained in it. After all the X's have been marked, we proceed to select a minimum number of prime implicants.

The completed prime-implicant table is inspected for columns containing only a single X. In this example, there are four minterms whose columns have a single X: 1, 4, 8, and 10. Minterm 1 is covered by prime implicant $x'y'z$, i.e., the selection of prime implicant $x'y'z$ guarantees that minterm 1 is included in the function. Similarly, minterm 4 is covered by prime implicant $w'xz'$, and minterms 8 and 10, by prime implicant wx'. Prime implicants that cover minterms with a single X in their column are called *essential prime implicants*. To enable the final simplified expression to contain all the minterms, we have no alternative but to include essential prime implicants. A check mark is placed in the table next to the essential prime implicants to indicate that they have been selected.

Next we check each column whose minterm is covered by the selected essential prime implicants. For example, the selected prime implicant $x'y'z$ covers minterms 1 and 9. A check is inserted in the bottom of the columns. Similarly, prime implicant $w'xz'$ covers minterms 4 and 6, and wx' covers minterms 8, 9, 10, and 11. Inspection of the prime-implicant table shows that the selection of the essential prime implicants

TABLE 3-8
Prime Implicant Table for Example 3-15

		1	4	6	7	8	9	10	11	15
√ $x'y'z$	1, 9	X					X			
√ $w'xz'$	4, 6		X	X						
$w'xy$	6, 7			X	X					
xyz	7, 15				X					X
wyz	11, 15								X	X
√ wx'	8, 9, 10, 11					X	X	X	X	
		√	√	√		√	√	√	√	

covers all the minterms of the function except 7 and 15. These two minterms must be included by the selection of one or more prime implicants. In this example, it is clear that prime implicant *xyz* covers both minterms and is therefore the one to be selected. We have thus found the minimum set of prime implicants whose sum gives the required minimized function:

$$F = x'y'z + w'xz' + wx' + xyz$$

■

The simplified expressions derived in the preceding examples were all in the sum of products form. The tabulation method can be adapted to give a simplified expression in product of sums. As in the map method, we have to start with the complement of the function by taking the 0's as the initial list of minterms. This list contains those minterms not included in the original function that are numerically equal to the maxterms of the function. The tabulation process is carried out with the 0's of the function and terminates with a simplified expression in sum of products of the complement of the function. By taking the complement again, we obtain the simplified product of sums expression.

A function with don't-care conditions can be simplified by the tabulation method after a slight modification. The don't-care terms are included in the list of minterms when the prime implicants are determined. This allows the derivation of prime implicants with the least number of literals. The don't-care terms are not included in the list of minterms when the prime implicant table is set up, because don't-care terms do not have to be covered by the selected prime implicants.

3-12   CONCLUDING REMARKS

Two methods of Boolean-function simplification were introduced in this chapter. The criterion for simplification was taken to be the minimization of the number of literals in sum of product or products of sums expressions. Both the map and the tabulation methods are restricted in their capabilities since they are useful for simplifying only Boolean functions expressed in the standard forms. Although this is a disadvantage of the methods, it is not very critical. Most applications prefer the standard forms over any other form. We have seen from Fig. 3-15 that the gate implementation of expressions in standard form consists of no more than two levels of gates. Expressions not in the standard form are implemented with more than two levels.

One should recognize that the Gray-code sequence chosen for the maps is not unique. It is possible to draw a map and assign a Gray-code sequence to the rows and columns different from the sequence employed here. As long as the binary sequence chosen produces a change in only one bit between adjacent squares, it will produce a valid and useful map.

Two alternate versions of the three-variable maps that are often found in the digital logic literature are shown in Fig. 3-29. The minterm numbers are written in each square for reference. In (a), the assignment of the variables to the rows and columns is different from the one used in this book. In (b), the map has been rotated in a vertical position. The minterm number assignment in all maps remains in the order xyz. For example, the square for minterm 6 is found by assigning to the ordered variables the binary number $xyz = 110$. The square for this minterm is found in (a) from the column marked $xy = 11$ and the row with $z = 0$. The corresponding square in (b) belongs in the column marked with $x = 1$ and the row with $yz = 10$. The simplification procedure with these maps is exactly the same as described in this chapter except, of course, for the variations in minterm and variable assignment.

Two other versions of the four-variable map are shown in Fig. 3-30. The map in (a) is very popular and is used quite often in the literature. Here again, the difference is

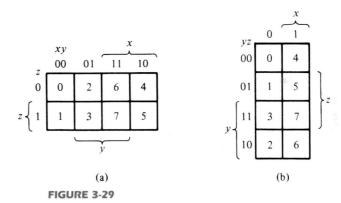

(a)   (b)

FIGURE 3-29
Variations of the three-variable map

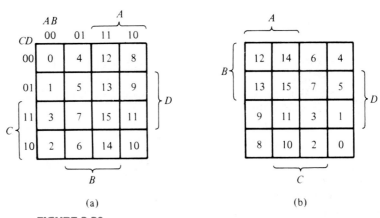

(a)   (b)

FIGURE 3-30
Variations of the four-variable map

slight and is manifested by a mere interchange of variable assignment from rows to columns and vice versa. The map in (b) is the original Veitch diagram that Karnaugh modified to the one shown in (a). Again, the simplification procedures do not change when these maps are used instead of the one employed in this book. There are also variations of the five-variable map. In any case, any map that looks different from the one used in this book, or is called by a different name, should be recognized merely as a variation of minterm assignment to the squares in the map.

As is evident from Examples 3-13 and 3-14, the tabulation method has the drawback that errors inevitably occur in trying to compare numbers over long lists. The map method would seem to be preferable, but for more than five variables, we cannot be certain that the best simplified expression has been found. The real advantage of the tabulation method lies in the fact that it consists of specific step-by-step procedures that guarantee an answer. Moreover, this formal procedure is suitable for computer mechanization.

In this chapter, we have considered the simplification of functions with many input variables and a single output variable. However, some digital circuits have more than one output. Such circuits are described by a set of Boolean functions, one for each output variable. A circuit with multiple outputs may sometimes have common terms among the various functions that can be utilized to form common gates during the implementation. This results in further simplification not taken into consideration when each function is simplified separately. There exists an extension of the tabulation method for multiple-output circuits. However, this method is too specialized and very tedious for human manipulation. It is of practical importance only if a computer program based on this method is available to the user.

REFERENCES

1. VEITCH, E. W., "A Chart Method for Simplifying Truth Functions." *Proc. ACM* (May 1952), 127–133.

2. KARNAUGH, M., "A Map Method for Synthesis of Combinational Logic Circuits." *Trans. AIEE, Comm. and Electron.,* **72,** Part I (November 1953), 593–599.

3. QUINE, W. V., "The Problem of Simplifying Truth Functions." *Am. Math. Monthly,* **59 (8)** (October 1952), 521–531.

4. McCLUSKEY, E. J., "Minimization of Boolean Functions." *Bell Syst. Tech. J.,* **35 (6)** (November 1956), 1417–1444.

5. McCLUSKEY, E. J., *Logic Design Principles.* Englewood Cliffs, NJ: Prentice-Hall, 1986.

6. KOHAVI, Z., *Switching and Automata Theory,* 2nd Ed. New York: McGraw-Hill, 1978.

7. HILL, F. J., and G. R. PETERSON, *Introduction to Switching Theory and Logical Design,* 3rd Ed. New York: John Wiley, 1981.

8. GIVONE, D. D., *Introduction to Switching Circuit Theory.* New York: McGraw-Hill, 1970.

PROBLEMS

3-1 Simplify the following Boolean functions using three-variable maps:
(a) $F(x, y, z) = \Sigma(0, 1, 5, 7)$
(b) $F(x, y, z) = \Sigma(1, 2, 3, 6, 7)$
(c) $F(x, y, z) = \Sigma(3, 5, 6, 7)$
(d) $F(A, B, C) = \Sigma(0, 2, 3, 4, 6)$

3-2 Simplify the following Boolean expressions using three-variable maps:
(a) $xy + x'y'z' + x'yz'$
(b) $x'y' + yz + x'yz'$
(c) $A'B + BC' + B'C'$

3-3 Simplify the following Boolean functions using four-variable maps:
(a) $F(A, B, C, D) = \Sigma(4, 6, 7, 15)$
(b) $F(w, x, y, z) = \Sigma(2, 3, 12, 13, 14, 15)$
(c) $F(A, B, C, D) = \Sigma(3, 7, 11, 13, 14, 15)$

3-4 Simplify the following Boolean functions using four-variable maps:
(a) $F(w, x, y, z) = \Sigma(1, 4, 5, 6, 12, 14, 15)$
(b) $F(A, B, C, D) = \Sigma(0, 1, 2, 4, 5, 7, 11, 15)$
(c) $F(w, x, y, z) = \Sigma(2, 3, 10, 11, 12, 13, 14, 15)$
(d) $F(A, B, C, D) = \Sigma(0, 2, 4, 5, 6, 7, 8, 10, 13, 15)$

3-5 Simplify the following Boolean expressions using four-variable maps:
(a) $w'z + xz + x'y + wx'z$
(b) $B'D + A'BC' + AB'C + ABC'$
(c) $AB'C + B'C'D' + BCD + ACD' + A'B'C + A'BC'D$
(d) $wxy + yz + xy'z + x'y$

3-6 Find the minterms of the following Boolean expressions by first plotting each function in a map:
(a) $xy + yz + xy'z$
(b) $C'D + ABC' + ABD' + A'B'D$
(c) $wxy + x'z' + w'xz$

3-7 Simplify the following Boolean functions by first finding the essential prime implicants:
(a) $F(w, x, y, z) = \Sigma(0, 2, 4, 5, 6, 7, 8, 10, 13, 15)$
(b) $F(A, B, C, D) = \Sigma(0, 2, 3, 5, 7, 8, 10, 11, 14, 15)$
(c) $F(A, B, C, D) = \Sigma(1, 3, 4, 5, 10, 11, 12, 13, 14, 15)$

3-8 Simplify the following Boolean functions using five-variable maps:
(a) $F(A, B, C, D, E) = \Sigma(0, 1, 4, 5, 16, 17, 21, 25, 29)$
(b) $F(A, B, C, D, E) = \Sigma(0, 2, 3, 4, 5, 6, 7, 11, 15, 16, 18, 19, 23, 27, 31)$
(c) $F = A'B'CE' + A'B'C'D' + B'D'E' + B'CD' + CDE' + BDE'$

3-9 Simplify the following Boolean functions in product of sums:
(a) $F(w, x, y, z) = \Sigma(0, 2, 5, 6, 7, 8, 10)$
(b) $F(A, B, C, D) = \Pi(1, 3, 5, 7, 13, 15)$
(c) $F(x, y, z) = \Sigma(2, 3, 6, 7)$
(d) $F(A, B, C, D) = \Pi(0, 1, 2, 3, 4, 10, 11)$

3-10 Simplify the following expressions in (i) sum of products and (ii) products of sums:
(a) $x'z' + y'z' + yz' + xy$
(b) $AC' + B'D + A'CD + ABCD$
(c) $(A' + B' + D')(A + B' + C')(A' + B + D')(B + C' + D')$

3-11 Draw the AND-OR gate implementation of the following function after simplifying it in (a) sum of products and (b) product of sums:

$$F = (A, B, C, D) = \Sigma(0, 2, 5, 6, 7, 8, 10)$$

3-12 Simplify the following expressions and implement them with two-level NAND gate circuits:
(a) $AB' + ABD + ABD' + A'C'D' + A'BC'$
(b) $BD + BCD' + AB'C'D'$

3-13 Draw a NAND logic diagram that implements the complement of the following function:

$$F(A, B, C, D) = \Sigma(0, 1, 2, 3, 4, 8, 9, 12)$$

3-14 Draw a logic diagram using only two-input NAND gates to implement the following expression:

$$(AB + A'B')(CD' + C'D)$$

3-15 Simplify the following functions and implement them with two-level NOR gate circuits:
(a) $F = wx' + y'z' + w'yz'$
(b) $F(w, x, y, z) = \Sigma(5, 6, 9, 10)$

3-16 Implement the functions of Problem 3-15 with three-level NOR gate circuits [similar to Fig. 3-21(b)].

3-17 Implement the expressions of Problem 3-12 with three-level NAND circuits [similar to Fig. 3-19(c)].

3-18 Give three possible ways to express the function F with eight or fewer literals.

$$F(A, B, C, D) = \Sigma(0, 2, 5, 7, 10, 13)$$

3-19 Find eight different two-level gate circuits to implement

$$F = xy'z + x'yz + w$$

3-20 Implement the function F with the following two-level forms: NAND-AND, AND-NOR, OR-NAND, and NOR-OR.

$$F(A,B,C,D) = \Sigma(0, 1, 2, 3, 4, 8, 9, 12)$$

3-21 List the eight degenerate two-level forms and show that they reduce to a single operation. Explain how the degenerate two-level forms can be used to extend the number of inputs to a gate.

3-22 Simplify the following Boolean function F together with the don't-care conditions d; then express the simplified function in sum of minterms.
(a) $F(x, y, z) = \Sigma(0, 1, 2, 4, 5)$
   $d(x, y, z) = \Sigma(3, 6, 7)$

(b) $F(A, B, C, D) = \Sigma(0, 6, 8, 13, 14)$
   $d(A, B, C, D) = \Sigma(2, 4, 10)$
(c) $F(A, B, C, D) = \Sigma(1, 3, 5, 7, 9, 15)$
   $d(A, B, C, D) = \Sigma(4, 6, 12, 13)$

3-23 Simplify the Boolean function F together with the don't-care conditions d in (i) sum of products and (ii) product of sums.
(a) $F(w, x, y, z) = \Sigma(0, 1, 2, 3, 7, 8, 10)$
   $d(w, x, y, z) = \Sigma(5, 6, 11, 15)$
(b) $F(A, B, C, D) = \Sigma(3, 4, 13, 15)$
   $d(A, B, C, D) = \Sigma(1, 2, 5, 6, 8, 10, 12, 14)$

3-24 A logic circuit implements the following Boolean function:

$$F = A'C + AC'D'$$

It is found that the circuit input combination $A = C = 1$ can never occur. Find a simpler expression for F using the proper don't-care conditions.

3-25 Implement the following Boolean function F together with the don't-care conditions d using no more than two NOR gates. Assume that both the normal and complement inputs are available.

$$F(A, B, C, D) = \Sigma(0, 1, 2, 9, 11)$$

$$d(A, B, C, D) = \Sigma(8, 10, 14, 15)$$

3-26 Simplify the following Boolean function using the map presented in Fig. 3-30(a). Repeat using the map of Fig. 3-30(b).

$$F(A, B, C, D) = \Sigma(1, 2, 3, 5, 7, 9, 10, 11, 13, 15)$$

3-27 Simplify the following Boolean functions by means of the tabulation method:
(a) $P(A, B, C, D, E, F, G) = \Sigma(20, 28, 52, 60)$
(b) $P(A, B, C, D, E, F, G) = \Sigma(20, 28, 38, 39, 52, 60, 102, 103, 127)$
(c) $P(A, B, C, D, E, F) = \Sigma(6, 9, 13, 18, 19, 25, 27, 29, 41, 45, 57, 61)$

Combinational Logic

4-1 INTRODUCTION

Logic circuits for digital systems may be combinational or sequential. A combinational circuit consists of logic gates whose outputs at any time are determined directly from the present combination of inputs without regard to previous inputs. A combinational circuit performs a specific information-processing operation fully specified logically by a set of Boolean functions. Sequential circuits employ memory elements (binary cells) in addition to logic gates. Their outputs are a function of the inputs and the state of the memory elements. The state of memory elements, in turn, is a function of previous inputs. As a consequence, the outputs of a sequential circuit depend not only on present inputs, but also on past inputs, and the circuit behavior must be specified by a time sequence of inputs and internal states. Sequential circuits are discussed in Chapter 6.

In Chapter 1, we learned to recognize binary numbers and binary codes that represent discrete quantities of information. These binary variables are represented by electric voltages or by some other signal. The signals can be manipulated in digital logic gates to perform required functions. In Chapter 2, we introduced Boolean algebra as a way to express logic functions algebraically. In Chapter 3, we learned how to simplify Boolean functions to achieve economical gate implementations. The purpose of this chapter is to use the knowledge acquired in previous chapters and formulate various systematic design and analysis procedures of combinational circuits. The solution of some typical examples will provide a useful catalog of elementary functions important for the understanding of digital computers and systems.

A combinational circuit consists of input variables, logic gates, and output variables. The logic gates accept signals from the inputs and generate signals to the outputs. This process transforms binary information from the given input data to the required output data. Obviously, both input and output data are represented by binary signals, i.e., they exist in two possible values, one representing logic-1 and the other logic-0. A block diagram of a combinational circuit is shown in Fig. 4-1. The n input binary variables come from an external source; the m output variables go to an external destination. In many applications, the source and/or destination are storage registers (Section 1-7) located either in the vicinity of the combinational circuit or in a remote external device. By definition, an external register does not influence the behavior of the combinational circuit because, if it does, the total system becomes a sequential circuit.

For n input variables, there are 2^n possible combinations of binary input values. For each possible input combination, there is one and only one possible output combination. A combinational circuit can be described by m Boolean functions, one for each output variable. Each output function is expressed in terms of the n input variables.

Each input variable to a combinational circuit may have one or two wires. When only one wire is available, it may represent the variable either in the normal form (unprimed) or in the complement form (primed). Since a variable in a Boolean expression may appear primed and/or unprimed, it is necessary to provide an inverter for each literal not available in the input wire. On the other hand, an input variable may appear in two wires, supplying both the normal and complement forms to the input of the circuit. If so, it is unnecessary to include inverters for the inputs. The type of binary cells used in most digital systems are flip-flop circuits (Chapter 6) that have outputs for both the normal and complement values of the stored binary variable. In our subsequent work, we shall assume that each input variable appears in two wires, supplying both the normal and complement values simultaneously. We must also realize that an inverter circuit can always supply the complement of the variable if only one wire is available.

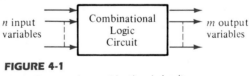

n input variables

Combinational Logic Circuit

m output variables

FIGURE 4-1
Block diagram of a combinational circuit

4-2 DESIGN PROCEDURE

The design of combinational circuits starts from the verbal outline of the problem and ends in a logic circuit diagram or a set of Boolean functions from which the logic diagram can be easily obtained. The procedure involves the following steps:

1. The problem is stated.

2. The number of available input variables and required output variables is determined.

3. The input and output variables are assigned letter symbols.
4. The truth table that defines the required relationships between inputs and outputs is derived.
5. The simplified Boolean function for each output is obtained.
6. The logic diagram is drawn.

A truth table for a combinational circuit consists of input columns and output columns. The 1's and 0's in the input columns are obtained from the 2^n binary combinations available for n input variables. The binary values for the outputs are determined from examination of the stated problem. An output can be equal to either 0 or 1 for every valid input combination. However, the specifications may indicate that some input combinations will not occur. These combinations become don't-care conditions.

The output functions specified in the truth table give the exact definition of the combinational circuit. It is important that the verbal specifications be interpreted correctly into a truth table. Sometimes the designer must use intuition and experience to arrive at the correct interpretation. Word specifications are very seldom complete and exact. Any wrong interpretation that results in an incorrect truth table produces a combinational circuit that will not fulfill the stated requirements.

The output Boolean functions from the truth table are simplified by any available method, such as algebraic manipulation, the map method, or the tabulation procedure. Usually, there will be a variety of simplified expressions from which to choose. However, in any particular application, certain restrictions, limitations, and criteria will serve as a guide in the process of choosing a particular algebraic expression. A practical design method would have to consider such constraints as (1) minimum number of gates, (2) minimum number of inputs to a gate, (3) minimum propagation time of the signal through the circuit, (4) minimum number of interconnections, and (5) limitations of the driving capabilities of each gate. Since all these criteria cannot be satisfied simultaneously, and since the importance of each constraint is dictated by the particular application, it is difficult to make a general statement as to what constitutes an acceptable simplification. In most cases, the simplification begins by satisfying an elementary objective, such as producing a simplified Boolean function in a standard form, and from that proceeds to meet any other performance criteria.

In practice, designers tend to go from the Boolean functions to a wiring list that shows the interconnections among various standard logic gates. In that case, the design need not go any further than the required simplified output Boolean functions. However, a logic diagram is helpful for visualizing the gate implementation of the expressions.

4-3 ADDERS

Digital computers perform a variety of information-processing tasks. Among the basic functions encountered are the various arithmetic operations. The most basic arithmetic operation, no doubt, is the addition of two binary digits. This simple addition consists of four possible elementary operations, namely, $0 + 0 = 0$, $0 + 1 = 1$, $1 + 0 = 1$, and $1 + 1 = 10$. The first three operations produce a sum whose length is one digit, but when both augend and addend bits are equal to 1, the binary sum consists of two digits. The higher significant bit of this result is called a *carry*. When the augend and addend numbers contain more significant digits, the carry obtained from the addition of two bits is added to the next higher-order pair of significant bits. A combinational circuit that performs the addition of two bits is called a *half-adder*. One that performs the addition of three bits (two significant bits and a previous carry) is a *full-adder*. The name of the former stems from the fact that two half-adders can be employed to implement a full-adder. The two adder circuits are the first combinational circuits we shall design.

Half-Adder

From the verbal explanation of a half-adder, we find that this circuit needs two binary inputs and two binary outputs. The input variables designate the augend and addend bits; the output variables produce the sum and carry. It is necessary to specify two output variables because the result may consist of two binary digits. We arbitrarily assign symbols x and y to the two inputs and S (for sum) and C (for carry) to the outputs.

Now that we have established the number and names of the input and output variables, we are ready to formulate a truth table to identify exactly the function of the half-adder. This truth table is

x	y	C	S
0	0	0	0
0	1	0	1
1	0	0	1
1	1	1	0

The carry output is 0 unless both inputs are 1. The S output represents the least significant bit of the sum.

The simplified Boolean functions for the two outputs can be obtained directly from the truth table. The simplified sum of products expressions are

$$S = x'y + xy'$$

$$C = xy$$

The logic diagram for this implementation is shown in Fig. 4-2(a), as are four other implementations for a half-adder. They all achieve the same result as far as the input–output behavior is concerned. They illustrate the flexibility available to the designer when implementing even a simple combinational logic function such as this.

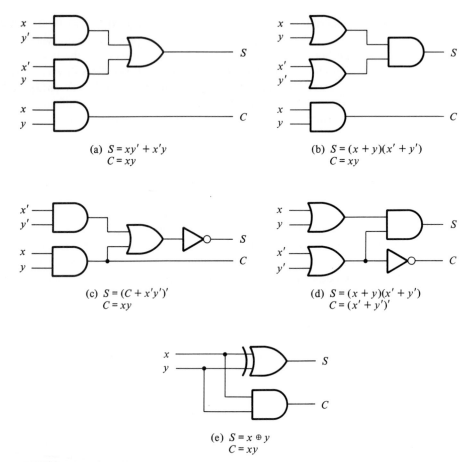

(a) $S = xy' + x'y$
    $C = xy$

(b) $S = (x + y)(x' + y')$
    $C = xy$

(c) $S = (C + x'y')'$
    $C = xy$

(d) $S = (x + y)(x' + y')$
    $C = (x' + y')'$

(e) $S = x \oplus y$
    $C = xy$

FIGURE 4-2
Various implementations of a half-adder

Figure 4-2(a), as mentioned before, is the implementation of the half-adder in sum of products. Figure 4-2(b) shows the implementation in product of sums:

$$S = (x + y)(x' + y')$$

$$C = xy$$

To obtain the implementation of Fig. 4-2(c), we note that S is the exclusive-OR of x and y. The complement of S is the equivalence of x and y (Section 2-6.):

$$S' = xy + x'y'$$

but $C = xy$, and, therefore, we have

$$S = (C + x'y')'$$

In Fig. 4-2(d), we use the product of sums implementation with C derived as follows:

$$C = xy = (x' + y')'$$

The half-adder can be implemented with an exclusive-OR and an AND gate, as shown in Fig. 4-2(e). This form is used later to show that two half-adder circuits are needed to construct a full-adder circuit.

Full-Adder

A full-adder is a combinational circuit that forms the arithmetic sum of three input bits. It consists of three inputs and two outputs. Two of the input variables, denoted by x and y, represent the two significant bits to be added. The third input, z, represents the carry from the previous lower significant position. Two outputs are necessary because the arithmetic sum of three binary digits ranges in value from 0 to 3, and binary 2 or 3 needs two digits. The two outputs are designated by the symbols S for sum and C for carry. The binary variable S gives the value of the least significant bit of the sum. The binary variable C gives the output carry. The truth table of the full-adder is

x	y	z	C	S
0	0	0	0	0
0	0	1	0	1
0	1	0	0	1
0	1	1	1	0
1	0	0	0	1
1	0	1	1	0
1	1	0	1	0
1	1	1	1	1

The eight rows under the input variables designate all possible combinations of 1's and 0's that these variables may have. The 1's and 0's for the output variables are determined from the arithmetic sum of the input bits. When all input bits are 0's, the output is 0. The S output is equal to 1 when only one input is equal to 1 or when all three inputs are equal to 1. The C output has a carry of 1 if two or three inputs are equal to 1.

The input and output bits of the combinational circuit have different interpretations at various stages of the problem. Physically, the binary signals of the input wires are considered binary digits added arithmetically to form a two-digit sum at the output wires. On the other hand, the same binary values are considered variables of Boolean functions when expressed in the truth table or when the circuit is implemented with logic gates. It is important to realize that two different interpretations are given to the values of the bits encountered in this circuit.

The input-output logical relationship of the full-adder circuit may be expressed in two Boolean functions, one for each output variable. Each output Boolean function re-

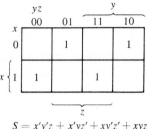

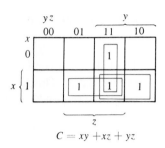

$$S = x'y'z + x'yz' + xy'z' + xyz$$

$$C = xy + xz + yz$$

FIGURE 4-3

Maps for a full-adder

quires a unique map for its simplification. Each map must have eight squares, since each output is a function of three input variables. The maps of Fig. 4-3 are used for simplifying the two output functions. The 1's in the squares for the maps of S and C are determined directly from the truth table. The squares with 1's for the S output do not combine in adjacent squares to give a simplified expression in sum of products. The C output can be simplified to a six-literal expression. The logic diagram for the full-adder implemented in sum of products is shown in Fig. 4-4. This implementation uses the following Boolean expressions:

$$S = x'y'z + x'yz' + xy'z' + xyz$$

$$C = xy + xz + yz$$

Other configurations for a full-adder may be developed. The product of sums implementation requires the same number of gates as in Fig. 4-4, with the number of AND and OR gates interchanged. A full-adder can be implemented with two half-adders and

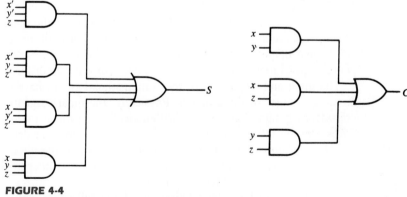

FIGURE 4-4

Implementation of a full-adder in sum of products

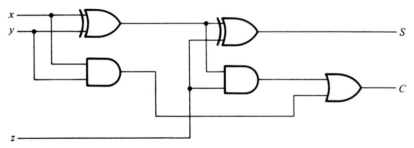

FIGURE 4-5
Implementation of a full-adder with two half-adders and an OR gate

one OR gate, as shown in Fig. 4-5. The S output from the second half-adder is the exclusive-OR of z and the output of the first half-adder, giving

$$S = z \oplus (x \oplus y)$$
$$= z'(xy' + x'y) + z(xy' + x'y)'$$
$$= z'(xy' + x'y) + z(xy + x'y')$$
$$= xy'z' + x'yz' + xyz + x'y'z$$

and the carry output is

$$C = z(xy' + x'y) + xy = xy'z + x'yz + xy$$

4-4 SUBTRACTORS

The subtraction of two binary numbers may be accomplished by taking the complement of the subtrahend and adding it to the minuend (Section 1-5). By this method, the subtraction operation becomes an addition operation requiring full-adders for its machine implementation. It is possible to implement subtraction with logic circuits in a direct manner, as done with paper and pencil. By this method, each subtrahend bit of the number is subtracted from its corresponding significant minuend bit to form a difference bit. If the minuend bit is smaller than the subtrahend bit, a 1 is borrowed from the next significant position. The fact that a 1 has been borrowed must be conveyed to the next higher pair of bits by means of a binary signal coming out (output) of a given stage and going into (input) the next higher stage. Just as there are half- and full-adders, there are half- and full-subtractors.

Half-Subtractor

A half-subtractor is a combinational circuit that subtracts two bits and produces their difference. It also has an output to specify if a 1 has been borrowed. Designate the minuend bit by x and the subtrahend bit by y. To perform $x - y$, we have to check the relative magnitudes of x and y. If $x \geqslant y$, we have three possibilities: $0 - 0 = 0$, $1 - 0 = 1$, and $1 - 1 = 0$. The result is called the *difference bit*. If $x < y$, we have $0 - 1$, and it is necessary to borrow a 1 from the next higher stage. The 1 borrowed from the next higher stage adds 2 to the minuend bit, just as in the decimal system a borrow adds 10 to a minuend digit. With the minuend equal to 2, the difference becomes $2 - 1 = 1$. The half-subtractor needs two outputs. One output generates the difference and will be designated by the symbol D. The second output, designated B for borrow, generates the binary signal that informs the next stage that a 1 has been borrowed. The truth table for the input–output relationships of a half-subtractor can now be derived as follows:

x	y	B	D
0	0	0	0
0	1	1	1
1	0	0	1
1	1	0	0

The output borrow B is a 0 as long as $x \geqslant y$. It is a 1 for $x = 0$ and $y = 1$. The D output is the result of the arithmetic operation $2B + x - y$.

The Boolean functions for the two outputs of the half-subtractor are derived directly from the truth table:

$$D = x'y + xy'$$

$$B = x'y$$

It is interesting to note that the logic for D is exactly the same as the logic for output S in the half-adder.

Full-Subtractor

A full-subtractor is a combinational circuit that performs a subtraction between two bits, taking into account that a 1 may have been borrowed by a lower significant stage. This circuit has three inputs and two outputs. The three inputs, x, y, and z, denote the minuend, subtrahend, and previous borrow, respectively. The two outputs, D and B, represent the difference and output borrow, respectively. The truth table for the circuit is

x	y	z	B	D
0	0	0	0	0
0	0	1	1	1
0	1	0	1	1
0	1	1	1	0
1	0	0	0	1
1	0	1	0	0
1	1	0	0	0
1	1	1	1	1

The eight rows under the input variables designate all possible combinations of 1's and 0's that the binary variables may take. The 1's and 0's for the output variables are determined from the subtraction of $x - y - z$. The combinations having input borrow $z = 0$ reduce to the same four conditions of the half-adder. For $x = 0$, $y = 0$, and $z = 1$, we have to borrow a 1 from the next stage, which makes $B = 1$ and adds 2 to x. Since $2 - 0 - 1 = 1$, $D = 1$. For $x = 0$ and $yz = 11$, we need to borrow again, making $B = 1$ and $x = 2$. Since $2 - 1 - 1 = 0$, $D = 0$. For $x = 1$ and $yz = 01$, we have $x - y - z = 0$, which makes $B = 0$ and $D = 0$. Finally, for $x = 1$, $y = 1$, $z = 1$, we have to borrow 1, making $B = 1$ and $x = 3$, and $3 - 1 - 1 = 1$, making $D = 1$.

The simplified Boolean functions for the two outputs of the full-subtractor are derived in the maps of Fig. 4-6. The simplified sum of products output functions are

$$D = x'y'z + x'yz' + xy'z' + xyz$$

$$B = x'y + x'z + yz$$

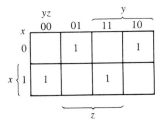

$$D = x'y'z + x'yz' + xy'z' + xyz$$

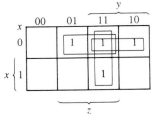

$$B = x'y + x'z + yz$$

FIGURE 4-6

Maps for a full-subtractor

4-5 CODE CONVERSION

The availability of a large variety of codes for the same discrete elements of information results in the use of different codes by different digital systems. It is sometimes necessary to use the output of one system as the input to another. A conversion circuit must be inserted between the two systems if each uses different codes for the same information. Thus, a code converter is a circuit that makes the two systems compatible even though each uses a different binary code.

To convert from binary code A to binary code B, the input lines must supply the bit combination of elements as specified by code A and the output lines must generate the corresponding bit combination of code B. A combinational circuit performs this transformation by means of logic gates. The design procedure of code converters will be illustrated by means of a specific example of conversion from the BCD to the excess-3 code.

The bit combinations for the BCD and excess-3 codes are listed in Table 1-2 (Section 1-7). Since each code uses four bits to represent a decimal digit, there must be four input variables and four output variables. Let us designate the four input binary variables by the symbols A, B, C, and D, and the four output variables by w, x, y, and z. The truth table relating the input and output variables is shown in Table 4-1. The bit combinations for the inputs and their corresponding outputs are obtained directly from Table 1-2. We note that four binary variables may have 16 bit combinations, only 10 of which are listed in the truth table. The six bit combinations not listed for the *input* variables are don't-care combinations. Since they will never occur, we are at liberty to assign to the output variables either a 1 or a 0, whichever gives a simpler circuit.

The maps in Fig. 4-7 are drawn to obtain a simplified Boolean function for each output. Each of the four maps of Fig. 4-7 represents one of the four outputs of this circuit as a function of the four input variables. The 1's marked inside the squares are obtained

TABLE 4-1
Truth Table for Code-Conversion Example

Input BCD				Output Excess-3 Code			
A	B	C	D	w	x	y	z
0	0	0	0	0	0	1	1
0	0	0	1	0	1	0	0
0	0	1	0	0	1	0	1
0	0	1	1	0	1	1	0
0	1	0	0	0	1	1	1
0	1	0	1	1	0	0	0
0	1	1	0	1	0	0	1
0	1	1	1	1	0	1	0
1	0	0	0	1	0	1	1
1	0	0	1	1	1	0	0

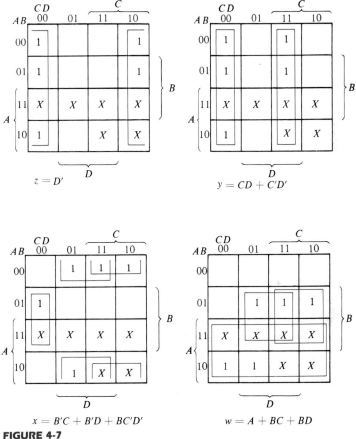

FIGURE 4-7

Maps for a BCD-to-excess-3-code converter

from the minterms that make the output equal to 1. The 1's are obtained from the truth table by going over the output columns one at a time. For example, the column under output z has five 1's; therefore, the map for z must have five 1's, each being in a square corresponding to the minterm that makes z equal to 1. The six don't-care combinations are marked by X's. One possible way to simplify the functions in sum of products is listed under the map of each variable.

A two-level logic diagram may be obtained directly from the Boolean expressions derived by the maps. There are various other possibilities for a logic diagram that implements this circuit. The expressions obtained in Fig. 4-7 may be manipulated algebraically for the purpose of using common gates for two or more outputs. This manipulation, shown below, illustrates the flexibility obtained with multiple-output systems when implemented with three or more levels of gates.

$$z = D'$$
$$y = CD + C'D' = CD + (C + D)'$$
$$x = B'C + B'D + BC'D' = B'(C + D) + BC'D'$$
$$= B'(C + D) + B(C + D)'$$
$$w = A + BC + BD = A + B(C + D)$$

The logic diagram that implements these expressions is shown in Fig. 4-8. In it we see that the OR gate whose output is $C + D$ has been used to implement partially each of three outputs.

Not counting input inverters, the implementation in sum of products requires seven AND gates and three OR gates. The implementation of Fig. 4-8 requires four AND gates, four OR gates, and one inverter. If only the normal inputs are available, the first implementation will require inverters for variables B, C, and D, whereas the second implementation requires inverters for variables B and D.

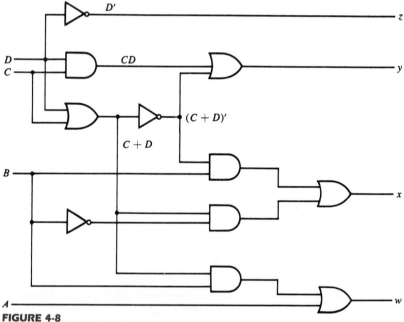

FIGURE 4-8

Logic diagram for a BCD-to-excess-3-code converter

4-6 ANALYSIS PROCEDURE

The design of a combinational circuit starts from the verbal specifications of a required function and culminates with a set of output Boolean functions or a logic diagram. The *analysis* of a combinational circuit is somewhat the reverse process. It starts with a given logic diagram and culminates with a set of Boolean functions, a truth table, or a verbal explanation of the circuit operation. If the logic diagram to be analyzed is accompanied by a function name or an explanation of what it is assumed to accomplish, then the analysis problem reduces to a verification of the stated function.

The first step in the analysis is to make sure that the given circuit is combinational and not sequential. The diagram of a combinational circuit has logic gates with no feedback paths or memory elements. A feedback path is a connection from the output of one gate to the input of a second gate that forms part of the input to the first gate. Feedback paths or memory elements in a digital circuit define a sequential circuit and must be analyzed according to procedures outlined in Chapter 6 or Chapter 9.

Once the logic diagram is verified as a combinational circuit, one can proceed to obtain the output Boolean functions and/or the truth table. If the circuit is accompanied by a verbal explanation of its function, then the Boolean functions or the truth table is sufficient for verification. If the function of the circuit is under investigation, then it is necessary to interpret the operation of the circuit from the derived truth table. The success of such investigation is enhanced if one has previous experience and familiarity with a wide variety of digital circuits. The ability to correlate a truth table with an information-processing task is an art one acquires with experience.

To obtain the output Boolean functions from a logic diagram, proceed as follows:

1. Label with arbitrary symbols all gate outputs that are a function of the input variables. Obtain the Boolean functions for each gate.
2. Label with other arbitrary symbols those gates that are a function of input variables and/or previously labeled gates. Find the Boolean functions for these gates.
3. Repeat the process outlined in step 2 until the outputs of the circuit are obtained.
4. By repeated substitution of previously defined functions, obtain the output Boolean functions in terms of input variables only.

Analysis of the combinational circuit in Fig. 4-9 illustrates the proposed procedure. We note that the circuit has three binary inputs, A, B, and C, and two binary outputs, F_1 and F_2. The outputs of various gates are labeled with intermediate symbols. The outputs of gates that are a function of input variables only are F_2, T_1, and T_2. The Boolean functions for these three outputs are

$$F_2 = AB + AC + BC$$

$$T_1 = A + B + C$$

$$T_2 = ABC$$

Next we consider outputs of gates that are a function of already defined symbols:

$$T_3 = F_2'T_1$$

$$F_1 = T_3 + T_2$$

The output Boolean function F_2 just expressed is already given as a function of the in-

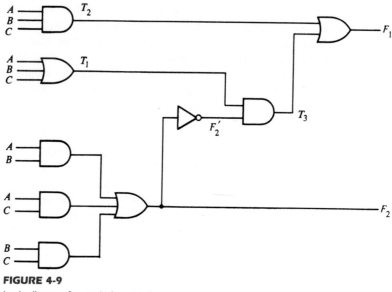

FIGURE 4-9
Logic diagram for analysis example

puts only. To obtain F_1 as a function of A, B, and C, form a series of substitutions as follows:

$$F_1 = T_3 + T_2 = F_2' T_1 + ABC = (AB + AC + BC)'(A + B + C) + ABC$$
$$= (A' + B')(A' + C')(B' + C')(A + B + C) + ABC$$
$$= (A' + B'C')(AB' + AC' + BC' + B'C) + ABC$$
$$= A'BC' + A'B'C + AB'C' + ABC$$

If we want to pursue the investigation and determine the information-transformation task achieved by this circuit, we can derive the truth table directly from the Boolean functions and try to recognize a familiar operation. For this example, we note that the circuit is a full-adder, with F_1 being the sum output and F_2 the carry output. A, B, and C are the three inputs added arithmetically.

The derivation of the truth table for the circuit is a straightforward process once the output Boolean functions are known. To obtain the truth table directly from the logic diagram without going through the derivations of the Boolean functions, proceed as follows:

1. Determine the number of input variables to the circuit. For n inputs, form the 2^n possible input combinations of 1's and 0's by listing the binary numbers from 0 to $2^n - 1$.

2. Label the outputs of selected gates with arbitrary symbols.
3. Obtain the truth table for the outputs of those gates that are a function of the input variables only.
4. Proceed to obtain the truth table for the outputs of those gates that are a function of previously defined values until the columns for all outputs are determined.

This process can be illustrated using the circuit of Fig. 4-9. In Table 4-2, we form the eight possible combinations for the three input variables. The truth table for F_2 is determined directly from the values of A, B, and C, with F_2 equal to 1 for any combination that has two or three inputs equal to 1. The truth table for F_2' is the complement of F_2. The truth tables for T_1 and T_2 are the OR and AND functions of the input variables, respectively. The values for T_3 are derived from T_1 and F_2': T_3 is equal to 1 when both T_1 and F_2' are equal to 1, and to 0 otherwise. Finally, F_1 is equal to 1 for those combinations in which either T_2 or T_3 or both are equal to 1. Inspection of the truth table combinations for A, B, C, F_1, and F_2 of Table 4-2 shows that it is identical to the truth table of the full-adder given in Section 4-3 for x, y, z, S, and C, respectively.

TABLE 4-2
Truth Table for the Logic Diagram of Fig. 4-9

A	B	C	F_2	F_2'	T_1	T_2	T_3	F_1
0	0	0	0	1	0	0	0	0
0	0	1	0	1	1	0	1	1
0	1	0	0	1	1	0	1	1
0	1	1	1	0	1	0	0	0
1	0	0	0	1	1	0	1	1
1	0	1	1	0	1	0	0	0
1	1	0	1	0	1	0	0	0
1	1	1	1	0	1	1	0	1

Consider now a combinational circuit that has don't-care input combinations. When such a circuit is designed, the don't-care combinations are marked by X's in the map and assigned an output of either 1 or 0, whichever is more convenient for the simplification of the output Boolean function. When a circuit with don't-care combinations is being analyzed, the situation is entirely different. Even though we assume that the don't-care input combinations will never occur, if any one of these combinations is applied to the inputs (intentionally or in error), a binary output will be present. The value of the output will depend on the choice for the X's taken during the design. Part of the analysis of such a circuit may involve the determination of the output values for the don't-care input combinations. As an example, consider the BCD-to-excess-3-code converter designed in Section 4-5. The outputs obtained when the six unused combinations of the BCD code are applied to the inputs are

Unused BCD Inputs				Outputs			
A	B	C	D	w	x	y	z
1	0	1	0	1	1	0	1
1	0	1	1	1	1	1	0
1	1	0	0	1	1	1	1
1	1	0	1	1	0	0	0
1	1	1	0	1	0	0	1
1	1	1	1	1	0	1	0

These outputs may be derived by means of the truth table analysis method as outlined in this section. In this particular case, the outputs may be obtained directly from the maps of Fig. 4-7. From inspection of the maps, we determine whether the X's in the corresponding minterm squares for each output have been included with the 1's or the 0's. For example, the square for minterm m_{10} (1010) has been included with the 1's for outputs w, x, and z, but not for y. Therefore, the outputs for m_{10} are $wxyz = 1101$, as listed in the previous table. We also note that the first three outputs in the table have no meaning in the excess-3 code, and the last three outputs correspond to decimal 5, 6, and 7, respectively. This coincidence is entirely a function of the choice for the X's taken during the design.

4-7   MULTILEVEL NAND CIRCUITS

Combinational circuits are more frequently constructed with NAND or NOR gates rather than AND and OR gates. NAND and NOR gates are more common from the hardware point of view because they are readily available in integrated-circuit form. Because of the prominence of NAND and NOR gates in the design of combinational circuits, it is important to be able to recognize the relationships that exist between circuits constructed with AND-OR gates and their equivalent NAND or NOR diagrams.

The implementation of two-level NAND and NOR logic diagrams was presented in Section 3-6. Here we consider the more general case of multilevel circuits. The procedure for obtaining NAND circuits is presented in this section, and for NOR circuits in the next section.

Universal Gate

The NAND gate is said to be a universal gate because any digital system can be implemented with it. Combinational circuits and sequential circuits as well can be constructed with this gate because the flip-flop circuit (the memory element most frequently used in sequential circuits) can be constructed from two NAND gates connected back to back, as shown in Section 6-2.

To show that any Boolean function can be implemented with NAND gates, we need only show that the logical operations AND, OR, and NOT can be implemented with

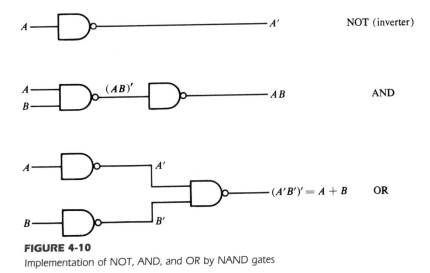

FIGURE 4-10
Implementation of NOT, AND, and OR by NAND gates

NAND gates. The implementation of the AND, OR, and NOT operations with NAND gates is shown in Fig. 4-10. The NOT operation is obtained from a one-input NAND gate, actually another symbol for an inverter circuit. The AND operation requires two NAND gates. The first produces the inverted AND and the second acts as an inverter to produce the normal output. The OR operation is achieved through a NAND gate with additional inverters in each input.

Boolean-Function Implementation

One possible way to implement a Boolean function with NAND gates is to obtain the simplified Boolean function in terms of Boolean operators and then convert the function to NAND logic. The conversion of an algebraic expression from AND, OR, and complement to NAND can be done by simple circuit-manipulation techniques that change AND-OR diagrams to NAND diagrams.

To facilitate the conversion to NAND logic, it is convenient to use the two alternate graphic symbols shown in Fig. 4-11. (These two graphic symbols for the NAND gate were introduced in Fig. 3-17(a) and are repeated here for convenience.) The AND-invert graphic symbol consists of an AND graphic symbol followed by a small circle. The invert-OR graphic symbol consists of an OR graphic symbol that is preceded by small circles in all the inputs. Either symbol can be used to represent a NAND gate.

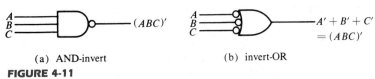

(a) AND-invert   (b) invert-OR

FIGURE 4-11
Two graphic symbols for a NAND gate

To obtain a multilevel NAND diagram from a Boolean expression, proceed as follows:

1. From the given Boolean expression, draw the logic diagram with AND, OR, and inverter gates. Assume that both the normal and complement inputs are available.
2. Convert all AND gates to NAND gates with AND-invert graphic symbols.
3. Convert all OR gates to NAND gates with invert-OR graphic symbols.
4. Check all small circles in the diagram. For every small circle that is not compensated by another small circle along the same line, insert an inverter (one-input NAND gate) or complement the input variable.

This procedure will be demonstrated with two examples. First, consider the Boolean function

$$F = A + (B' + C)(D' + BE')$$

Although it is possible to remove the parentheses and convert the expression into a standard sum of products form, we choose to implement it as a multilevel circuit for illustration. The AND-OR implementation is shown in Fig. 4-12(a). There are four levels of gating in the circuit. The first level has an AND and an OR gate. The second level has an OR gate followed by an AND gate in the third level and an OR gate in the fourth level. A logic diagram with a pattern of alternate levels of AND and OR gates

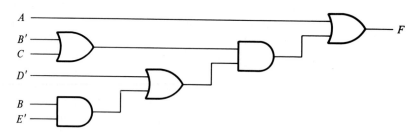

(a) AND-OR diagram

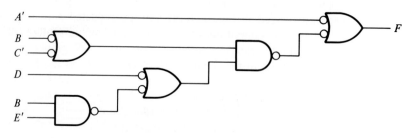

(b) NAND diagram using two graphic symbols

FIGURE 4-12

Implementing $F = A + (B' + C)(D' + BE')$ with NAND gates (continued on next page)

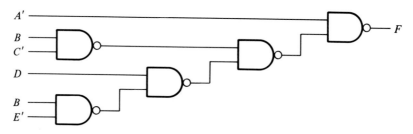

(c) NAND diagram using one graphic symbol

FIGURE 4-12 (continued)

can be easily converted into a NAND circuit. This is shown in Fig. 4-12(b). The procedure requires that we change every AND gate to an AND-invert graphic symbol and every OR gate to an invert-OR graphic symbol. The NAND circuit performs the same logic as the AND-OR circuit as long as the complementing small circles do not change the value of the function. Any connection between an output of a gate that has a complementing circle and the input of another gate that also has a complementing circle represents double complementation and does not change the logic of the circuit. However, the small circles associated with inputs A, B', C, and D' cause extra complementations that are not compensated with other small circles along the same line. We can insert inverters after each of these inputs or, as shown in the figure, complement the literals to obtain A', B, C', and D.

Because it does not matter whether we use the AND-invert or the invert-OR graphic symbol to represent a NAND gate, the diagram of Fig. 4-12(c) is identical to the NAND diagram of part (b). In fact, the diagram of Fig. 4-12(b) is preferable because it represents a clearer picture of the Boolean expression it implements.

As another example, consider the multilevel Boolean expression

$$F = (CD + E)(A + B')$$

The AND-OR implementation is shown in Fig. 4-13(a) with three levels of gating. The conversion into a NAND circuit is presented in part (b) of the diagram. The three additional small circles associated with inputs E, A, and B' cause these three literals to be complemented to E', A', and B. The small circle in the last NAND gate complements the output, so we need to insert an inverter gate at the output in order to complement the signal again and obtain the original value.

The number of NAND gates required to implement the expression of the first example is the same as the number of AND and OR gates in the AND-OR diagram. The number of NAND gates in the second example is equal to the number of AND-OR gates plus an additional inverter in the output. In general, the number of NAND gates required to implement a Boolean expression is equal to the number of AND-OR gates except for an occasional inverter. This is true provided both the normal and complement inputs are available, because the conversion forces certain input variables to be complemented.

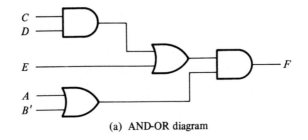

(a) AND-OR diagram

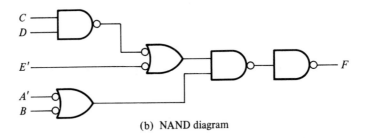

(b) NAND diagram

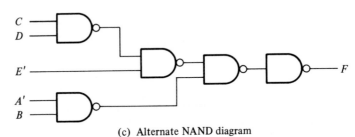

(c) Alternate NAND diagram

FIGURE 4-13
Implementing $F = (CD + E)(A + B')$ with NAND gates

Analysis Procedure

The foregoing procedure considered the problem of deriving a NAND logic diagram from a given Boolean function. The reverse process is the analysis problem that starts with a given NAND logic diagram and culminates with a Boolean expression or a truth table. The analysis of NAND logic diagrams follows the same procedures presented in Section 4-6 for the analysis of combinational circuits. The only difference is that NAND logic requires a repeated application of DeMorgan's theorem. We shall now demonstrate the derivation of the Boolean function from a logic diagram. Then we will show the derivation of the truth table directly from the NAND logic diagram. Finally, a method will be presented for converting a NAND logic diagram to AND-OR logic diagram.

Derivation of the Boolean Function by Algebraic Manipulation

The procedure for deriving the Boolean function from a logic diagram is outlined in Section 4-6. This procedure is demonstrated for the NAND logic diagram shown in Fig. 4-14. First, all gate outputs are labeled with arbitrary symbols. Second, the Boolean functions for the outputs of gates that receive only external inputs are derived:

$$T_1 = (CD)' = C' + D'$$
$$T_2 = (BC')' = B' + C$$

The second form follows directly from DeMorgan's theorem and may, at times, be more convenient to use. Third, Boolean functions of gates that have inputs from previously derived functions are determined in consecutive order until the output is expressed in terms of input variables:

$$T_3 = (B'T_1)' = (B'C' + B'D')'$$
$$= (B + C)(B + D) = B + CD$$
$$T_4 = (AT_3)' = [A(B + CD)]'$$
$$F = (T_2 T_4)' = \{(BC')'[A(B + CD)]'\}'$$
$$= BC' + A(B + CD)$$

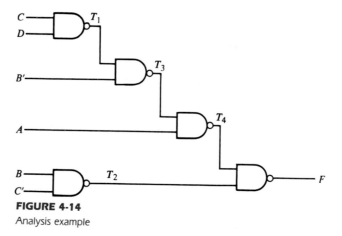

FIGURE 4-14
Analysis example

Derivation of the Truth Table

The procedure for obtaining the truth table directly from a logic diagram is also outlined in Section 4-6. This procedure is demonstrated for the NAND logic diagram of Fig. 4-14. First, the four input variables, together with their 16 combinations of 1's and 0's, are listed as in Table 4-3. Second, the outputs of all gates are labeled with arbitrary symbols as in Fig. 4-14. Third, we obtain the truth table for the outputs of those

TABLE 4-3
Truth Table for the Circuit of Figure 4-14

A	B	C	D	T_1	T_2	T_3	T_4	F
0	0	0	0	1	1	0	1	0
0	0	0	1	1	1	0	1	0
0	0	1	0	1	1	0	1	0
0	0	1	1	0	1	1	1	0
0	1	0	0	1	0	1	1	1
0	1	0	1	1	0	1	1	1
0	1	1	0	1	1	1	1	0
0	1	1	1	0	1	1	1	0
1	0	0	0	1	1	0	1	0
1	0	0	1	1	1	0	1	0
1	0	1	0	1	1	0	1	0
1	0	1	1	0	1	1	0	1
1	1	0	0	1	0	1	0	1
1	1	0	1	1	0	1	0	1
1	1	1	0	1	1	1	0	1
1	1	1	1	0	1	1	0	1

gates that are a function of the input variables only. These are T_1 and T_2. $T_1 = (CD)'$; so we mark 0's in those rows where both C and D are equal to 1 and fill the rest of the rows of T_1 with 1's. Also, $T_2 = (BC')'$; so we mark 0's in those rows where $B = 1$ and $C = 0$, and fill the rest of the rows of T_2 with 1's. We then proceed to obtain the truth table for the outputs of those gates that are a function of previously defined outputs until the column for the output F is determined. It is now possible to obtain an algebraic expression for the output from the derived truth table. The map shown in Fig. 4-15 is ob-

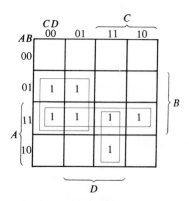

$$F = AB + BC' + ACD$$

FIGURE 4-15
Derivation of F from Table 4-3

tained directly from Table 4-3 and has 1's in the squares of those minterms for which F is equal to 1. The simplified expression obtained from the map is

$$F = AB + ACD + BC' = A(B + CD) + BC'$$

Transformation to AND-OR Diagram

It is sometimes convenient to convert a NAND logic diagram to its equivalent AND-OR logic diagram to facilitate the analysis procedure. By doing so, the Boolean expression can be derived more easily from the diagram without employing DeMorgan's the-

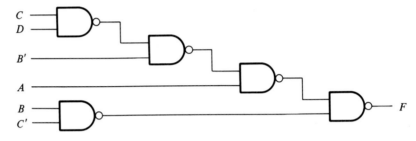

(a) NAND logic diagram

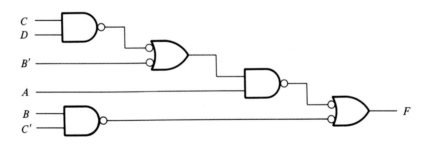

(b) Substitution of invert-OR symbols in alternate levels

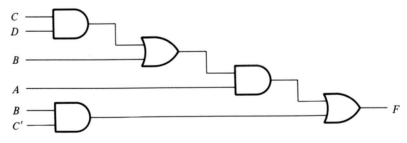

(c) AND-OR logic diagram

FIGURE 4-16
Conversion of NAND logic diagram to AND-OR

orem. The conversion is achieved through a change in graphic symbols from AND-invert to invert-OR in *alternate* levels in the gate structure. The first level to be changed to an invert-OR symbol should be the last level. These changes produce pairs of small circles along the same line, which are then removed since they represent double complementation. Any small circle associated with an input can be removed provided the input variable is complemented. A one-input AND or OR gate with a small circle in the input or output represents an inverter circuit.

The procedure is demonstrated in Fig. 4-16. The NAND logic diagram of Fig. 4-16(a) is to be converted to an equivalent AND-OR diagram. The graphic symbol of the NAND gate in the last level is changed to an invert-OR symbol. Looking for alternate levels, we find one more gate requiring a change of symbol, as shown in Fig. 4-16(b). Any two small circles along the same line are removed. The small circle connected to input B' is removed and the input variable is complemented. The required AND-OR logic diagram is shown in Fig. 4-16(c). The Boolean expression for F can be easily determined from the AND-OR diagram to be

$$F = BC' + A(B + CD)$$

4-8 MULTILEVEL NOR CIRCUITS

The NOR function is the dual of the NAND function. For this reason, all procedures and rules for NOR logic form a dual of the corresponding procedures and rules developed for NAND logic. This section enumerates various methods for NOR logic implementation and analysis by following the same list of topics used for NAND logic. However, less detailed explanation is included so as to avoid excessive repetition of the material in Section 4-7.

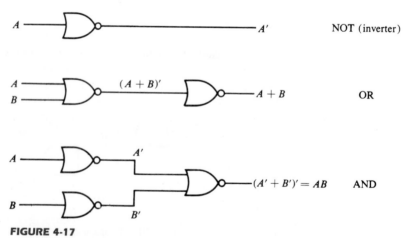

FIGURE 4-17
Implementation of NOT, OR, and AND by NOR gates

Universal Gate

The NOR gate is universal because any Boolean function can be implemented with it, including a flip-flop circuit, as shown in Section 6-2. The conversion of AND, OR, and NOT to NOR is shown in Fig. 4-17. The NOT operation is obtained from a one-input NOR gate, yet another symbol for an inverter circuit. The OR operation requires two NOR gates. The first produces the inverted-OR and the second acts as an inverter to obtain the normal output. The AND operation is achieved through a NOR gate with additional inverters at each input.

Boolean-Function Implementation

The two graphic symbols for the NOR gate are shown in Fig. 4-18. The OR-invert symbol defines the NOR operation as an OR followed by a complement. The invert-AND symbol complements each input and then performs an AND operation. The two symbols designate the same NOR operation and are logically identical because of DeMorgan's theorem.

The procedure for implementing a Boolean function with NOR gates is similar to the procedure outlined in the previous section for NAND gates.

1. Draw the AND-OR logic diagram from the given algebraic expression. Assume that both the normal and complement inputs are available.
2. Convert all OR gates to NOR gates with OR-invert graphic symbols.
3. Convert all AND gates to NOR gates with invert-AND graphic symbols.
4. Any small circle that is not compensated by another small circle along the same line needs an inverter or the complementation of the input variable.

The procedure is illustrated in Fig. 4-19 for the Boolean function

$$F = (AB + E)(C + D)$$

The AND-OR implementation of the expression is shown in the logic diagram of Fig. 4-19(a). For each OR gate, we substitute a NOR gate with the OR-invert graphic symbol. For each AND gate, we substitute a NOR gate with the invert-AND graphic symbol. The two small circles associated with inputs A and B cause these two variables to be complemented to A' and B', respectively. The NOR diagram is shown in Fig. 4-19(b). The diagram of Fig. 4-19(c) is an alternate way of drawing the diagram using only one type of graphic symbol for the NOR gate.

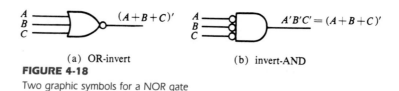

(a) OR-invert   (b) invert-AND

FIGURE 4-18

Two graphic symbols for a NOR gate

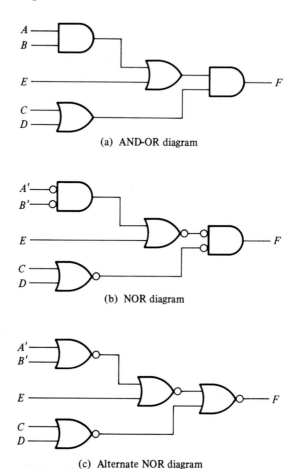

(a) AND-OR diagram

(b) NOR diagram

(c) Alternate NOR diagram

FIGURE 4-19
Implementing $F = (AB + E)(C + D)$ with NOR gates

In general, the number of NOR gates required to implement a Boolean function will be the same as the number of gates in the AND-OR diagram. This is true provided both the normal and complement inputs are available, because the conversion may require that certain input variables be complemented.

Analysis Procedure

The analysis of NOR logic diagrams follows the same procedure presented in Section 4-6 for the analysis of combinational circuits. To derive the Boolean expression from a logic diagram, we mark the outputs of various gates with arbitrary symbols. By repetitive substitutions, we obtain the output variable as a function of the input variables.

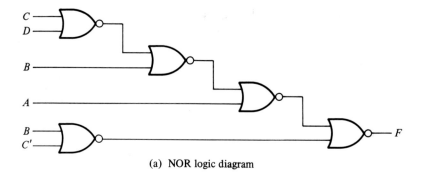

(a) NOR logic diagram

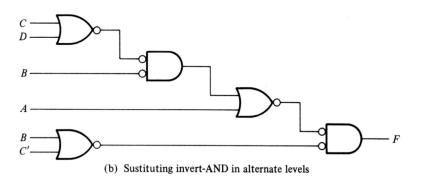

(b) Sustituting invert-AND in alternate levels

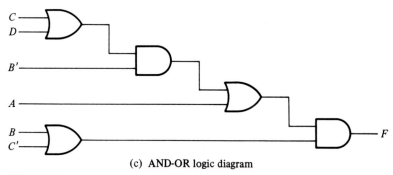

(c) AND-OR logic diagram

FIGURE 4-20

Conversion of NOR diagram to AND-OR

To obtain the truth table from a logic diagram without first deriving the Boolean expression, we form a table with the n variables by listing the 2^n binary combinations. The truth table of selected NOR gate outputs is derived in succession until the output truth table is obtained. The output expression of a typical NOR gate is of the form $T = (A + B' + C)'$. By using DeMorgan's theorem, this can be expressed as $T = A'BC'$. The truth table for T is marked with 1's for those combinations where $ABC = 010$ and the rest of the rows are filled with 0's.

The conversion of a NOR logic diagram to an AND-OR diagram is achieved through a change of graphic symbols from OR-invert to invert-AND starting from the last logic level and in alternate levels. Pairs of small circles along the same line are removed. A one-input AND or OR gate is removed, but if it has a small circle at the input or output, it is converted to an inverter. Any small circle associated with an input is removed and the input variable is complemented.

This procedure is demonstrated in Fig. 4-20, where the NOR logic diagram in part (a) is converted to an AND-OR diagram. The graphic symbol of the gate in the last (fourth) logic level is changed to an invert-AND. Looking for alternate levels, we find a gate in level two that needs to undergo a symbol change, as shown in part (b). Any two circles along the same line are removed. The circle associated with external input B is removed and the input variable is changed to B'. The required AND-OR logic diagram is drawn in part (c). The Boolean expression for the circuit can be obtained by inspection and then manipulated into a product of sums form:

$$F = [(C + D)B' + A](B + C')$$
$$= (A + C + D)(A + B')(B + C')$$

4-9 EXCLUSIVE-OR FUNCTION

The exclusive-OR (XOR) denoted by the symbol $\oplus$ is a logical operation that performs the following Boolean operation:

$$x \oplus y = xy' + x'y$$

It is equal to 1 if only x is equal to 1 or if only y is equal to 1 but not when both are equal to 1. The exclusive-NOR, also known as equivalence, performs the following Boolean operation:

$$(x \oplus y)' = xy + x'y'$$

It is equal to 1 if both x and y are equal to 1 or if both are equal to 0. The exclusive-NOR can be shown to be the complement of the exclusive-OR by means of a truth table or by algebraic manipulation.

$$(x \oplus y)' = (xy' + x'y)' = (x' + y)(x + y') = xy + x'y'$$

The following identities apply to the exclusive-OR operation:

$$x \oplus 0 = x \qquad\qquad x \oplus 1 = x'$$
$$x \oplus x = 0 \qquad\qquad x \oplus x' = 1$$
$$x \oplus y' = (x \oplus y)' \qquad x' \oplus y = (x \oplus y)'$$

Any of these identities can be proven by using a truth table or by replacing the $\oplus$ operation by its equivalent Boolean expression. It can be shown also that the exclusive-OR operation is both commutative and associative.

$$A \oplus B = B \oplus A$$

$$(A \oplus B) \oplus C = A \oplus (B \oplus C) = A \oplus B \oplus C$$

This means that the two inputs to an exclusive-OR gate can be interchanged without affecting the operation. It also means that we can evaluate a three-variable exclusive-OR operation in any order and for this reason, three or more variables can be expressed without parentheses. This would imply the possibility of using exclusive-OR gates with three or more inputs. However, multiple-input exclusive-OR gates are difficult to fabricate with hardware. In fact even a two-input function is usually constructed with other types of gates. A two-input exclusive-OR function is constructed with conventional gates using two inverters, two AND gates, and an OR gate, as shown in Fig. 4-21(a). Figure 4-21(b) shows the implementation of the exclusive-OR with four NAND gates.

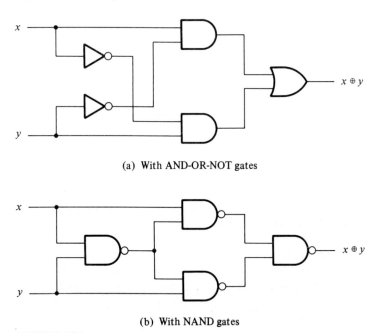

(a) With AND-OR-NOT gates

(b) With NAND gates

FIGURE 4-21
Exclusive-OR implementations

The first NAND gate performs the operation $(xy)' = (x' + y')$. The other two-level NAND circuit produces the sum of products of its inputs:

$$(x' + y')x + (x' + y')y = xy' + x'y = x \oplus y$$

Only a limited number of Boolean functions can be expressed in terms of exclusive-OR operations. Nevertheless, this function emerges quite often during the design of digital systems. It is particularly useful in arithmetic operations and error-detection and correction circuits.

Odd Function

The exclusive-OR operation with three or more variables can be converted into an ordinary Boolean function by replacing the $\oplus$ symbol with its equivalent Boolean expression. In particular, the three-variable case can be converted to a Boolean expression as follows:

$$
\begin{aligned}
A \oplus B \oplus C &= (AB' + A'B)C' + (AB + A'B')C \\
&= AB'C' + A'BC' + ABC + A'B'C \\
&= \Sigma\,(1, 2, 4, 7)
\end{aligned}
$$

The Boolean expression clearly indicates that the three-variable exclusive-OR function is equal to 1 if only one variable is equal to 1 or if all three variables are equal to 1. Contrary to the two-variable case, where only one variable must be equal to 1, in the three or more variable case, the requirement is that an odd number of variables be equal to 1. As a consequence, the multiple-variable exclusive-OR operation is defined as an *odd function*.

The Boolean function derived from the three-variable exclusive-OR operation is expressed as the logical sum of four minterms whose binary numerical values are 001, 010, 100, and 111. Each of these binary numbers has an odd number of 1's. The other four minterms not included in the function are 000, 011, 101, and 110, and they have an even number of 1's in their binary numerical values. In general, an n-variable exclusive-OR function is an odd function defined as the logical sum of the $2^n/2$ minterms whose binary numerical values have an odd number of 1's.

The definition of an odd function can be clarified by plotting it in a map. Figure 4-22(a) shows the map for the three-variable exclusive-OR function. The four minterms of the function are a unit distance apart from each other. The odd function is identified from the four minterms whose binary values have an odd number of 1's. The complement of an odd function is an even function. As shown in Fig. 4-22(b), the three-variable even function is equal to 1 when an even number of variables is equal to 1 (including the condition that none of the variables is equal to 1).

The 3-input odd function is implemented by means of 2-input exclusive-OR gates, as shown in Fig. 4-23(a). The complement of an odd function is obtained by replacing the output gate with an exclusive-NOR gate, as shown in Fig. 4-23(b).

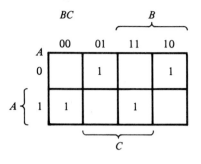

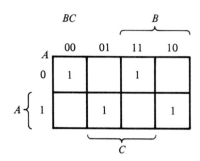

(a) Odd function
$F = A \oplus B \oplus C$

(b) Even function
$F = (A \oplus B \oplus C)'$

FIGURE 4-22
Map for a three-variable exclusive-OR function

Consider now the the four-variable exclusive-OR operation. By algebraic manipulation, we can obtain the sum of minterms for this function:

$$A \oplus B \oplus C \oplus D = (AB' + A'B) \oplus (CD' + C'D)$$
$$= (AB' + A'B)(CD + C'D') + (AB + A'B')(CD' + C'D)$$
$$= \Sigma (1, 2, 4, 7, 8, 11, 13, 14)$$

There are 16 minterms for a four-variable Boolean function. Half of the minterms have binary numerical values with an odd number of 1's; the other half of the minterms have binary numerical values with an even number of 1's. When plotting the function in the map, the binary numerical value for a minterm is determined from the row and column numbers of the square that represents the minterm. The map of Fig. 4-24(a) is a plot of the four-variable exclusive-OR function. This is an odd function because the binary values of all the minterms have an odd number of 1's. The complement of an odd function is an even function. As shown in Fig. 4-24(b), the four-variable even function is equal to 1 when an even number of variables is equal to 1.

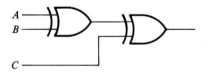

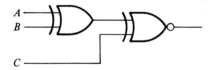

(a) 3-input odd function

(b) 3-input even function

FIGURE 4-23
Logic diagram of odd and even functions

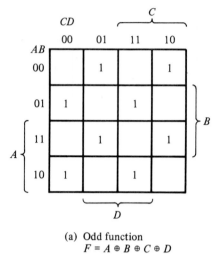

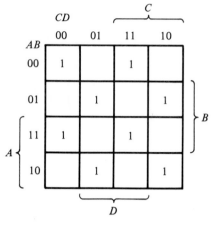

(a) Odd function
$F = A \oplus B \oplus C \oplus D$

(b) Even function
$F = (A \oplus B \oplus C \oplus D)'$

FIGURE 4-24
Map for a four-variable exclusive-OR function

Parity Generation and Checking

Exclusive-OR functions are very useful in systems requiring error-detection and correction codes. As discussed in Section 1-7, a parity bit is used for the purpose of detecting errors during transmission of binary information. A parity bit is an extra bit included with a binary message to make the number of 1's either odd or even. The message, including the parity bit, is transmitted and then checked at the receiving end for errors. An error is detected if the checked parity does not correspond with the one transmitted. The circuit that generates the parity bit in the transmitter is called a *parity generator*. The circuit that checks the parity in the receiver is called a *parity checker*.

As an example, consider a 3-bit message to be transmitted together with an even parity bit. Table 4-4 shows the truth table for the parity generator. The three bits, x, y, and z, constitute the message and are the inputs to the circuit. The parity bit P is the output. For even parity, the bit P must be generated to make the total number of 1's even (including P). From the truth table, we see that P constitutes an odd function be-cause it is equal to 1 for those minterms whose numerical values have an odd number of 1's. Therefore, P can be expressed as a three-variable exclusive-OR function:

$$P = x \oplus y \oplus z$$

The logic diagram for the parity generator is shown in Fig. 4-25(a).

The three bits in the message together with the parity bit are transmitted to their destination, where they are applied to a parity-checker circuit to check for possible er-rors in the transmission. Since the information was transmitted with even parity, the

TABLE 4-4
Even-Parity-Generator Truth Table

Three-Bit Message			Parity Bit
x	y	z	P
0	0	0	0
0	0	1	1
0	1	0	1
0	1	1	0
1	0	0	1
1	0	1	0
1	1	0	0
1	1	1	1

four bits received must have an even number of 1's. An error occurs during the transmission if the four bits received have an odd number of 1's, indicating that one bit has changed in value during transmission. The output of the parity checker, denoted by C, will be equal to 1 if an error occurs, that is, if the four bits received have an odd number of 1's. Table 4-5 is the truth table for the even-parity checker. From it we see that the function C consists of the eight minterms with binary numerical values having an odd number of 1's. This corresponds to the map of Fig. 4-24(a), which represents an odd function. The parity checker can be implemented with exclusive-OR gates:

$$C = x \oplus y \oplus z \oplus P$$

The logic diagram of the parity checker is shown in Fig. 4-25(b).

It is worth noting that the parity generator can be implemented with the circuit of Fig. 4-25(b) if the input P is connected to logic-0 and the output is marked with P. This is because $z \oplus 0 = z$, causing the value of z to pass through the gate unchanged. The advantage of this is that the same circuit can be used for both parity generation and checking.

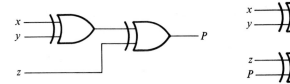

      (a) 3-bit even parity generator   (b) 4-bit even parity checker

FIGURE 4-25
Logic diagram of a parity generator and checker

TABLE 4-5
Even-Parity-Checker Truth Table

Four Bits Received				Parity Error Check
x	y	z	P	C
0	0	0	0	0
0	0	0	1	1
0	0	1	0	1
0	0	1	1	0
0	1	0	0	1
0	1	0	1	0
0	1	1	0	0
0	1	1	1	1
1	0	0	0	1
1	0	0	1	0
1	0	1	0	0
1	0	1	1	1
1	1	0	0	0
1	1	0	1	1
1	1	1	0	1
1	1	1	1	0

It is obvious from the foregoing example that parity-generation and checking circuits always have an output function that includes half of the minterms whose numerical values have either an odd or even number of 1's. As a consequence, they can be implemented with exclusive-OR gates. A function with an even number of 1's is the complement of an odd function. It is implemented with exclusive-OR gates except that the gate associated with the output must be an exclusive-NOR to provide the required complementation.

REFERENCES

1. HILL, F. J., and G. R. PETERSON, *Introduction to Switching Theory and Logical Design,* 3rd Ed. New York: John Wiley, 1981.

2. KOHAVI, Z., *Switching and Automata Theory,* 2nd Ed. New York: McGraw-Hill, 1978.

3. ROTH, C. H., *Fundamentals of Logic Design,* 3rd Ed. St. Paul, Minnesota: West Publishing Co., 1985.

4. BOOTH, T. L., *Introduction to Computer Engineering,* 3rd Ed. New York: John Wiley, 1984.

5. MANO, M. M., *Computer Engineering: Hardware Design.* Englewood Cliffs, NJ: Prentice-Hall, 1988.

6. FLETCHER, W. I., *An Engineering Approach to Digital Design.* Englewood Cliffs, NJ: Prentice-Hall, 1979.

7. ERCEGOVAC, M. D., and T. LANG, *Digital Systems and Hardware/Firmware Algorithms*. New York: John Wiley, 1985.

8. MANGE, D., *Analysis and Synthesis of Logic Systems*. Norwood, MA: Artech House, 1986.

9. SHIVA, S. G., *Introduction to Logic Design*. Glenview, IL: Scott, Foresman, 1988.

10. McCLUSKEY, E. J., *Logic Design Principles*. Englewood Cliffs, NJ: Prentice-Hall, 1986.

PROBLEMS

4-1 Design a combinational circuit with three inputs and one output. The output is equal to logic-1 when the binary value of the input is less than 3. The output is logic-0 otherwise.

4-2 Design a combinational circuit with three inputs, x, y, and z, and three outputs, A, B, and C. When the binary input is 0, 1, 2, or 3, the binary output is one greater than the input. When the binary input is 4, 5, 6, or 7, the binary output is one less than the input.

4-3 A majority function is generated in a combinational circuit when the output is equal to 1 if the input variables have more 1's than 0's. The output is 0 otherwise. Design a 3-input majority function.

4-4 Design a combinational circuit that adds one to a 4-bit binary number, $A_3 A_2 A_1 A_0$. For example, if the input of the circuit is $A_3 A_2 A_1 A_0 = 1101$, the output is 1110. The circuit can be designed using four half-adders.

4-5 A combinational circuit produces the binary sum of two 2-bit numbers, $x_1 x_0$ and $y_1 y_0$. The outputs are C, S_1, and S_0. Provide a truth table of the combinational circuit.

4-6 Design the circuit of Problem 4-5 using two full-adders.

4-7 Design a combinational circuit that multiplies two 2-bit numbers, $a_1 a_0$ and $b_1 b_0$, to produce a 4-bit product, $c_3 c_2 c_1 c_0$. Use AND gates and half-adders.

4-8 Show that a full-subtractor can be constructed with two half-subtractors and an OR gate.

4-9 Design a combinational circuit with three inputs and six outputs. The output binary number should be the square of the input binary number.

4-10 Design a combinational circuit with four inputs that represent a decimal digit in BCD and four outputs that produce the 9's complement of the input digit. The six unused combinations can be treated as don't-care conditions.

4-11 Design a combinational circuit with four inputs and four outputs. The output generates the 2's complement of the input binary number.

4-12 Design a combinational circuit that detects an error in the representation of a decimal digit in BCD. The output of the circuit must be equal to logic-1 when the inputs contain any one of the six unused bit combinations in the BCD code.

4-13 Design a code converter that converts a decimal digit from the 8 4 −2 −1 code to BCD (see Table 1-2.)

4-14 Design a combinational circuit that converts a decimal digit from the 2 4 2 1 code to the 8 4 −2 −1 code (see Table 1-2.)

4-15 Design a combinational circuit that converts a binary number of four bits to a decimal number in BCD. Note that the BCD number is the same as the binary number as long as the input is less than or equal to 9. The binary number from 1010 to 1111 converts into BCD numbers from 1 0000 to 1 0101.

4-16 A BCD-to-seven-segment decoder is a combinational circuit that converts a decimal digit in BCD to an appropriate code for the selection of segments in a display indicator used for displaying the decimal digit in a familiar form. The seven outputs of the decoder (a, b, c, d, e, f, g) select the corresponding segments in the display, as shown in Fig. P4-16(a). The numeric designation chosen to represent the decimal digit is shown in Fig. P4-16(b). Design the BCD-to-seven-segment decoder using a minimum number of NAND gates. The six invalid combinations should result in a blank display.

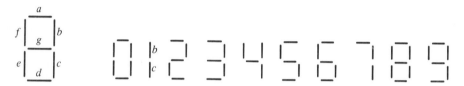

(a) Segment designation   (b) Numerical designation for display

FIGURE P4-16

4-17 Analyze the two-output combinational circuit shown in Fig. P4-17. Find the Boolean functions for the two outputs as a function of the three inputs and explain the circuit operation.

4-18 Derive the truth table of the circuit shown in Fig. P4-17.

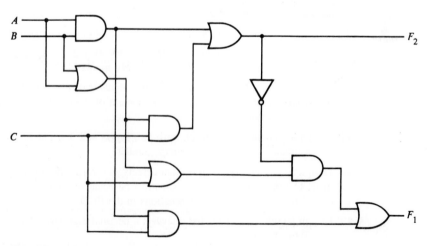

FIGURE P4-17

4-19 Draw the NAND logic diagram for each of the following expressions using multiple-level NAND gate circuits:
(a) $(AB' + CD')E + BC(A + B)$
(b) $w(x + y + z) + xyz$

4-20 Convert the logic diagram of the code converter shown in Fig. 4-8 to a multiple-level NAND circuit.

4-21 Determine the Boolean functions for outputs F and G as a function of four inputs, A, B, C, and D, in Fig. P4-21.

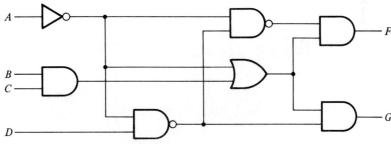

FIGURE P4-21

4-22 Verify that the circuit of Fig. P4-22 generates the exclusive-NOR function.

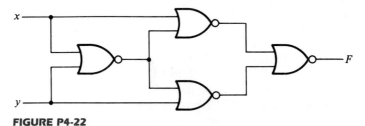

FIGURE P4-22

4-23 Convert the logic diagram of the code converter shown in Fig. 4-8 to a multiple-level NOR circuit.

4-24 Derive the truth table for the output of each NOR gate in Fig. 4-20(a).

4-25 Prove that $x' \oplus y = x \oplus y' = (x \oplus y)' = xy + x'y'$.

4-26 Prove that $x \oplus 1 = x'$ and $x \oplus 0 = x$.

4-27 Show that if $xy = 0$, then $x \oplus y = x + y$.

4-28 Design a combinational circuit that converts a 4-bit Gray code number (Table 1-4) to a 4-bit straight binary number. Implement the circuit with exclusive-OR gates.

4-29 Design the circuit of a 3-bit parity generator and the circuit of a 4-bit parity checker using an odd parity bit.

4-30 Manipulate the following Boolean expression in such a way so that it can be implemented using exclusive-OR and AND gates only.

$$AB'CD' + A'BCD' + AB'C'D + A'BC'D$$

5 MSI and PLD Components

5-1 INTRODUCTION

The purpose of Boolean-algebra simplification is to obtain an algebraic expression that, when implemented, results in a low-cost circuit. However, the criteria that determine a low-cost circuit must be defined if we are to evaluate the success of the achieved simplification. The design procedure for combinational circuits presented in Section 4-2 minimizes the number of gates required to implement a given function. This procedure assumes that given two circuits that perform the same function, the one that requires fewer gates is preferable because it will cost less. This is not necessarily true when integrated circuits are used.

The circuit complexity of integrated circuits (ICs) has been classified in Section 2-8 as having four levels of integration: small- (SSI), medium- (MSI), large- (LSI), and very large- (VLSI) scale integration. A combinational circuit designed with individual gates can be implemented with SSI circuits that contain several independent gates. The number of gates in an SSI circuit is limited by the number of pins in the package, typically 14 or 16. Since several gates are included in a single IC package, it becomes economical to use as many of the gates from an already used package even if, by doing so, we increase the total number of gates. Moreover, some interconnections among the gates in many ICs are internal to the chip and it is more economical to use as many internal interconnections as possible in order to minimize the number of wires between package pins. With integrated circuits, it is not the count of gates that determines the cost, but the number and types of ICs employed and the number of interconnections needed to implement the given digital circuit.

There are several combinational circuits that are employed extensively in the design of digital systems. These circuits are available in integrated circuits and are classified as MSI components. MSI components perform specific digital functions commonly needed in the design of digital systems. In this chapter we introduce the most important combinational circuit-type MSI components that are readily available in IC packages. These are adders, subtractors, comparators, decoders, encoders, and multiplexers. These components are also used as standard modules within more complex LSI and VLSI circuits. The MSI components presented here provide a catalog of elementary digital modules used extensively as basic building blocks in the design of digital computers and systems.

The components of a digital system can be classified as being specific to an application or as being standard circuits. Standard components are taken from a set that has been used in other systems. MSI components are standard circuits and their use results in a significant reduction in the total cost as compared to the cost of using SSI circuits. In contrast, specific components are particular to the system being implemented and are not commonly found among the standard components. The implementation of specific circuits with LSI chips can be done by means of ICs that can be programmed to provide the required logic.

A programmable logic device (PLD) is an integrated circuit with internal logic gates that are connected through electronic fuses. Programming the device involves the blowing of fuses along the paths that must be disconnected so as to obtain a particular configuration. The word "programming" here refers to a hardware procedure that specifies the internal configuration of the device. The gates in a PLD are divided into an AND array and an OR array that are connected together to provide an AND-OR sum of product implementation. The initial state of a PLD has all the fuses intact. Programming the device involves the blowing of internal fuses to achieve a desired logic function.

In this chapter we introduce three programmable logic devices and establish procedures for their use in the design of digital systems. The three types of PLDs differ in the placement of fuses in the AND-OR array. Figure 5-1 shows the fuse locations of the three PLDs. The programmable read-only memory (PROM) has a fixed AND array and programmable fuses for the output OR gates. The PROM implements Boolean functions in sum of minterms, as explained in Section 5-7. The programmable array logic (PAL) has a fused programmable AND array and a fixed OR array. The AND gates are programmed to provide the product terms for the Boolean functions that are logically summed in each OR gate. PALs are presented in Section 5-9. The most flexible PLD is the programmable logic array (PLA), where both the AND and OR arrays can be programmed. The product terms in the AND array may be shared by any OR gate to provide the required sum of products implementation. The operation of the PLA is explained in Section 5-8.

The advantage of using PLDs in the design of digital systems is that they can be programmed to incorporate complex logic functions within one LSI circuit. The use of programmable logic devices is an alternative to another design technology called VLSI design. VLSI design refers to the design of digital systems that contain thousands of

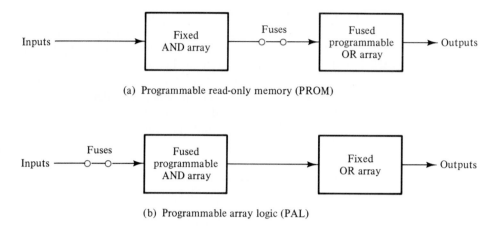

(a) Programmable read-only memory (PROM)

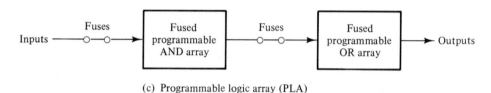

(b) Programmable array logic (PAL)

(c) Programmable logic array (PLA)

FIGURE 5-1

Basic configuration of three PLDs

gates within a single integrated-circuit chip. The basic component used in VLSI design is the *gate array*. A gate array consists of a pattern of gates fabricated in an area of silicon that is repeated thousands of times until the entire chip is covered with identical gates. Arrays of 1000 to 10,000 gates can be fabricated within a single integrated-circuit chip, depending on the technology used. The design with gate arrays requires that the designer specify the layout of the chip and the way that the gates are routed and connected. The first few levels of the fabrication process are common and independent of the final logic function. Additional fabrication levels are required to interconnect the gates in order to realize the desired function. This is usually done by means of computer-aided design methods. Both the gate array and the programmable logic device require extensive computer software tools to facilitate the design procedure.

5-2 BINARY ADDER AND SUBTRACTOR

The full-adder introduced in Section 4-3 forms the sum of two bits and a previous carry. Two binary numbers of n bits each can be added by means of this circuit. To demonstrate with a specific example, consider two binary numbers, $A = 1011$ and $B = 0011$, whose sum is $S = 1110$. When a pair of bits are added through a full-adder, the circuit produces a carry to be used with the pair of bits one significant position higher. This is shown in the following table:

Subscript i	4 3 2 1		Full-adder of Fig. 4-5
Input carry	0 1 1 0	C_i	z
Augend	1 0 1 1	A_i	x
Addend	0 0 1 1	B_i	y
Sum	1 1 1 0	S_i	S
Output carry	0 0 1 1	C_{i+1}	C

The bits are added with full-adders, starting from the least significant position (subscript 1), to form the sum bit and carry bit. The inputs and outputs of the full-adder circuit of Fig. 4-5 are also indicated. The input carry C_1 in the least significant position must be 0. The value of C_{i+1} in a given significant position is the output carry of the full-adder. This value is transferred into the input carry of the full-adder that adds the bits one higher significant position to the left. The sum bits are thus generated starting from the rightmost position and are available as soon as the corresponding previous carry bit is generated.

The sum of two n-bit binary numbers, A and B, can be generated in two ways: either in a serial fashion or in parallel. The serial addition method uses only one full-adder circuit and a storage device to hold the generated output carry. The pair of bits in A and B are transferred serially, one at a time, through the single full-adder to produce a string of output bits for the sum. The stored output carry from one pair of bits is used as an input carry for the next pair of bits. The parallel method uses n full-adder circuits, and all bits of A and B are applied simultaneously. The output carry from one full-adder is connected to the input carry of the full-adder one position to its left. As soon as the carries are generated, the correct sum bits emerge from the sum outputs of all full-adders.

Binary Parallel Adder

A binary parallel adder is a digital circuit that produces the arithmetic sum of two binary numbers in parallel. It consists of full-adders connected in a chain, with the output carry from each full-adder connected to the input carry of the next full-adder in the chain.

Figure 5-2(a) shows the interconnection of four full-adder (FA) circuits to provide a 4-bit binary parallel adder. The augend bits of A and the addend bits of B are designated by subscript numbers from right to left, with subscript 1 denoting the low-order bit. The carries are connected in a chain through the full-adders. The input carry to the adder is C_1 and the output carry is C_5. The S outputs generate the required sum bits. When the 4-bit full-adder circuit is enclosed within an IC package, it has four terminals for the augend bits, four terminals for the addend bits, four terminals for the sum bits, and two terminals for the input and output carries.

An n-bit parallel adder requires n full-adders. It can be constructed from 4-bit, 2-bit, and 1-bit full-adders ICs by cascading several packages. The output carry from one

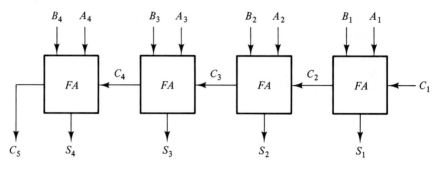

(a) 4-bit parallel adder

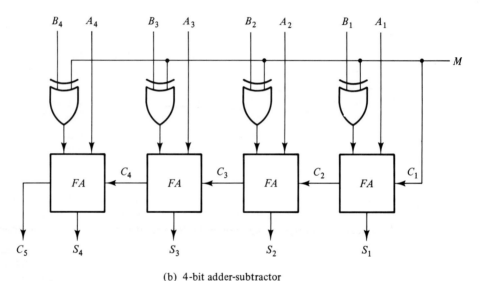

(b) 4-bit adder-subtractor

FIGURE 5-2
Adder and subtractor circuits

package must be connected to the input carry of the one with the next higher-order bits.

The 4-bit full-adder is a typical example of an MSI function. It can be used in many applications involving arithmetic operations. Observe that the design of this circuit by the classical method would require a truth table with $2^9 = 512$ entries, since there are nine inputs to the circuit. By using an iterative method of cascading an already known function, we were able to obtain a simple and well-organized implementation.

Binary Adder-Subtractor

The subtraction of binary numbers can be done most conveniently by means of complements, as discussed in Section 1-5. Remember that the subtraction $A - B$ can be done by taking the 2's complement of B and adding it to A. The 2's complement can be obtained by taking the 1's complement and adding one to the least significant pair of bits. The 1's complement can be implemented with inverters and a one can be added to the sum through the input carry.

The circuit for subtracting $A - B$ consists of a parallel adder with inverters placed between each data input B and the corresponding input of the full-adder. The input carry C_1 must be equal to 1 when performing subtraction. The operation thus performed becomes A plus the 1's complement of B plus 1. This is equal to A plus the 2's complement of B. For unsigned numbers, this gives $A - B$ if $A \geq B$ or the 2's complement of $B - A$ if $A < B$ (see Section 1-5). For signed numbers, the result is $A - B$ provided there is no overflow. (See Section 1-6.)

The addition and subtraction operations can be combined into one circuit with one common binary adder. This is done by including an exclusive-OR gate with each full-adder. A 4-bit adder-subtractor circuit is shown in Fig. 5-2(b). The mode input M controls the operation. When $M = 0$, the circuit is an adder, and when $M = 1$, the circuit becomes a subtractor. Each exclusive-OR gate receives input M and one of the inputs of B. When $M = 0$, we have $B \oplus 0 = B$. The full-adders receive the value of B, the input carry is 0, and the circuit performs A plus B. When $M = 1$, we have $B \oplus 1 = B'$ and $C_1 = 1$. The B inputs are all complemented and a 1 is added through the input carry. The circuit performs the operation A plus the 2's complement of B.

Carry Propagation

The addition of two binary numbers in parallel implies that all the bits of the augend and the addend are available for computation at the same time. As in any combinational circuit, the signal must propagate through the gates before the correct output sum is available in the output terminals. The total propagation time is equal to the propagation delay of a typical gate times the number of gate levels in the circuit. The longest propagation delay time in a parallel adder is the time it takes the carry to propagate through the full-adders. Since each bit of the sum output depends on the value of the input carry, the value of S_i in any given stage in the adder will be in its steady-state final value only after the input carry to that stage has been propagated. Consider output S_4 in Fig. 5-2(a). Inputs A_4 and B_4 reach a steady value as soon as input signals are applied to the adder. But input carry C_4 does not settle to its final steady-state value until C_3 is available in its steady-state value. Similarly, C_3 has to wait for C_2, and so on down to C_1. Thus, only after the carry propagates through all stages will the last output S_4 and carry C_5 settle to their final steady-state value.

The number of gate levels for the carry propagation can be found from the circuit of the full-adder. This circuit was derived in Fig. 4-5 and is redrawn in Fig. 5-3 for convenience. The input and output variables use the subscript i to denote a typical stage in the parallel adder. The signals at P_i and G_i settle to their steady-state values after the propagation through their respective gates. These two signals are common to all full-adders and depend only on the input augend and addend bits. The signal from the input carry, C_i, to the output carry, C_{i+1}, propagates through an AND gate and an OR gate, which constitute two gate levels. If there are four full-adders in the parallel adder, the

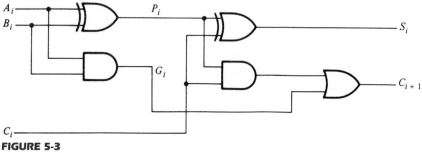

FIGURE 5-3
Full-adder circuit

output carry C_5 would have $2 \times 4 = 8$ gate levels from C_1 to C_5. The total propagation time in the adder would be the propagation time in one half-adder plus eight gate levels. For an n-bit parallel adder, there are $2n$ gate levels for the carry to propagate through.

The carry propagation time is a limiting factor on the speed with which two numbers are added in parallel. Although a parallel adder, or any combinational circuit, will always have some value at its output terminals, the outputs will not be correct unless the signals are given enough time to propagate through the gates connected from the inputs to the outputs. Since all other arithmetic operations are implemented by successive additions, the time consumed during the addition process is very critical. An obvious solution for reducing the carry propagation delay time is to employ faster gates with reduced delays. But physical circuits have a limit to their capability. Another solution is to increase the equipment complexity in such a way that the carry delay time is reduced. There are several techniques for reducing the carry propagation time in a parallel adder. The most widely used technique employs the principle of *look-ahead* carry and is described below.

Consider the circuit of the full-adder shown in Fig. 5-3. If we define two new binary variables:

$$P_i = A_i \oplus B_i$$
$$G_i = A_i B_i$$

the output sum and carry can be expressed as

$$S_i = P_i \oplus C_i$$
$$C_{i+1} = G_i + P_i C_i$$

G_i is called a *carry generate* and it produces an output carry when both A_i and B_i are one, regardless of the input carry. P_i is called a *carry propagate* because it is the term associated with the propagation of the carry from C_i to C_{i+1}.

We now write the Boolean function for the carry output of each stage and substitute for each C_i its value from the previous equations:

$$C_2 = G_1 + P_1 C_1$$
$$C_3 = G_2 + P_2 C_2 = G_2 + P_2(G_1 + P_1 C_1) = G_2 + P_2 G_1 + P_2 P_1 C_1$$
$$C_4 = G_3 + P_3 C_3 = G_3 + P_3 G_2 + P_3 P_2 G_1 + P_3 P_2 P_1 C_1$$

Since the Boolean function for each output carry is expressed in sum of products, each function can be implemented with one level of AND gates followed by an OR gate (or by a two-level NAND). The three Boolean functions for C_2, C_3, and C_4 are implemented in the look-ahead carry generator shown in Fig. 5-4. Note that C_4 does not have to wait for C_3 and C_2 to propagate; in fact, C_4 is propagated at the same time as C_2 and C_3.

The construction of a 4-bit parallel adder with a look-ahead carry scheme is shown in Fig. 5-5. Each sum output requires two exclusive-OR gates. The output of the first exclusive-OR gate generates the P_i variable, and the AND gate generates the G_i variable. All the P's and G's are generated in two gate levels. The carries are propagated through the look-ahead carry generator (similar to that in Fig. 5-4) and applied as in-

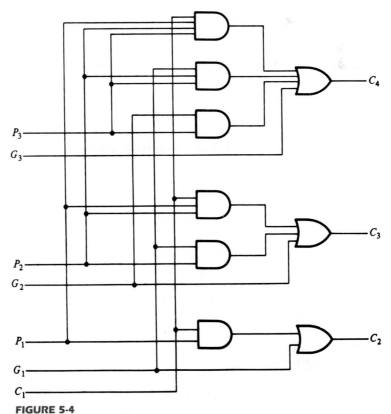

FIGURE 5-4
Logic diagram of a look-ahead carry generator

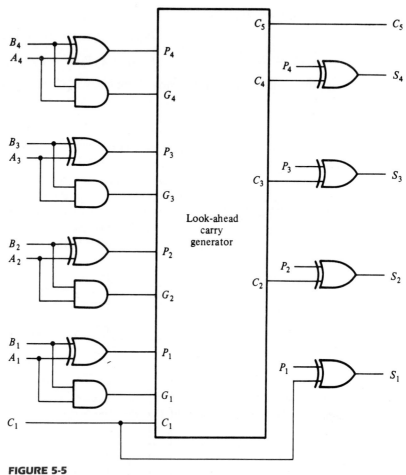

FIGURE 5-5
4-bit full-adders with look-ahead carry

puts to the second exclusive-OR gate. After the P and G signals settle into their steady-state values, all output carries are generated after a delay of two levels of gates. Thus, outputs S_2 through S_4 have equal propagation delay times. The two-level circuit for the output carry C_5 is not shown in Fig. 5-4. This circuit can be easily derived by the equation-substitution method, as done above (see Problem 5-8).

5-3   DECIMAL ADDER

Computers or calculators that perform arithmetic operations directly in the decimal number system represent decimal numbers in binary-coded form. An adder for such a computer must employ arithmetic circuits that accept coded decimal numbers and present results in the accepted code. For binary addition, it was sufficient to consider a pair of significant bits at a time, together with a previous carry. A decimal adder requires a minimum of nine inputs and five outputs, since four bits are required to code each decimal digit and the circuit must have an input carry and output carry. Of course, there is a wide variety of possible decimal adder circuits, dependent upon the code used to represent the decimal digits.

The design of a nine-input, five-output combinational circuit by the classical method requires a truth table with $2^9 = 512$ entries. Many of the input combinations are don't-care conditions, since each binary code input has six combinations that are invalid. The simplified Boolean functions for the circuit may be obtained by a computer-generated tabular method, and the result would probably be a connection of gates forming an irregular pattern. An alternate procedure is to add the numbers with full-adder circuits, taking into consideration the fact that six combinations in each 4-bit input are not used. The output must be modified so that only those binary combinations that are valid combinations of the decimal code are generated.

BCD Adder

Consider the arithmetic addition of two decimal digits in BCD, together with a possible carry from a previous stage. Since each input digit does not exceed 9, the output sum cannot be greater than $9 + 9 + 1 = 19$, the 1 in the sum being an input carry. Suppose we apply two BCD digits to a 4-bit binary adder. The adder will form the sum in *binary* and produce a result that may range from 0 to 19. These binary numbers are listed in Table 5-1 and are labeled by symbols K, Z_8, Z_4, Z_2, and Z_1. K is the carry, and the subscripts under the letter Z represent the weights 8, 4, 2, and 1 that can be assigned to the four bits in the BCD code. The first column in the table lists the binary sums as they appear in the outputs of a 4-bit *binary* adder. The output sum of two *decimal digits* must be represented in BCD and should appear in the form listed in the second column of the table. The problem is to find a simple rule by which the binary number in the first column can be converted to the correct BCD-digit representation of the number in the second column.

In examining the contents of the table, it is apparent that when the binary sum is equal to or less than 1001, the corresponding BCD number is identical, and therefore no conversion is needed. When the binary sum is greater than 1001, we obtain a nonvalid BCD representation. The addition of binary 6 (0110) to the binary sum converts it to the correct BCD representation and also produces an output carry as required.

The logic circuit that detects the necessary correction can be derived from the table entries. It is obvious that a correction is needed when the binary sum has an output carry $K = 1$. The other six combinations from 1010 to 1111 that need a correction have a 1 in position Z_8. To distinguish them from binary 1000 and 1001, which also have a 1 in position Z_8, we specify further that either Z_4 or Z_2 must have a 1. The condition for a correction and an output carry can be expressed by the Boolean function

$$C = K + Z_8 Z_4 + Z_8 Z_2$$

TABLE 5-1
Derivation of a BCD Adder

Binary Sum					BCD Sum					Decimal
K	Z_8	Z_4	Z_2	Z_1	C	S_8	S_4	S_2	S_1	
0	0	0	0	0	0	0	0	0	0	0
0	0	0	0	1	0	0	0	0	1	1
0	0	0	1	0	0	0	0	1	0	2
0	0	0	1	1	0	0	0	1	1	3
0	0	1	0	0	0	0	1	0	0	4
0	0	1	0	1	0	0	1	0	1	5
0	0	1	1	0	0	0	1	1	0	6
0	0	1	1	1	0	0	1	1	1	7
0	1	0	0	0	0	1	0	0	0	8
0	1	0	0	1	0	1	0	0	1	9
0	1	0	1	0	1	0	0	0	0	10
0	1	0	1	1	1	0	0	0	1	11
0	1	1	0	0	1	0	0	1	0	12
0	1	1	0	1	1	0	0	1	1	13
0	1	1	1	0	1	0	1	0	0	14
0	1	1	1	1	1	0	1	0	1	15
1	0	0	0	0	1	0	1	1	0	16
1	0	0	0	1	1	0	1	1	1	17
1	0	0	1	0	1	1	0	0	0	18
1	0	0	1	1	1	1	0	0	1	19

When $C = 1$, it is necessary to add 0110 to the binary sum and provide an output carry for the next stage.

A *BCD adder* is a circuit that adds two BCD digits in parallel and produces a sum digit also in BCD. A BCD adder must include the correction logic in its internal construction. To add 0110 to the binary sum, we use a second 4-bit binary adder, as shown in Fig. 5-6. The two decimal digits, together with the input carry, are first added in the top 4-bit binary adder to produce the binary sum. When the output carry is equal to zero, nothing is added to the binary sum. When it is equal to one, binary 0110 is added to the binary sum through the bottom 4-bit binary adder. The output carry generated from the bottom binary adder can be ignored, since it supplies information already available at the output-carry terminal.

The BCD adder can be constructed with three IC packages. Each of the 4-bit adders is an MSI function and the three gates for the correction logic need one SSI package. However, the BCD adder is available in one MSI circuit. To achieve shorter propagation delays, an MSI BCD adder includes the necessary circuits for look-ahead carries. The adder circuit for the correction does not need all four full-adders, and this circuit can be optimized within the IC package.

A decimal parallel adder that adds n decimal digits needs n BCD adder stages. The

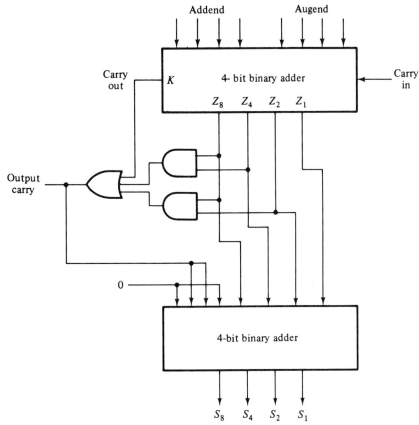

FIGURE 5-6
Block diagram of a BCD adder

output carry from one stage must be connected to the input carry of the next higher-order stage.

5-4 MAGNITUDE COMPARATOR

The comparison of two numbers is an operation that determines if one number is greater than, less than, or equal to the other number. A *magnitude comparator* is a combinational circuit that compares two numbers, A and B, and determines their relative magnitudes. The outcome of the comparison is specified by three binary variables that indicate whether $A > B$, $A = B$, or $A < B$.

The circuit for comparing two n-bit numbers has 2^{2n} entries in the truth table and becomes too cumbersome even with $n = 3$. On the other hand, as one may suspect, a comparator circuit possesses a certain amount of regularity. Digital functions that possess an inherent well-defined regularity can usually be designed by means of an algorithmic procedure if one is found to exist. An *algorithm* is a procedure that specifies a finite set of steps that, if followed, give the solution to a problem. We illustrate this method here by deriving an algorithm for the design of a 4-bit magnitude comparator.

The algorithm is a direct application of the procedure a person uses to compare the relative magnitudes of two numbers. Consider two numbers, A and B, with four digits each. Write the coefficients of the numbers with descending significance as follows:

$$A = A_3 A_2 A_1 A_0$$

$$B = B_3 B_2 B_1 B_0$$

where each subscripted letter represents one of the digits in the number. The two numbers are equal if all pairs of significant digits are equal, i.e., if $A_3 = B_3$ and $A_2 = B_2$ and $A_1 = B_1$ and $A_0 = B_0$. When the numbers are binary, the digits are either 1 or 0 and the equality relation of each pair of bits can be expressed logically with an equivalence function:

$$x_i = A_i B_i + A_i' B_i' \qquad i = 0, 1, 2, 3$$

where $x_i = 1$ only if the pair of bits in position i are equal, i.e., if both are 1's or both are 0's.

The equality of the two numbers, A and B, is displayed in a combinational circuit by an output binary variable that we designate by the symbol $(A = B)$. This binary variable is equal to 1 if the input numbers, A and B, are equal, and it is equal to 0 otherwise. For the equality condition to exist, all x_i variables must be equal to 1. This dictates an AND operation of all variables:

$$(A = B) = x_3 x_2 x_1 x_0$$

The *binary* variable $(A = B)$ is equal to 1 only if all pairs of digits of the two numbers are equal.

To determine if A is greater than or less than B, we inspect the relative magnitudes of pairs of significant digits starting from the most significant position. If the two digits are equal, we compare the next lower significant pair of digits. This comparison continues until a pair of unequal digits is reached. If the corresponding digit of A is 1 and that of B is 0, we conclude that $A > B$. If the corresponding digit of A is 0 and that of B is 1, we have that $A < B$. The sequential comparison can be expressed logically by the following two Boolean functions:

$$(A > B) = A_3 B_3' + x_3 A_2 B_2' + x_3 x_2 A_1 B_1' + x_3 x_2 x_1 A_0 B_0'$$

$$(A < B) = A_3' B_3 + x_3 A_2' B_2 + x_3 x_2 A_1' B_1 + x_3 x_2 x_1 A_0' B_0$$

The symbols $(A > B)$ and $(A < B)$ are *binary* output variables that are equal to 1 when $A > B$ or $A < B$, respectively.

The gate implementation of the three output variables just derived is simpler than it seems because it involves a certain amount of repetition. The "unequal" outputs can use the same gates that are needed to generate the "equal" output. The logic diagram of the 4-bit magnitude comparator is shown in Fig. 5-7. The four x outputs are generated with

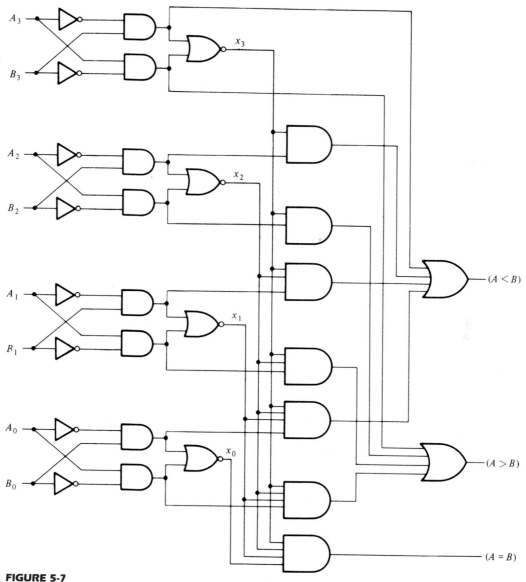

FIGURE 5-7

4-bit magnitude comparator

equivalence (exclusive-NOR) circuits and applied to an AND gate to give the output binary variable $(A = B)$. The other two outputs use the x variables to generate the Boolean functions listed before. This is a multilevel implementation and, as clearly seen, it has a regular pattern. The procedure for obtaining magnitude-comparator circuits for binary numbers with more than four bits should be obvious from this example. The same circuit can be used to compare the relative magnitudes of two BCD digits.

5-5 DECODERS AND ENCODERS

Discrete quantities of information are represented in digital systems with binary codes. A binary code of n bits is capable of representing up to 2^n distinct elements of the coded information. A *decoder* is a combinational circuit that converts binary information from n input lines to a maximum of 2^n unique output lines. If the n-bit decoded information has unused or don't-care combinations, the decoder output will have fewer than 2^n outputs.

The decoders presented here are called n-to-m-line decoders, where $m \leq 2^n$. Their purpose is to generate the 2^n (or fewer) minterms of n input variables. The name *decoder* is also used in conjunction with some code converters such as a BCD-to-seven-segment decoder.

As an example, consider the 3-to-8-line decoder circuit of Fig. 5-8. The three inputs are decoded into eight outputs, each output representing one of the minterms of the 3-input variables. The three inverters provide the complement of the inputs, and each one of the eight AND gates generates one of the minterms. A particular application of this decoder would be a binary-to-octal conversion. The input variables may represent a binary number, and the outputs will then represent the eight digits in the octal number

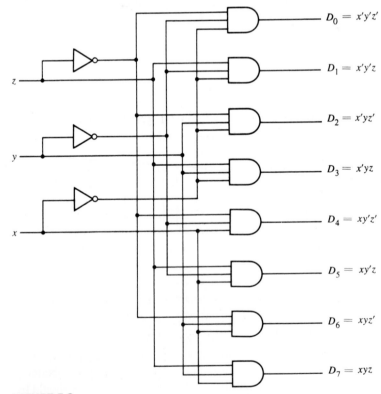

FIGURE 5-8
A 3-to-8 line decoder

TABLE 5-2
Truth Table of a 3-to-8-Line Decoder

Inputs			Outputs							
x	y	z	D_0	D_1	D_2	D_3	D_4	D_5	D_6	D_7
0	0	0	1	0	0	0	0	0	0	0
0	0	1	0	1	0	0	0	0	0	0
0	1	0	0	0	1	0	0	0	0	0
0	1	1	0	0	0	1	0	0	0	0
1	0	0	0	0	0	0	1	0	0	0
1	0	1	0	0	0	0	0	1	0	0
1	1	0	0	0	0	0	0	0	1	0
1	1	1	0	0	0	0	0	0	0	1

system. However, a 3-to-8-line decoder can be used for decoding any 3-bit code to provide eight outputs, one for each element of the code.

The operation of the decoder may be further clarified from its input–output relationship, listed in Table 5-2. Observe that the output variables are mutually exclusive because only one output can be equal to 1 at any one time. The output line whose value is equal to 1 represents the minterm equivalent of the binary number presently available in the input lines.

Combinational Logic Implementation

A decoder provides the 2^n minterm of n input variables. Since any Boolean function can be expressed in sum of minterms canonical form, one can use a decoder to generate the minterms and an external OR gate to form the sum. In this way, any combinational circuit with n inputs and m outputs can be implemented with an n-to-2^n-line decoder and m OR gates.

The procedure for implementing a combinational circuit by means of a decoder and OR gates requires that the Boolean functions for the circuit be expressed in sum of minterms. This form can be easily obtained from the truth table or by expanding the functions to their sum of minterms (see Section 2-5). A decoder is then chosen that generates all the minterms of the n input variables. The inputs to each OR gate are selected from the decoder outputs according to the minterm list in each function.

Example
5-1

Implement a full-adder circuit with a decoder and two OR gates.

From the truth table of the full-adder (Section 4-3), we obtain the functions for this combinational circuit in sum of minterms:

$$S(x, y, z) = \Sigma(1, 2, 4, 7)$$
$$C(x, y, z) = \Sigma(3, 5, 6, 7)$$

Since there are three inputs and a total of eight minterms, we need a 3-to-8-line decoder. The implementation is shown in Fig. 5-9. The decoder generates the eight

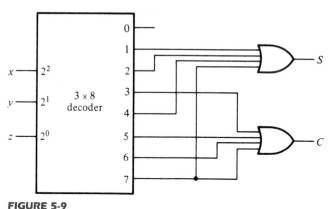

FIGURE 5-9
Implementation of a full-adder with a decoder

minterms for x, y, z. The OR gate for output S forms the sum of minterms 1, 2, 4, and 7. The OR gate for output C forms the sum of minterms 3, 5, 6, and 7.   ■

A function with a long list of minterms requires an OR gate with a large number of inputs. A function F having a list of k minterms can be expressed in its complemented form F' with $2^n - k$ minterms. If the number of minterms in a function is greater than $2^n/2$, then F' can be expressed with fewer minterms than required for F. In such a case, it is advantageous to use a NOR gate to sum the minterms of F'. The output of the NOR gate will generate the normal output F.

The decoder method can be used to implement any combinational circuit. However, its implementation must be compared with all other possible implementations to determine the best solution. In some cases, this method may provide the best implementation, especially if the combinational circuit has many outputs and if each output function (or its complement) is expressed with a small number of minterms.

Demultiplexers

Some IC decoders are constructed with NAND gates. Since a NAND gate produces the AND operation with an inverted output, it becomes more economical to generate the decoder minterms in their complemented form. Most, if not all, IC decoders include one or more *enable* inputs to control the circuit operation. A 2-to-4-line decoder with an enable input constructed with NAND gates is shown in Fig. 5-10. All outputs are equal to 1 if enable input E is 1, regardless of the values of inputs A and B. When the enable input is 0, the circuit operates as a decoder with complemented outputs. The truth table lists these conditions. The X's under A and B are don't-care conditions. Normal decoder operation occurs only with $E = 0$, and the outputs are selected when they are in the 0 state.

The block diagram of the decoder is shown in Fig. 5-11(a). The small circle at input E indicates that the decoder is enabled when $E = 0$. The small circles at the outputs indicate that all outputs are complemented.

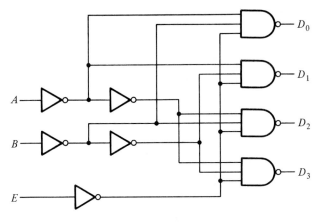

E	A	B	D_0	D_1	D_2	D_3
1	X	X	1	1	1	1
0	0	0	0	1	1	1
0	0	1	1	0	1	1
0	1	0	1	1	0	1
0	1	1	1	1	1	0

(a) Logic diagram   (b) Truth table

FIGURE 5-10
A 2-to-4-line decoder with enable (E) input

A decoder with an enable input can function as a demultiplexer. A *demultiplexer* is a circuit that receives information on a single line and transmits this information on one of 2^n possible output lines. The selection of a specific output line is controlled by the bit values of n selection lines. The decoder of Fig. 5-10 can function as a demultiplexer if the E line is taken as a data input line and lines A and B are taken as the selection lines. This is shown in Fig. 5-11(b). The single input variable E has a path to all four outputs, but the input information is directed to only one of the output lines, as specified by the binary value of the two selection lines, A and B. This can be verified from the truth table of this circuit, shown in Fig. 5-10(b). For example, if the selection lines $AB = 10$, output D_2 will be the same as the input value E, while all other outputs are maintained at 1. Because decoder and demultiplexer operations are obtained from the same circuit, a decoder with an enable input is referred to as a *decoder/demultiplexer*. It is the enable input that makes the circuit a demultiplexer; the decoder itself can use AND, NAND, or NOR gates.

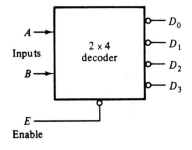

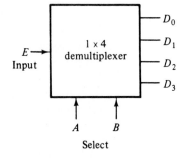

(a) Decoder with enable   (b) Demultiplexer

FIGURE 5-11
Block diagrams for the circuit of Fig. 5-10

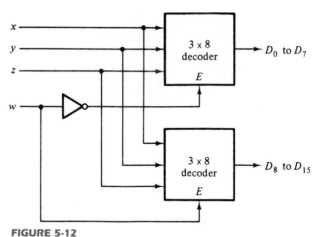

FIGURE 5-12
A 4 × 16 decoder constructed with two 3 × 8 decoders

Decoder/demultiplexer circuits can be connected together to form a larger decoder circuit. Figure 5-12 shows two 3×8 decoders with enable inputs connected to form a 4×16 decoder. When $w = 0$, the top decoder is enabled and the other is disabled. The bottom decoder outputs are all 0's, and the top eight outputs generate minterms 0000 to 0111. When $w = 1$, the enable conditions are reversed; the bottom decoder outputs generate minterms 1000 to 1111, while the outputs of the top decoder are all 0's. This example demonstrates the usefulness of enable inputs in ICs. In general, enable lines are a convenient feature for connecting two or more IC packages for the purpose of expanding the digital function into a similar function with more inputs and outputs.

Encoders

An encoder is a digital circuit that performs the inverse operation of a decoder. An encoder has 2^n (or fewer) input lines and n output lines. The output lines generate the binary code corresponding to the input value. An example of an encoder is the octal-to-binary encoder whose truth table is given in Table 5-3. It has eight inputs, one for each of the octal digits, and three outputs that generate the corresponding binary number. It is assumed that only one input has a value of 1 at any given time; otherwise the circuit has no meaning.

The encoder can be implemented with OR gates whose inputs are determined directly from the truth table. Output z is equal to 1 when the input octal digit is 1 or 3 or 5 or 7. Output y is 1 for octal digits 2, 3, 6, or 7, and output x is 1 for digits 4, 5, 6, or 7. These conditions can be expressed by the following output Boolean functions:

$$z = D_1 + D_3 + D_5 + D_7$$

$$y = D_2 + D_3 + D_6 + D_7$$

$$x = D_4 + D_5 + D_6 + D_7$$

TABLE 5-3
Truth Table of Octal-to-Binary Encoder

			Inputs					Outputs		
D_0	D_1	D_2	D_3	D_4	D_5	D_6	D_7	x	y	z
1	0	0	0	0	0	0	0	0	0	0
0	1	0	0	0	0	0	0	0	0	1
0	0	1	0	0	0	0	0	0	1	0
0	0	0	1	0	0	0	0	0	1	1
0	0	0	0	1	0	0	0	1	0	0
0	0	0	0	0	1	0	0	1	0	1
0	0	0	0	0	0	1	0	1	1	0
0	0	0	0	0	0	0	1	1	1	1

The encoder is implemented with three OR gates, as shown in Fig. 5-13.

The encoder defined in Table 5-3 has the limitation that only one input can be active at any given time. If two inputs are active simultaneously, the output produces an undefined combination. For example, if D_3 and D_6 are 1 simultaneously, the output of the encoder will be 111 because all three outputs are equal to 1. This does not represent binary 3 nor binary 6. To resolve this ambiguity, encoder circuits must establish a priority to ensure that only one input is encoded. If we establish a higher priority for inputs with higher subscript numbers, and if both D_3 and D_6 are 1 at the same time, the output will be 110 because D_6 has higher priority than D_3.

Another ambiguity in the octal-to-binary encoder is that an output with all 0's is generated when all the inputs are 0. The problem is that an output with all 0's is also generated when D_0 is equal to 1. This ambiguity can be resolved by providing an additional output that specifies the condition that none of the inputs are active.

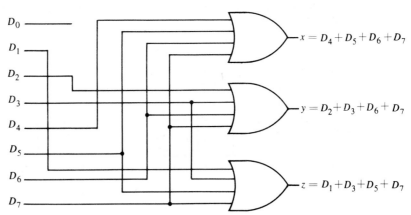

FIGURE 5-13
Octal-to-binary encoder

TABLE 5-4
Truth Table of a Priority Encoder

Inputs				Outputs		
D_0	D_1	D_2	D_3	x	y	V
0	0	0	0	X	X	0
1	0	0	0	0	0	1
X	1	0	0	0	1	1
X	X	1	0	1	0	1
X	X	X	1	1	1	1

Priority Encoder

A priority encoder is an encoder circuit that includes the priority function. The operation of the priority encoder is such that if two or more inputs are equal to 1 at the same time, the input having the highest priority will take precedence. The truth table of a four-input priority encoder is given in Table 5-4. The X's are don't-care conditions that designate the fact that the binary value may be equal either to 0 or 1. Input D_3 has the highest priority; so regardless of the values of the other inputs, when this input is 1, the output for xy is 11 (binary 3). D_2 has the next priority level. The output is 10 if $D_2 = 1$ provided that $D_3 = 0$, regardless of the values of the other two lower-priority inputs. The output for D_1 is generated only if higher-priority inputs are 0, and so on down the priority level. A *valid*-output indicator, designated by V, is set to 1 only when one or more of the inputs are equal to 1. If all inputs are 0, V is equal to 0, and the other two outputs of the circuit are not used.

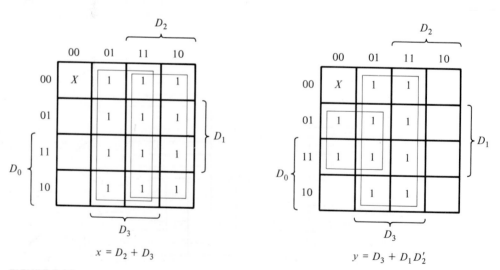

$$x = D_2 + D_3$$

$$y = D_3 + D_1 D_2'$$

FIGURE 5-14
Maps for a priority encoder

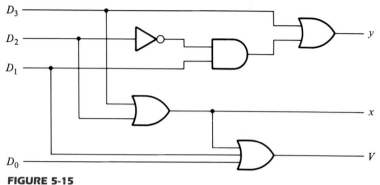

FIGURE 5-15
4-input priority encoder

The maps for simplifying outputs x and y are shown in Fig. 5-14. The minterms for the two functions are derived from Table 5-4. Although the table has only five rows, when each don't-care condition is replaced first by 0 and then by 1, we obtain all 16 possible input combinations. For example, the third row in the table with $X100$ represents minterms 0100 and 1100 since X can be assigned either 0 or 1. The simplified Boolean expressions for the priority encoder are obtained from the maps. The condition for output V is an OR function of all the input variables. The priority encoder is implemented in Fig. 5-15 according to the following Boolean functions:

$$x = D_2 + D_3$$

$$y = D_3 + D_1 D_2'$$

$$V = D_0 + D_1 + D_2 + D_3$$

5-6 MULTIPLEXERS

Multiplexing means transmitting a large number of information units over a smaller number of channels or lines. A *digital multiplexer* is a combinational circuit that selects binary information from one of many input lines and directs it to a single output line. The selection of a particular input line is controlled by a set of selection lines. Normally, there are 2^n input lines and n selection lines whose bit combinations determine which input is selected.

A 4-to-1-line multiplexer is shown in Fig. 5-16. Each of the four input lines, I_0 to I_3, is applied to one input of an AND gate. Selection lines s_1 and s_0 are decoded to select a particular AND gate. The function table, Fig. 5-16(b), lists the input-to-output path for each possible bit combination of the selection lines. When this MSI function is used in the design of a digital system, it is represented in block diagram form, as shown in Fig. 5-16(c). To demonstrate the circuit operation, consider the case when $s_1 s_0 = 10$. The AND gate associated with input I_2 has two of its inputs equal to 1 and the third input connected to I_2. The other three AND gates have at least one input equal to 0, which

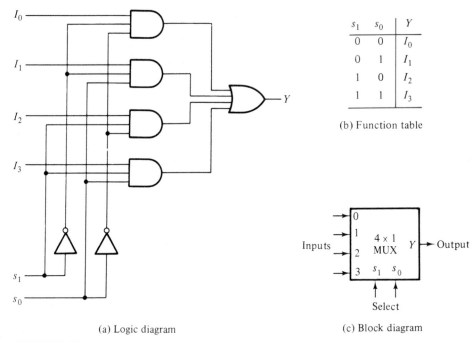

s_1	s_0	Y
0	0	I_0
0	1	I_1
1	0	I_2
1	1	I_3

(b) Function table

(a) Logic diagram   (c) Block diagram

FIGURE 5-16

A 4-to-1-line multiplexer

makes their outputs equal to 0. The OR gate output is now equal to the value of I_2, thus providing a path from the selected input to the output. A multiplexer is also called a *data selector*, since it selects one of many inputs and steers the binary information to the output line.

The AND gates and inverters in the multiplexer resemble a decoder circuit and, indeed, they decode the input-selection lines. In general, a 2^n-to-1-line multiplexer is constructed from an n-to-2^n decoder by adding to it 2^n input lines, one to each AND gate. The outputs of the AND gates are applied to a single OR gate to provide the 1-line output. The size of a multiplexer is specified by the number 2^n of its input lines and the single output line. It is then implied that it also contains n selection lines. A multiplexer is often abbreviated as MUX.

As in decoders, multiplexer ICs may have an *enable* input to control the operation of the unit. When the enable input is in a given binary state, the outputs are disabled, and when it is in the other state (the enable state), the circuit functions as a normal multiplexer. The enable input (sometimes called *strobe*) can be used to expand two or more multiplexer ICs to a digital multiplexer with a larger number of inputs.

In some cases, two or more multiplexers are enclosed within one IC package. The selection and enable inputs in multiple-unit ICs may be common to all multiplexers. As an illustration, a quadruple 2-to-1-line multiplexer IC is shown in Fig. 5-17. It has four multiplexers, each capable of selecting one of two input lines. Output Y_1 can be selected

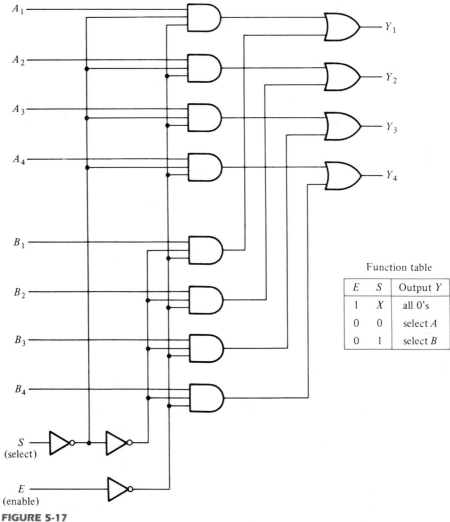

FIGURE 5-17
Quadruple 2-to-1-line multiplexer

to be equal to either A_1 or B_1. Similarly, output Y_2 may have the value of A_2 or B_2, and so on. One input selection line, S, suffices to select one of two lines in all four multiplexers. The control input E enables the multiplexers in the 0 state and disables them in the 1 state. Although the circuit contains four multiplexers, we may think of it as a circuit that selects one in a pair of 4-input lines. As shown in the function table, the unit is selected when $E = 0$. Then, if $S = 0$, the four A inputs have a path to the outputs. On the other hand, if $S = 1$, the four B inputs are selected. The outputs have all 0's when $E = 1$, regardless of the value of S.

Boolean-Function Implementation

It was shown in the previous section that a decoder can be used to implement a Boolean function by employing an external OR gate. A quick reference to the multiplexer of Fig. 5-16 reveals that it is essentially a decoder with the OR gate already available. The minterms out of the decoder to be chosen can be controlled with the input lines. The minterms to be included with the function being implemented are chosen by making their corresponding input lines equal to 1; those minterms not included in the function are disabled by making their input lines equal to 0. This gives a method for implementing any Boolean function of n variables with a 2^n-to-1 multiplexer. However, it is possible to do better than that.

If we have a Boolen function of $n + 1$ variables, we take n of these variables and connect them to the selection lines of a multiplexer. The remaining single variable of the function is used for the inputs of the multiplexer. If A is this single variable, the inputs of the multiplexer are chosen to be either A or A' or 1 or 0. By judicious use of these four values for the inputs and by connecting the other variables to the selection lines, one can implement any Boolean function with a multiplexer. In this way, it is possible to generate any function of $n + 1$ variables with a 2^n-to-1 multiplexer.

To demonstrate this procedure with a concrete example, consider the function of three variables:

$$F(A, B, C) = \Sigma(1, 3, 5, 6)$$

The function can be implemented with a 4-to-1 multiplexer, as shown in Fig. 5-18. Two of the variables, B and C, are applied to the selection lines in that order, i.e., B is connected to s_1 and C to s_0. The inputs of the multiplexer are 0, 1, A, and A'. When $BC = 00$, output $F = 0$ since $I_0 = 0$. Therefore, both minterms $m_0 = A'B'C'$ and $m_4 = AB'C'$ produce a 0 output, since the output is 0 when $BC = 00$ regardless of the value of A. When $BC = 01$, output $F = 1$, since $I_1 = 1$. Therefore, both minterms $m_1 = A'B'C$ and $m_5 = AB'C$ produce a 1 output, since the output is 1 when $BC = 01$ regardless of the value of A. When $BC = 10$, input I_2 is selected. Since A is connected to this input, the output will be equal to 1 only for minterm $m_6 = ABC'$, but not for minterm $m_2 = A'BC'$, because when $A' = 1$, then $A = 0$, and since $I_2 = 0$, we have $F = 0$. Finally, when $BC = 11$, input I_3 is selected. Since A' is connected to this input, the output will be equal to 1 only for minterm $m_3 = A'BC$, but not for $m_7 = ABC$. This information is summarized in Fig. 5-18(b), which is the truth table of the function we want to implement.

This discussion shows by analysis that the multiplexer implements the required function. We now present a general procedure for implementing any Boolean function of n variables with a 2^{n-1}-to-1 multiplexer.

First, express the function in its sum of minterms form. Assume that the ordered sequence of variables chosen for the minterms is $ABCD$ . . . , where A is the leftmost variable in the ordered sequence of n variables and BCD . . . are the remaining $n - 1$ variables. Connect the $n - 1$ variables to the selection lines of the multiplexer, with B connected to the high-order selection line, C to the next lower selection line, and so on

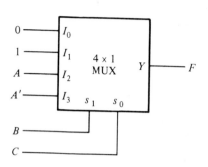

Minterm	A	B	C	F
0	0	0	0	0
1	0	0	1	1
2	0	1	0	0
3	0	1	1	1
4	1	0	0	0
5	1	0	1	1
6	1	1	0	1
7	1	1	1	0

(a) Multiplexer implementation

(b) Truth table

	I_0	I_1	I_2	I_3
A'	0	①	2	③
A	4	⑤	⑥	7
	0	1	A	A'

(c) Implementation table

FIGURE 5-18
Implementing $F(A, B, C) = \sum (1, 3, 5, 6)$ with a multiplexer

down to the last variable, which is connected to the lowest-order selection line s_0. Consider now the single variable A. Since this variable is in the highest-order position in the sequence of variables, it will be complemented in minterms 0 to $(2^n/2) - 1$, which comprise the first half in the list of minterms. The second half of the minterms will have their A variable uncomplemented. For a three-variable function, A, B, C, we have eight minterms. Variable A is complemented in minterms 0 to 3 and uncomplemented in minterms 4 to 7.

List the inputs of the multiplexer and under them list all the minterms in two rows. The first row lists all those minterms where A is complemented, and the second row all the minterms with A uncomplemented, as shown in Fig. 5-18(c). Circle all the minterms of the function and inspect each column separately.

If the two minterms in a column are not circled, apply 0 to the corresponding multiplexer input.

If the two minterms are circled, apply 1 to the corresponding multiplexer input.

If the bottom minterm is circled and the top is not circled, apply A to the corresponding multiplexer input.

If the top minterm is circled and the bottom is not circled, apply A' to the corresponding multiplexer input.

This procedure follows from the conditions established during the previous analysis. Figure 5-18(c) shows the implementation table for the Boolean function

$$F(A, B, C) = \Sigma(1, 3, 5, 6)$$

from which we obtain the multiplexer connections of Fig. 5-18(a). Note that B must be connected to s_1 and C to s_0.

It is not necessary to choose the leftmost variable in the ordered sequence of a variable list for the data inputs of the multiplexer. In fact, any one of the variables can be chosen for the inputs, provided we modify the multiplexer implementation table. Moreover, it is possible to derive the multiplexer circuit directly from the truth table. Consider, for example, the following three-variable Boolean function:

$$F(A, B, C) = \Sigma(1, 2, 4, 5)$$

We wish to implement the function with a multiplexer, but in this case, we will connect variables A and B to selection inputs s_1 and s_0, respectively, and use the rightmost vari-

A	B	C	F	
0	0	0	0	
				$F = C$
0	0	1	1	
0	1	0	1	
				$F = C'$
0	1	1	0	
1	0	0	1	
				$F = 1$
1	0	1	1	
1	1	0	0	
				$F = 0$
1	1	1	0	

(a) Truth table

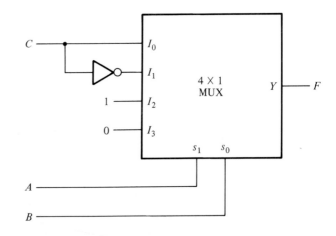

(b) Multiplexer implementation

	I_0	I_1	I_2	I_3
C'	0	②	④	6
C	①	3	⑤	7
	C	C'	1	0

(c) Implementation table

FIGURE 5-19
Implementing $F(A, B, C) = \sum (1, 2, 4, 5)$ with a multiplexer

able C for the data inputs of the multiplexer. Figure 5-19(a) is the truth table of the function. The table is divided into sections, with each section having identical values for variables A and B. We note that when $AB = 00$, output F is the same as input C. When $AB = 01$, F is the same as C'. When $AB = 10$, $F = 1$, and when $AB = 11$, $F = 0$. The multiplexer circuit of Fig. 5-19(b) can be derived directly from the truth table without the need of an implementation table. However, if an implementation table is desired, it must be modified to take into account the relationship between the minterms and the inputs of the multiplexer. As seen from the truth table, variable C is complemented in the even-numbered minterms 0, 2, 4, and 6, and uncomplemented in the odd-numbered minterms 1, 3, 5, and 7. The arrangement of the two rows in the implementation table must be as shown in Fig. 5-19(c). By circling the minterms of the function and using the rules stated before, we obtain the multiplexer inputs for imple-menting the function.

In a similar fashion, it is possible to choose any other variable of the function for the multiplexer data inputs. In any case, all input variables except one are applied to the se-lection inputs of the multiplexer. The remaining single variable, or its complement, or 0, or 1, is then applied to the data inputs of the multiplexer.

Example 5-2

Implement the following function with a multiplexer:

$$F(A, B, C, D) = \Sigma(0, 1, 3, 4, 8, 9, 15)$$

This is a four-variable function and, therefore, we need a multiplexer with three selec-tion lines and eight inputs. We choose to apply variables B, C, and D to the selection lines. The implementation table is then as shown in Fig. 5-20. The first half of the

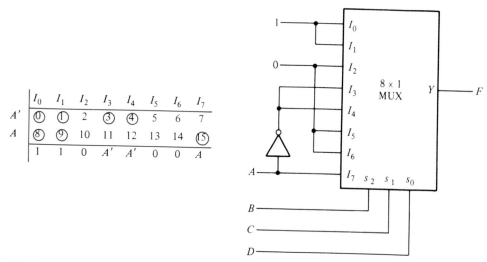

FIGURE 5-20

Implementing $F(A, B, C, D) = \Sigma (0, 1, 3, 4, 8, 9, 15)$

minterms are associated with A' and the second half with A. By circling the minterms of the function and applying the rules for finding values for the multiplexer inputs, we obtain the implementation shown.   ∎

Let us now compare the multiplexer method with the decoder method for implementing combinational circuits. The decoder method requires an OR gate for each output function, but only one decoder is needed to generate all minterms. The multiplexer method uses smaller-size units but requires one multiplexer for each output function. It would seem reasonable to assume that combinational circuits with a small number of outputs should be implemented with multiplexers. Combinational circuits with many output functions would probably use fewer ICs with the decoder method.

Although multiplexers and decoders may be used in the implementation of combinational circuits, it must be realized that decoders are mostly used for decoding binary information and multiplexers are mostly used to form a selected path between multiple sources and a single destination.

5-7   READ-ONLY MEMORY (ROM)

We saw in Section 5-5 that a decoder generates the 2^n minterms of the n input variables. By inserting OR gates to sum the minterms of Boolean functions, we were able to generate any desired combinational circuit. A read-only memory (ROM) is a device that includes both the decoder and the OR gates within a single IC package. The connections between the outputs of the decoder and the inputs of the OR gates can be specified for each particular configuration. The ROM is used to implement complex combinational circuits within one IC package or as permanent storage for binary information.

A ROM is essentially a memory (or storage) device in which permanent binary information is stored. The binary information must be specified by the designer and is then embedded in the unit to form the required interconnection pattern. ROMs come with special internal electronic fuses that can be "programmed" for a specific configuration. Once the pattern is established, it stays within the unit even when power is turned off and on again.

A block diagram of a ROM is shown in Fig. 5-21. It consists of n input lines and m output lines. Each bit combination of the input variables is called an *address*. Each bit combination that comes out of the output lines is called a *word*. The number of bits per word is equal to the number of output lines, m. An address is essentially a binary number that denotes one of the minterms of n variables. The number of distinct addresses possible with n input variables is 2^n. An output word can be selected by a unique address, and since there are 2^n distinct addresses in a ROM, there are 2^n distinct words that are said to be stored in the unit. The word available on the output lines at any given time depends on the address value applied to the input lines. A ROM is charac-

n inputs

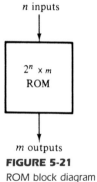

m outputs

FIGURE 5-21

ROM block diagram

terized by the number of words 2^n and the number of bits per word m. This terminology is used because of the similarity between the read-only memory and the random-access memory, which is presented in Section 7-7.

Consider a 32×8 ROM. The unit consists of 32 words of 8 bits each. This means that there are eight output lines and that there are 32 distinct words stored in the unit, each of which may be applied to the output lines. The particular word selected that is presently available on the output lines is determined from the five input lines. There are only five inputs in a 32×8 ROM because $2^5 = 32$, and with five variables, we can specify 32 addresses or minterms. For each address input, there is a unique selected word. Thus, if the input address is 00000, word number 0 is selected and it appears on the output lines. If the input address is 11111, word number 31 is selected and applied to the output lines. In between, there are 30 other addresses that can select the other 30 words.

The number of addressed words in a ROM is determined from the fact that n input lines are needed to specify 2^n words. A ROM is sometimes specified by the total number of bits it contains, which is $2^n \times m$. For example, a 2048-bit ROM may be organized as 512 words of 4 bits each. This means that the unit has four output lines and nine input lines to specify $2^9 = 512$ words. The total number of bits stored in the unit is $512 \times 4 = 2048$.

Internally, the ROM is a combinational circuit with AND gates connected as a decoder and a number of OR gates equal to the number of outputs in the unit. Figure 5-22 shows the internal logic construction of a 32×4 ROM. The five input variables are decoded into 32 lines by means of 32 AND gates and 5 inverters. Each output of the decoder represents one of the minterms of a function of five variables. Each one of the 32 addresses selects one and only one output from the decoder. The address is a 5-bit number applied to the inputs, and the selected minterm out of the decoder is the one marked with the equivalent decimal number. The 32 outputs of the decoder are connected through fuses to each OR gate. Only four of these fuses are shown in the diagram, but actually each OR gate has 32 inputs and each input goes through a fuse that can be blown as desired.

The ROM is a two-level implementation in sum of minterms form. It does not have to be an AND-OR implementation, but it can be any other possible two-level minterm

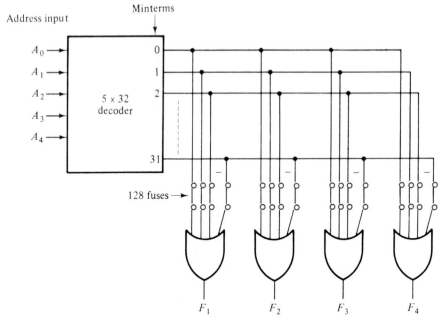

FIGURE 5-22

Logic construction of a 32 × 4 ROM

implementation. The second level is usually a wired-logic connection (see Section 3-7) to facilitate the blowing of fuses.

ROMs have many important applications in the design of digital computer systems. Their use for implementing complex combinational circuits is just one of these applications. Other uses of ROMs are presented in other parts of the book in conjunction with their particular applications.

Combinational Logic Implementation

From the logic diagram of the ROM, it is clear that each output provides the sum of all the minterms of the n input variables. Remember that any Boolean function can be expressed in sum of minterms form. By breaking the links of those minterms not included in the function, each ROM output can be made to represent the Boolean function of one of the output variables in the combinational circuit. For an n-input, m-output combinational circuit, we need a $2^n \times m$ ROM. The blowing of the fuses is referred to as *programming* the ROM. The designer need only specify a ROM program table that gives the information for the required paths in the ROM. The actual programming is a hardware procedure that follows the specifications listed in the program table.

Let us clarify the process with a specific example. The truth table in Fig. 5-23(a) specifies a combinational circuit with two inputs and two outputs. The Boolean functions can be expressed in sum of minterms:

A_1	A_0	F_1	F_2
0	0	0	1
0	1	1	0
1	0	1	1
1	1	1	0

(a) Truth table

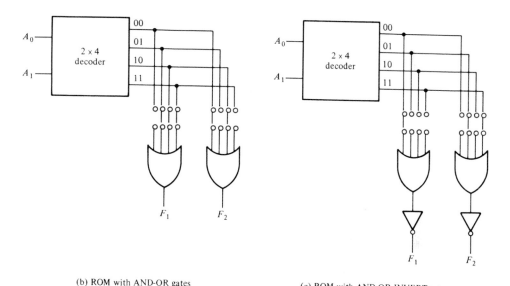

(b) ROM with AND-OR gates   (c) ROM with AND-OR-INVERT gates

FIGURE 5-23

Combinational-circuit implementation with a 4 × 2 ROM

$$F_1(A_1, A_0) = \Sigma(1, 2, 3)$$

$$F_2(A_1, A_0) = \Sigma(0, 2)$$

When a combinational circuit is implemented by means of a ROM, the functions must be expressed in sum of minterms or, better yet, by a truth table. If the output functions are simplified, we find that the circuit needs only one OR gate and an inverter. Obviously, this is too simple a combinational circuit to be implemented with a ROM. The advantage of a ROM is in complex combinational circuits. This example merely demonstrates the procedure and should not be considered in a practical situation.

The ROM that implements the combinational circuit must have two inputs and two outputs; so its size must be 4 ×.2. Figure 5-23(b) shows the internal construction of such a ROM. It is now necessary to determine which of the eight available fuses must be blown and which should be left intact. This can be easily done from the output functions listed in the truth table. Those minterms that specify an output of 0 should not have a path to the output through the OR gate. Thus, for this particular case, the truth table shows three 0's, and their corresponding fuses to the OR gates must be blown. It is obvious that we must assume here that an open input to an OR gate behaves as a 0 input.

Some ROM units come with an inverter after each of the OR gates and, as a consequence, they are specified as having initially all 0's at their outputs. The programming procedure in such ROMs requires that we open the paths of the minterms (or addresses) that specify an output of 1 in the truth table. The output of the OR gate will then generate the complement of the function, but the inverter placed after the OR gate complements the function once more to provide the normal output. This is shown in the ROM of Fig. 5-23(c).

The previous example demonstrates the general procedure for implementing any combinational circuit with a ROM. From the number of inputs and outputs in the combinational circuit, we first determine the size of ROM required. Then we must obtain the programming truth table of the ROM; no other manipulation or simplification is required. The 0's (or 1's) in the output functions of the truth table directly specify those fuses that must be blown to provide the required combinational circuit in sum of minterms form.

In practice, when one designs a circuit by means of a ROM, it is not necessary to show the internal gate connections of fuses inside the unit, as was done in Fig. 5-23. This was shown there for demonstration purposes only. All the designer has to do is specify the particular ROM (or its designation number) and provide the ROM truth table, as in Fig. 5-23(a). The truth table gives all the information for programming the ROM. No internal logic diagram is needed to accompany the truth table.

Example **5-3**	Design a combinational circuit using a ROM. The circuit accepts a 3-bit number and generates an output binary number equal to the square of the input number. The first step is to derive the truth table for the combinational circuit. In most cases, this is all that is needed. In some cases, we can fit a smaller truth table for the ROM by using certain properties in the truth table of the combinational circuit. Table 5-5 is the

TABLE 5-5
Truth Table for Circuit of Example 5-3

Inputs			Outputs						
A_2	A_1	A_0	B_5	B_4	B_3	B_2	B_1	B_0	Decimal
0	0	0	0	0	0	0	0	0	0
0	0	1	0	0	0	0	0	1	1
0	1	0	0	0	0	1	0	0	4
0	1	1	0	0	1	0	0	1	9
1	0	0	0	1	0	0	0	0	16
1	0	1	0	1	1	0	0	1	25
1	1	0	1	0	0	1	0	0	36
1	1	1	1	1	0	0	0	1	49

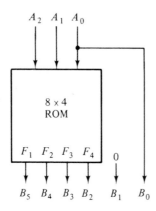

A_2 A_1 A_0

8 × 4
ROM

F_1 F_2 F_3 F_4 0

B_5 B_4 B_3 B_2 B_1 B_0

(a) Block diagram

A_2	A_1	A_0	F_1	F_2	F_3	F_4
0	0	0	0	0	0	0
0	0	1	0	0	0	0
0	1	0	0	0	0	1
0	1	1	0	0	1	0
1	0	0	0	1	0	0
1	0	1	0	1	1	0
1	1	0	1	0	0	1
1	1	1	1	1	0	0

(b) ROM truth table

FIGURE 5-24
ROM implementation of Example 5-3

truth table for the combinational circuit. Three inputs and six outputs are needed to accommodate all possible numbers. We note that output B_0 is always equal to input A_0; so there is no need to generate B_0 with a ROM since it is equal to an input variable. Moreover, output B_1 is always 0, so this output is always known. We actually need to generate only four outputs with the ROM; the other two are easily obtained. The minimum-size ROM needed must have three inputs and four outputs. Three inputs specify eight words, so the ROM size must be 8 × 4. The ROM implementation is shown in Fig. 5-24. The three inputs specify eight words of four bits each. The other two outputs of the combinational circuit are equal to 0 and A_0. The truth table in Fig. 5-24 specifies all the information needed for programming the ROM, and the block diagram shows the required connections.   ∎

Types of ROMs

The required paths in a ROM may be programmed in two different ways. The first is called *mask programming* and is done by the manufacturer during the last fabrication process of the unit. The procedure for fabricating a ROM requires that the customer fill out the truth table the ROM is to satisfy. The truth table may be submitted on a special form provided by the manufacturer. More often, it is submitted in a computer input medium in the format specified on the data sheet of the particular ROM. The manufacturer makes the corresponding mask for the paths to produce the 1's and 0's according to the customer's truth table. This procedure is costly because the vendor charges the customer a special fee for custom masking a ROM. For this reason, mask programming is economical only if large quantities of the same ROM configuration are to be manufactured.

For small quantities, it is more economical to use a second type of ROM called a *programmable read-only memory,* or PROM. When ordered, PROM units contain all 0's (or all 1's) in every bit of the stored words. The fuses in the PROM are blown by application of current pulses through the output terminals. A blown fuse defines one binary state and an unbroken link represents the other state. This allows the user to program the unit in the laboratory to achieve the desired relationship between input addresses and stored words. Special units called *PROM programmers* are available commercially to facilitate this procedure. In any case, all procedures for programming ROMs are *hardware* procedures even though the word *programming* is used.

The hardware procedure for programming ROMs or PROMs is irreversible and, once programmed, the fixed pattern is permanent and cannot be altered. Once a bit pattern has been established, the unit must be discarded if the bit pattern is to be changed. A third type of unit available is called *erasable PROM,* or EPROM. EPROMs can be restructured to the initial value (all 0's or all 1's) even though they have been changed previously. When an EPROM is placed under a special ultraviolet light for a given period of time, the shortwave radiation discharges the internal gates that serve as contacts. After erasure, the ROM returns to its initial state and can be reprogrammed. Certain ROMs can be erased with electrical signals instead of ultraviolet light, and these are called *electrically erasable PROMs,* or EEPROMs.

The function of a ROM can be interpreted in two different ways. The first interpretation is of a unit that implements any combinational circuit. From this point of view, each output terminal is considered separately as the output of a Boolean function expressed in sum of minterms. The second interpretation considers the ROM to be a storage unit having a fixed pattern of bit strings called *words.* From this point of view, the inputs specify an *address* to a specific stored word, which is then applied to the outputs. For example, the ROM of Fig. 5-24 has three address lines, which specify eight stored words as given by the truth table. Each word is four bits long. This is the reason why the unit is given the name *read-only memory. Memory* is commonly used to designate a storage unit. *Read* is commonly used to signify that the contents of a word specified by an address in a storage unit is placed at the output terminals. Thus, a ROM is a memory unit with a fixed word pattern that can be read out upon application of a given address. The bit pattern in the ROM is permanent and cannot be changed during normal operation.

ROMs are widely used to implement complex combinational circuits directly from their truth tables. They are useful for converting from one binary code to another (such as ASCII to EBCDIC and vice versa), for arithmetic functions such as multipliers, for display of characters in a cathode-ray tube, and in many other applications requiring a large number of inputs and outputs. They are also employed in the design of control units of digital systems. As such, they are used to store fixed bit patterns that represent the sequence of control variables needed to enable the various operations in the system. A control unit that utilizes a ROM to store binary control information is called a *microprogrammed control unit.*

5-8 PROGRAMMABLE LOGIC ARRAY (PLA)

A combinational circuit may occasionally have don't-care conditions. When implemented with a ROM, a don't-care condition becomes an address input that will never occur. The words at the don't-care addresses need not be programmed and may be left in their original state (all 0's or all 1's). The result is that not all the bit patterns available in the ROM are used, which may be considered a waste of available equipment.

Consider, for example, a combinational circuit that converts a 12-bit card code to a 6-bit internal alphanumeric code (see end of Section 1-7). The input card code consists of 12 lines designated by 0, 1, 2, . . . , 9, 11, 12. The size of the ROM for implementing the code converter must be 4096 × 6, since there are 12 inputs and 6 outputs. There are only 47 valid entries for the card code; all other input combinations are don't-care conditions. Thus, only 47 words of the 4096 available are used. The remaining 4049 words of ROM are not used and are thus wasted.

For cases where the number of don't-care conditions is excessive, it is more economical to use a second type of LSI component called a *programmable logic array,* or PLA. A PLA is similar to a ROM in concept; however, the PLA does not provide full decoding of the variables and does not generate all the minterms as in the ROM. In the PLA, the decoder is replaced by a group of AND gates, each of which can be programmed to generate a product term of the input variables. The AND and OR gates inside the PLA are initially fabricated with fuses among them. The specific Boolean functions are implemented in sum of products form by blowing appropriate fuses and leaving the desired connections.

A block diagram of the PLA is shown in Fig. 5-25. It consists of n inputs, m outputs, k product terms, and m sum terms. The product terms constitute a group of k AND gates and the sum terms constitute a group of m OR gates. Fuses are inserted between all n inputs and their complement values to each of the AND gates. Fuses are also provided between the outputs of the AND gates and the inputs of the OR gates. Another set of fuses in the output inverters allows the output function to be generated either in the AND-OR form or in the AND-OR-INVERT form. With the inverter fuse in place, the inverter is bypassed, giving an AND-OR implementation. With the fuse blown, the inverter becomes part of the circuit and the function is implemented in the AND-OR-INVERT form.

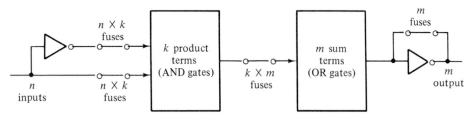

FIGURE 5-25
PLA block diagram

The size of the PLA is specified by the number of inputs, the number of product terms, and the number of outputs (the number of sum terms is equal to the number of outputs). A typical PLA has 16 inputs, 48 product terms, and 8 outputs. The number of programmed fuses is $2n \times k + k \times m + m$, whereas that of a ROM is $2^n \times m$.

Figure 5-26 shows the internal construction of a specific PLA. It has three inputs, three product terms, and two outputs. Such a PLA is too small to be available commerically; it is presented here merely for demonstration purposes. Each input and its complement are connected through fuses to the inputs of all AND gates. The outputs of the AND gates are connected through fuses to each input of the OR gates. Two more fuses are provided with the output inverters. By blowing selected fuses and leaving others intact, it is possible to implement Boolean functions in their sum of products form.

As with a ROM, the PLA may be mask-programmable or field-programmable. With a mask-programmable PLA, the customer must submit a PLA program table to the manufacturer. This table is used by the vendor to produce a custom-made PLA that has the required internal paths between inputs and outputs. A second type of PLA available is called a *field-programmable logic array*, or FPLA. The FPLA can be programmed by the user by means of certain recommended procedures. Commercial hardware programmer units are available for use in conjunction with certain FPLAs.

PLA Program Table

The use of a PLA must be considered for combinational circuits that have a large number of inputs and outputs. It is superior to a ROM for circuits that have a large number of don't-care conditions. The example to be presented demonstrates how a PLA is programmed. Bear in mind when going through the example that such a simple circuit will not require a PLA because it can be implemented more economically with SSI gates.

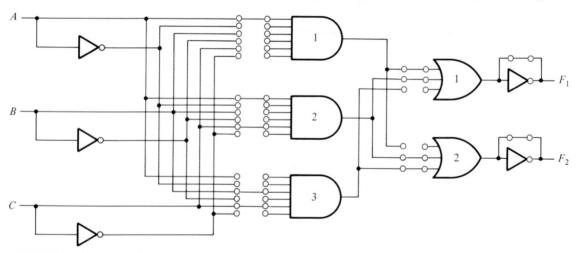

FIGURE 5-26

PLA with three inputs, three product terms, and two outputs; it implements the combinational circuit specified in Fig. 5-27

$$F_1 = AB' + AC$$

$$F_2 = AC + BC$$

(b) Map simplification

A	B	C	F_1	F_2
0	0	0	0	0
0	0	1	0	0
0	1	0	0	0
0	1	1	0	1
1	0	0	1	0
1	0	1	1	1
1	1	0	0	0
1	1	1	1	1

(a) Truth table

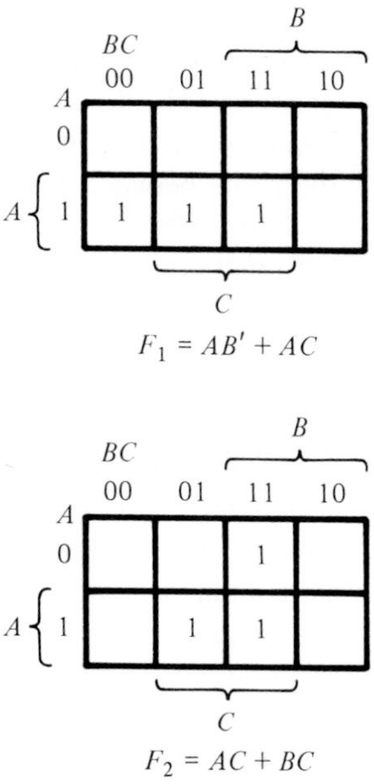

Product term	Inputs A	B	C	Outputs F_1	F_2
AB'	1	1	0	—	1	—
AC	2	1	—	1	1	1
BC	3	—	1	1	—	1
					T	T	T/C

(c) PLA program table

FIGURE 5-27
Steps required in PLA implementation

Consider the truth table of the combinational circuit, shown in Fig. 5-27(a). Although a ROM implements a combinational circuit in its sum of minterms form, a PLA implements the functions in their sum of products form. Each product term in the expression requires an AND gate. Since the number of AND gates in a PLA is finite, it is necessary to simplify the function to a minimum number of product terms in order to minimize the number of AND gates used. The simplified functions in sum of products are obtained from the maps of Fig. 5-27(b):

$$F_1 = AB' + AC$$

$$F_2 = AC + BC$$

There are three distinct product terms in this combinational circuit: AB', AC, and BC. The circuit has three inputs and two outputs; so the PLA of Fig. 5-26 can be used to implement this combinational circuit.

Programming the PLA means that we specify the paths in its AND-OR-NOT pattern. A typical PLA program table is shown in Fig. 5-27(c). It consists of three columns. The first column lists the product terms numerically. The second column specifies the required paths between inputs and AND gates. The third column specifies the paths between the AND gates and the OR gates. Under each output variable, we write a T (for true) if the output inverter is to be bypassed, and C (for complement) if the function is to be complemented with the output inverter. The Boolean terms listed at the left are not part of the table; they are included for reference only.

For each product term, the inputs are marked with 1, 0, or – (dash). If a variable in the product term appears in its normal form (unprimed), the corresponding input variable is marked with a 1. If it appears complemented (primed), the corresponding input variable is marked with a 0. If the variable is absent in the product term, it is marked with a dash. Each product term is associated with an AND gate. The paths between the inputs and the AND gates are specified under the column heading *inputs*. A 1 in the input column specifies a path from the corresponding input to the input of the AND gate that forms the product term. A 0 in the input column specifies a path from the corresponding complemented input to the input of the AND gate. A dash specifies no connection. The appropriate fuses are blown and the ones left intact form the desired paths, as shown in Fig. 5-26. It is assumed that the open terminals in the AND gate behave like a 1 input.

The paths between the AND and OR gates are specified under the column heading *outputs*. The output variables are marked with 1's for all those product terms that formulate the function. In the example of Fig. 5-27, we have

$$F_1 = AB' + AC$$

so F_1 is marked with 1's for product terms 1 and 2 and with a dash for product term 3. Each product term that has a 1 in the output column requires a path from the corresponding AND gate to the output OR gate. Those marked with a dash specify no connection. Finally, a T (true) output dictates that the fuse across the output inverter remains intact, and a C (complement) specifies that the corresponding fuse be blown. The internal paths of the PLA for this circuit are shown in Fig. 5-26. It is assumed that an open terminal in an OR gate behaves like a 0, and that a short circuit across the output inverter does not damage the circuit.

When designing a digital system with a PLA, there is no need to show the internal connections of the unit, as was done in Fig. 5-26. All that is needed is a PLA program table from which the PLA can be programmed to supply the appropriate paths.

When implementing a combinational circuit with PLA, careful investigation must be undertaken in order to reduce the total number of distinct product terms, since a given PLA would have a finite number of AND terms. This can be done by simplifying each function to a minimum number of terms. The number of literals in a term is not important since we have available all input variables. Both the true value and the complement of the function should be simplified to see which one can be expressed with fewer product terms and which one provides product terms that are common to other functions.

Example
5-4

A combinational circuit is defined by the functions

$$F_1(A, B, C) = \Sigma(3, 5, 6, 7)$$

$$F_2(A, B, C) = \Sigma(0, 2, 4, 7)$$

Implement the circuit with a PLA having three inputs, four product terms, and two outputs.

The two functions are simplified in the maps of Fig. 5-28. Both the true values and the complements of the functions are simplified. The combinations that give a minimum number of product terms are

$$F_1 = (B'C' + A'C' + A'B')'$$

$$F_2 = B'C' + A'C' + ABC$$

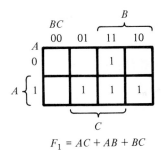

$$F_1 = AC + AB + BC$$

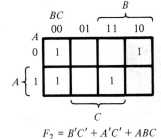

$$F_2 = B'C' + A'C' + ABC$$

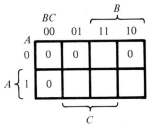

$$F_1' = B'C' + A'C' + A'B'$$

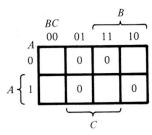

$$F_2' = B'C + A'C + ABC$$

PLA program table

Product term		Inputs			Outputs	
		A	B	C	F_1	F_2
$B'C'$	1	−	0	0	1	1
$A'C'$	2	0	−	0	1	1
$A'B'$	3	0	0	−	1	−
ABC	4	1	1	1	−	1
					C	T T/C

FIGURE 5-28
Solution to Example 5-4

This gives only four distinct product terms: $B'C'$, $A'C'$, $A'B'$, and ABC. The PLA program table for this combination is shown in Fig. 5-28. Note that output F_1 is the normal (or true) output even though a C is marked under it. This is because F_1' is generated *prior* to the output inverter. The inverter complements the function to produce F_1 in the output.

∎

The combinational circuit for this example is too small for practical implementation with a PLA. It was presented here merely for demonstration purposes. A typical commercial PLA would have over 10 inputs and about 50 product terms. The simplification of Boolean functions with so many variables should be carried out by means of a tabulation method or other computer-assisted simplification method. This is where a computer program may aid in the design of complex digital systems. The computer program should simplify each function of the combinational circuit and its complement to a minimum number of terms. The program then selects a minimum number of distinct terms that cover all functions in their true or complement form.

5-9 PROGRAMMABLE ARRAY LOGIC (PAL)

Programmable logic devices have hundreds of gates interconnected through hundreds of electronic fuses. It is sometimes convenient to draw the internal logic of such devices in a compact form referred to as *array logic*. Figure 5-29 shows the conventional and array logic symbols for a multiple-input AND gate. The conventional symbol is drawn with multiple lines showing the fuses connected to the inputs of the gate. The corresponding array logic symbol uses a single horizontal line connected to the gate input and multiple vertical lines to indicate the individual inputs. Each intersection between a vertical line and the common horizontal line has a fused connection. Thus, in Fig. 5-29(b), the AND gate has four inputs connected through fuses. In a similar fashion, we can draw the array logic for the OR gate or any other type of multiple-input gate.

The programmable array logic (PAL) is a programmable logic device with a fixed OR array and a programmable AND array. Because only the AND gates are programmable, the PAL is easier to program, but is not as flexible as the PLA. Figure 5-30 shows the array logic configuration of a typical PAL. It has four inputs and four out-

(a) Conventional symbol   (b) Array logic symbol

FIGURE 5-29
Two graphic symbols for an AND gate

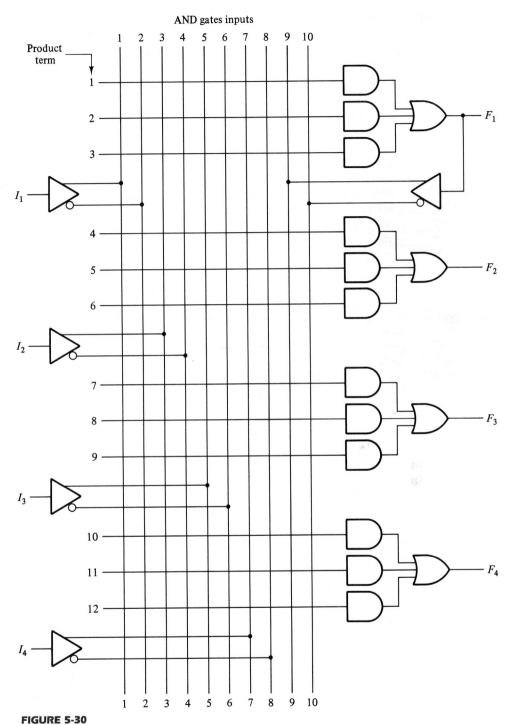

FIGURE 5-30
PAL with four inputs, four outputs, and three-wide AND-OR structure

puts. Each input has a buffer and an inverter gate. Note that the two gates are shown with one composite graphic symbol with normal and complement outputs. There are four sections in the unit, each being composed of a threewide AND-OR array. This is the term used to indicate that there are three programmable AND gates in each section and one fixed OR gate. Each AND gate has 10 fused programmable inputs. This is shown in the diagram by 10 vertical lines intersecting each horizontal line. The horizontal line symbolizes the multiple-input configuration of the AND gate. One of the outputs is connected to a buffer–inverter gate and then fed back into the inputs of the AND gates through fuses.

Commercial PAL devices contain more gates than the one shown in Fig. 5-30. A typical PAL integrated circuit may have eight inputs, eight outputs, and eight sections, each consisting of an eightwide AND-OR array. The output terminals are sometimes bidirectional, which means that they can be programmed as inputs instead of outputs if desired.

When designing with a PAL, the Boolean functions must be simplified to fit into each section. Unlike the PLA, a product term cannot be shared among two or more OR gates. Therefore, each function can be simplified by itself without regard to common product terms. The number of product terms in each section is fixed, and if the number of terms in the function is too large, it may be necessary to use two sections to implement one Boolean function.

As an example of using a PAL in the design of a combinational circuit, consider the following Boolean functions given in sum of minterms:

$$w(A, B, C, D) = \Sigma(2, 12, 13)$$

$$x(A, B, C, D) = \Sigma(7, 8, 9, 10, 11, 12, 13, 14, 15)$$

$$y(A, B, C, D) = \Sigma(0, 2, 3, 4, 5, 6, 7, 8, 10, 11, 15)$$

$$z(A, B, C, D) = \Sigma(1, 2, 8, 12, 13)$$

Simplifying the four functions to a minimum number of terms results in the following Boolean functions:

$$w = ABC' + A'B'CD'$$

$$x = A + BCD$$

$$y = A'B + CD + B'D'$$

$$z = ABC' + A'B'CD' + AC'D' + A'B'C'D$$

$$= w + AC'D' + A'B'C'D$$

Note that the function for z has four product terms. The logical sum of two of these terms is equal to w. By using w, it is possible to reduce the number of terms for z from four to three.

The PAL programming table is similar to the one used for the PLA except that only the inputs of the AND gates need to be programmed. Table 5-6 lists the PAL programming table for the four Boolean functions. The table is divided into four sections with three product terms in each to conform with the PAL of Fig. 5-30. The first two sections need only two product terms to implement the Boolean function. The last section for output z needs four product terms. Using the output from w, we can reduce the function to three terms.

The fuse map for the PAL as specified in the programming table is shown in Fig. 5-31. For each 1 or 0 in the table, we mark the corresponding intersection in the diagram with the symbol for an intact fuse. For each dash, we mark the diagram with blown fuses in both the true and complement inputs. If the AND gate is not used, we leave all its input fuses intact. Since the corresponding input receives both the true and complement of each input variable, we have $AA' = 0$ and the output of the AND gate is always 0.

As with all PLDs, the design with PALs is facilitated by using computer-aided design techniques. The blowing of internal fuses is a hardware procedure done with the help of special electronic instruments.

TABLE 5-6
PAL Programming Table

Product Term	AND Inputs A	B	C	D	W	Outputs
1	1	1	0	–	–	$w = ABC'$
2	0	0	1	0	–	$+ A'B'CD'$
3	–	–	–	–	–	
4	1	–	–	–	–	$x = A$
5	–	1	1	1	–	$+ BCD$
6	–	–	–	–	–	
7	0	1	–	–	–	$y = A'B$
8	–	–	1	1	–	$+ CD$
9	–	0	–	0	–	$+ B'D'$
10	–	–	–	–	1	$z = w$
11	1	–	0	0	–	$+ AC'D'$
12	0	0	0	1	–	$+ A'B'C'D$

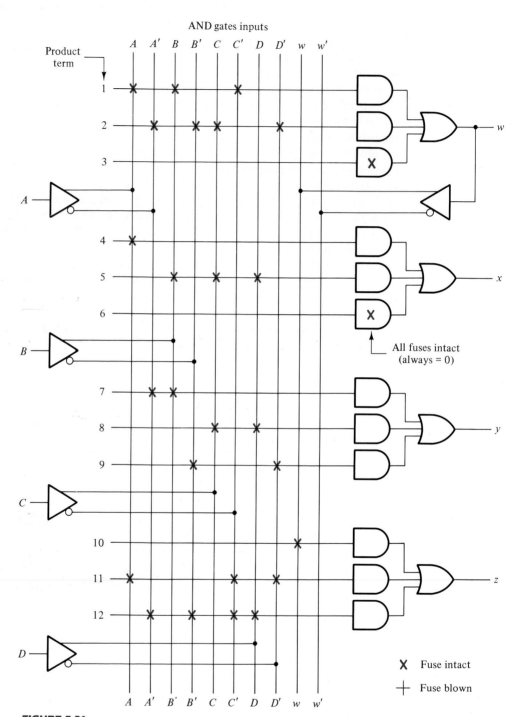

FIGURE 5-31
Fuse map for PAL as specified in Table 5-6

REFERENCES

1. Blakeslee, T. R., *Digital Design with Standard MSI and LSI*, 2nd Ed. New York: John Wiley, 1979.

2. Sandige, R. S., *Digital Concepts Using Standard Integrated Circuits*. New York: Mc-Graw-Hill, 1978.

3. Mano, M. M., *Computer Engineering: Hardware Design*. Englewood Cliffs, NJ: Prentice-Hall, 1988.

4. Mano, M. M., *Computer System Architecture*, 2nd Ed. Englewood Cliffs, NJ: Prentice-Hall, 1982.

5. Shiva, S. G., *Introduction to Logic Design*. Glenview, IL: Scott, Foresman, 1988.

6. *The TTL Logic Data Book*. Dallas: Texas Instruments, 1988.

7. Tocci, R. J., *Digital Systems Principles and Applications*, 4th Ed. Englewood Cliffs, NJ: Prentice-Hall, 1988.

8. Fletcher, W. I., *An Engineering Approach to Digital Design*. Englewood Cliffs, NJ: Prentice-Hall, 1979.

9. Kitson, B., *Programmable Array Logic Handbook*. Sunnyvale, CA: Advanced Micro Devices, 1983.

10. *Programmable Logic Data Manual*. Sunnyvale, CA: Signetics, 1986.

11. *Programmable Logic Data Book*. Dallas: Texas Instruments, 1988.

PROBLEMS

5-1 Construct a 16-bit parallel adder with four MSI circuits, each containing a 4-bit parallel adder. Use a block diagram with nine inputs and five outputs for each 4-bit adder. Show how the carries are connected between the MSI circuits.

5-2 Construct a BCD-to-excess-3-code converter with a 4-bit adder. Remember that the excess-3 code digit is obtained by adding three to the corresponding BCD digit. What must be done to change the circuit to an excess-3-to-BCD-code converter?

5-3 The adder-subtractor of Fig. 5-2(b) is used to subtract the following unsigned 4-bit numbers: 0110 − 1001 (6 − 9).
(a) What are the binary values in the nine inputs of the circuit?
(b) What are the binary values of the five outputs of the circuit? Explain how the output is related to the operation of 6 − 9.

5-4 The adder-subtractor circuit of Fig. 5-2(b) has the following values for mode input M and data inputs A and B. In each case, determine the values of the outputs: S_4, S_3, S_2, S_1, and C_5.

	M	A	B
(a)	0	0111	0110
(b)	0	1000	1001
(c)	1	1100	1000
(d)	1	0101	1010
(e)	1	0000	0001

5-5 (a) Using the AND-OR-INVERT implementation procedure described in Section 3-7, show that the output carry in a full-adder circuit can be expressed as

$$C_{i+1} = G_i + P_i C_i = (G_i' P_i + G_i' C_i')'$$

(b) IC type 74182 is a look-ahead carry generator MSI circuit that generates the carries with AND-OR-INVERT gates. The MSI circuit assumes that the input terminals have the complements of the G's, the P's, and of C_1. Derive the Boolean functions for the look-ahead carries C_2, C_3, and C_4 in this IC. (*Hint:* Use the equation-substitution method to derive the carries in terms of C_1'.)

5-6 (a) Redefine the carry propagate and carry generate as follows:

$$P_i = A_i + B_i$$

$$G_i = A_i B_i$$

Show that the output carry and output sum of a full-adder becomes

$$C_{i+1} = (C_i' G_i + P_i')' = G_i + P_i C_i$$

$$S_i = (P_i G_i') \oplus C_i$$

(b) The logic diagram of the first stage of a 4-bit parallel adder as implemented in IC type 74283 is shown in Fig. P5-6. Identify the P_i' and G_i' terminals as defined in part (a) and show that the circuit implements a full-adder circuit.

(c) Obtain the output carries C_3 and C_4 as functions of P_1', P_2', P_3', G_1', G_2', G_3', and C_1' in AND-OR-INVERT form, and draw the two-level look-ahead circuit for this IC. [*Hint:* Use the equation-substitution method as done in the text when deriving Fig. 5-4, but use the AND-OR-INVERT function given in part (a) for C_{i+1}.]

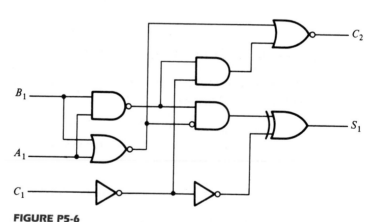

FIGURE P5-6
First stage of a parallel adder

5-7 Assume that the exclusive-OR gate has a propagation delay of 20 ns and that the AND or OR gates have a propagation delay of 10 ns. What is the total propagation delay time in the 4-bit adder of Fig. 5-5?

5-8 Derive the two-level Boolean expression for the output carry C_5 shown in the look-ahead carry generator of Fig. 5-5.

5-9 How many unused input combinations are there in a BCD adder?

5-10 Design a combinational circuit that generated the 9's complement of a BCD digit.

5-11 Construct a 4-digit BCD adder-subtractor using four BCD adders, as shown in Fig. 5-6, and four 9's complement circuits from Problem 5-10. Use block diagrams for each component, showing only inputs and outputs.

5-12 It is necessary to design a decimal adder for two digits represented in the excess-3 code. Show that the correction after adding the two digits with a 4-bit binary adder is as follows:
(a) The output carry is equal to the carry from the binary adder.
(b) If the output carry = 1, then add 0011.
(c) If the output carry = 0, then add 1101.
Construct the decimal adder with two 4-bit adders and an inverter.

5-13 Design a combinational circuit that compares two 4-bit numbers A and B to check if they are equal. The circuit has one output x, so that $x = 1$ if $A = B$ and $x = 0$ if A is not equal to B.

5-14 Design a BCD-to-decimal decoder using the unused combinations of the BCD code as don't-care conditions.

5-15 A combinational circuit is defined by the following three Boolean functions. Design the circuit with a decoder and external gates.

$$F_1 = x'y'z' + xz$$
$$F_2 = xy'z' + x'y$$
$$F_3 = x'y'z + xy$$

5-16 A combinational circuit is specified by the following three Boolean functions. Implement the circuit with a 3×8 decoder constructed with NAND gates (similar to Fig. 5-10) and three external NAND or AND gates. Use a block diagram for the decoder. Minimize the number of inputs in the external gates.

$$F_1(A, B, C) = \Sigma(2, 4, 7)$$
$$F_2(A, B, C) = \Sigma(0, 3)$$
$$F_3(A, B, C) = \Sigma(0, 2, 3, 4, 7)$$

5-17 Draw the logic diagram of a 2-to-4-line decoder with only NOR gates. Include an enable input.

5-18 Construct a 5×32 decoder with four 3×8 decoders with enable and one 2×4 decoder. Use block diagrams similar to Fig. 5-12.

5-19 Rearrange the truth table for the circuit of Fig. 5-10 and verify that it can function as a demultiplexer.

5-20 Design a 4-input priority encoder with inputs as in Table 5-4, but with input D_0 having the highest priority and input D_3 the lowest priority.

5-21 Specify the truth table of an octal-to-binary priority encoder. Provide an output V to indicate that at least one of the inputs is a 1. The input with the highest subscript number has the highest priority. What will be the value of the four outputs if inputs D_5 and D_3 are 1 and the other inputs are all 0's?

5-22 Draw the logic diagram of a dual 4-to-1-line multiplexer with common selection inputs and a common enable input.

5-23 Construct a 16×1 multiplexer with two 8×1 and one 2×1 multiplexers. Use block diagrams for the three multiplexers.

5-24 Implement the following Boolean function with an 8×1 multiplexer.

$$F(A, B, C, D) = \Sigma(0, 3, 5, 6, 8, 9, 14, 15)$$

5-25 Implement a full-adder with two 4×1 multiplexers.

5-26 Implement the Boolean function of Example 5-2 with an 8×1 multiplexer, but with inputs A, B, and C connected to selection inputs s_2, s_1, and s_0, respectively.

5-27 An 8×1 multiplexer has inputs A, B, and C connected to the selection inputs s_2, s_1, and s_0, respectively. The data inputs, I_0 through I_7, are as follows: $I_1 = I_2 = I_7 = 0$; $I_3 = I_5 = 1$; $I_0 = I_4 = D$; and $I_6 = D'$. Determine the Boolean function that the multiplexer implements.

5-28 Implement the following Boolean function with a 4×1 multiplexer and external gates. Connect inputs A and B to the selection lines. The input requirements for the four data lines will be a function of variables C and D. These values are obtained by expressing F as a function of C and D for each of the four cases when $AB = 00, 01, 10,$ and 11. These functions may have to be implemented with external gates.

$$F(A, B, C, D) = \Sigma(1, 3, 4, 11, 12, 13, 14, 15)$$

5-29 Given a 32×8 ROM chip with an enable input, show the external connections necessary to construct a 128×8 ROM with four chips and a decoder.

5-30 A ROM chip of 4096×8 bits has two enable inputs and operates from a 5-volt power supply. How many pins are needed for the integrated-circuit package? Draw a block diagram and label all input and output terminals in the ROM.

5-31 Specify the size of a ROM (number of words and number of bits per word) that will accommodate the truth table for the following combinational circuit components:
(a) A binary multiplier that multiplies two 4-bit numbers.
(b) A 4-bit adder-subtractor; see Fig. 5-2(b).
(c) A quadruple 2-to-1-line multiplexers with common select and enable inputs; see Fig. 5-17.
(d) A BCD-to-seven-segment decoder with an enable input; see Problem 4-16.

5-32 Tabulate the truth table for an 8×4 ROM that implements the following four Boolean functions:

$$A(x, y, z) = \Sigma(1, 2, 4, 6,)$$

$$B(x, y, z) = \Sigma(0, 1, 6, 7)$$

$$C(x, y, z) = \Sigma(2, 6)$$

$$D(x, y, z) = \Sigma(1, 2, 3, 5, 7)$$

5-33 Tabulate the PLA programming table for the four Boolean functions listed in Problem 5-32. Minimize the number of product terms.

5-34 Derive the PLA programming table for the combinational circuit that squares a 3-bit number. Minimize the number of product terms. (See Fig. 5-24 for the equivalent ROM implementation.)

5-35 List the PLA programming table for the BCD-to-excess-3-code converter whose Boolean functions are simplified in Fig. 4-7.

5-36 Repeat Problem 5-35 using a PAL.

5-37 The following is a truth table of a 3-input, 4-output combinational circuit. Tabulate the PAL programming table for the circuit and mark the fuses to be blown in a PAL diagram similar to the one shown in Fig. 5-30.

Inputs			Outputs			
x	y	z	A	B	C	D
0	0	0	0	1	0	0
0	0	1	1	1	1	1
0	1	0	1	0	1	1
0	1	1	0	1	0	1
1	0	0	1	0	1	0
1	0	1	0	0	0	1
1	1	0	1	1	1	0
1	1	1	0	1	1	1

6 Synchronous Sequential Logic

6-1 INTRODUCTION

The digital circuits considered thus far have been combinational, i.e., the outputs at any instant of time are entirely dependent upon the inputs present at that time. Although every digital system is likely to have combinational circuits, most systems encountered in practice also include memory elements, which require that the system be described in terms of *sequential logic*.

A block diagram of a sequential circuit is shown in Fig. 6-1. It consists of a combinational circuit to which memory elements are connected to form a feedback path. The memory elements are devices capable of storing binary information within them. The binary information stored in the memory elements at any given time defines the *state* of the sequential circuit. The sequential circuit receives binary information from external inputs. These inputs, together with the present state of the memory elements, determine the binary value at the output terminals. They also determine the condition for changing the state in the memory elements. The block diagram demonstrates that the external outputs in a sequential circuit are a function not only of external inputs, but also of the present state of the memory elements. The next state of the memory elements is also a function of external inputs and the present state. Thus, a sequential circuit is specified by a time sequence of inputs, outputs, and internal states.

There are two main types of sequential circuits. Their classification depends on the timing of their signals. A *synchronous* sequential circuit is a system whose behavior can be defined from the knowledge of its signals at discrete instants of time. The behavior of an *asynchronous* sequential circuit depends upon the order in which its input signals change and can be affected at any instant of time. The memory elements commonly

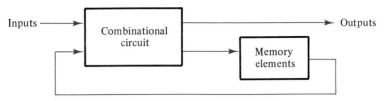

FIGURE 6-1
Block diagram of a sequential circuit

used in asynchronous sequential circuits are time-delay devices. The memory capability of a time-delay device is due to the finite time it takes for the signal to propagate through the device. In practice, the internal propagation delay of logic gates is of sufficient duration to produce the needed delay, so that physical time-delay units may be unnecessary. In gate-type asynchronous systems, the memory elements of Fig. 6-1 consist of logic gates whose propagation delays constitute the required memory. Thus, an asynchronous sequential circuit may be regarded as a combinational circuit with feedback. Because of the feedback among logic gates, an asynchronous sequential circuit may, at times, become unstable. The instability problem imposes many difficulties on the designer. Asynchronous sequential circuits are presented in Chapter 9.

A synchronous sequential logic system, by definition, must employ signals that affect the memory elements only at discrete instants of time. One way of achieving this goal is to use pulses of limited duration throughout the system so that one pulse amplitude represents logic-1 and another pulse amplitude (or the absence of a pulse) represents logic-0. The difficulty with a system of pulses is that any two pulses arriving from separate independent sources to the inputs of the same gate will exhibit unpredictable delays, will separate the pulses slightly, and will result in unreliable operation.

Practical synchronous sequential logic systems use fixed amplitudes such as voltage levels for the binary signals. Synchronization is achieved by a timing device called a *master-clock generator,* which generates a periodic train of *clock pulses.* The clock pulses are distributed throughout the system in such a way that memory elements are affected only with the arrival of the synchronization pulse. In practice, the clock pulses are applied into AND gates together with the signals that specify the required change in memory elements. The AND-gate outputs can transmit signals only at instants that coincide with the arrival of clock pulses. Synchronous sequential circuits that use clock pulses in the inputs of memory elements are called *clocked sequential circuits.* Clocked sequential circuits are the type encountered most frequently. They do not manifest instability problems and their timing is easily divided into independent discrete steps, each of which is considered separately. The sequential circuits discussed in this chapter are exclusively of the clocked type.

The memory elements used in clocked sequential circuits are called *flip-flops.* These circuits are binary cells capable of storing one bit of information. A flip-flop circuit has two outputs, one for the normal value and one for the complement value of the bit stored in it. Binary information can enter a flip-flop in a variety of ways, a fact that gives rise to different types of flip-flops. In the next section, we examine the various types of flip-flops and define their logical properties.

6-2 FLIP-FLOPS

A flip-flop circuit can maintain a binary state indefinitely (as long as power is delivered to the circuit) until directed by an input signal to switch states. The major differences among various types of flip-flops are in the number of inputs they possess and in the manner in which the inputs affect the binary state. The most common types of flip-flops are discussed in what follows.

Basic Flip-Flop Circuit

It was mentioned in Sections 4-7 and 4-8 that a flip-flop circuit can be constructed from two NAND gates or two NOR gates. These constructions are shown in the logic diagrams of Figs. 6-2 and 6-3. Each circuit forms a basic flip-flop upon which other more complicated types can be built. The cross-coupled connection from the output of one gate to the input of the other gate constitutes a feedback path. For this reason, the circuits are classified as asynchronous sequential circuits. Each flip-flop has two outputs, Q and Q', and two inputs, *set* and *reset*. This type of flip-flop is sometimes called a *direct-coupled RS* flip-flop, or *SR latch*. The R and S are the first letters of the two input names.

To analyze the operation of the circuit of Fig. 6-2, we must remember that the output of a NOR gate is 0 if any input is 1, and that the output is 1 only when all inputs are 0. As a starting point, assume that the set input is 1 and the reset input is 0. Since gate 2 has an input of 1, its output Q' must be 0, which puts both inputs of gate 1 at 0, so that output Q is 1. When the set input is returned to 0, the outputs remain the same, because output Q remains a 1, leaving one input of gate 2 at 1. That causes output Q' to stay at 0, which leaves both inputs of gate number 1 at 0, so that output Q is a 1. In the same manner, it is possible to show that a 1 in the reset input changes output Q to 0 and Q' to 1. When the reset input returns to 0, the outputs do not change.

When a 1 is applied to both the set and the reset inputs, both Q and Q' outputs go to 0. This condition violates the fact that outputs Q and Q' are the complements of each other. In normal operation, this condition must be avoided by making sure that 1's are not applied to both inputs simultaneously.

A flip-flop has two useful states. When $Q = 1$ and $Q' = 0$, it is in the *set state* (or

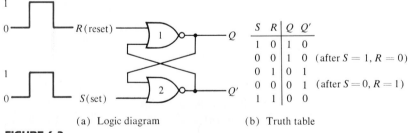

S	R	Q	Q'	
1	0	1	0	
0	0	1	0	(after $S = 1, R = 0$)
0	1	0	1	
0	0	0	1	(after $S = 0, R = 1$)
1	1	0	0	

(a) Logic diagram   (b) Truth table

FIGURE 6-2
Basic flip-flop circuit with NOR gates

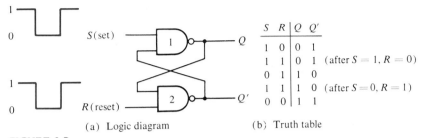

(a) Logic diagram | (b) Truth table

S	R	Q	Q'	
1	0	0	1	
1	1	0	1	(after $S = 1, R = 0$)
0	1	1	0	
1	1	1	0	(after $S = 0, R = 1$)
0	0	1	1	

FIGURE 6-3

Basic flip-flop circuit with NAND gates

1-state). When $Q = 0$ and $Q' = 1$, it is in the *clear state* (or 0-state). The outputs Q and Q' are complements of each other and are referred to as the normal and complement outputs, respectively. The binary state of the flip-flop is taken to be the value of the normal output.

Under normal operation, both inputs remain at 0 unless the state of the flip-flop has to be changed. The application of a momentary 1 to the set input causes the flip-flop to go to the set state. The set input must go back to 0 before a 1 is applied to the reset input. A momentary 1 applied to the reset input causes the flip-flop to go the clear state. When both inputs are initially 0, a 1 applied to the set input while the flip-flop is in the set state or a 1 applied to the reset input while the flip-flop is in the clear state leaves the outputs unchanged. When a 1 is applied to both the set and the reset inputs, both outputs go to 0. This state is undefined and is usually avoided. If both inputs now go to 0, the state of the flip-flop is indeterminate and depends on which input remains a 1 longer before the transition to 0.

The NAND basic flip-flop circuit of Fig. 6-3 operates with both inputs normally at 1 unless the state of the flip-flop has to be changed. The application of a momentary 0 to the set input causes output Q to go to 1 and Q' to go to 0, thus putting the flip-flop into the set state. After the set input returns to 1, a momentary 0 to the reset input causes a transition to the clear state. When both inputs go to 0, both outputs go to 1—a condition avoided in normal flip-flop operation.

RS Flip-flop

The operation of the basic flip-flop can be modified by providing an additional control input that determines when the state of the circuit is to be changed. An *RS* flip-flop with a clock pulse (*CP*) input is shown in Fig. 6-4(a). It consists of a basic flip-flop circuit and two additional NAND gates. The pulse input acts as an enable signal for the other two inputs. The outputs of NAND gates 3 and 4 stay at the logic 1 level as long as the *CP* input remains at 0. This is the quiescent condition for the basic flip-flop. When the pulse input goes to 1, information from the S or R input is allowed to reach the output. The set state is reached with $S = 1$, $R = 0$, and $CP = 1$. This causes the output of gate 3 to go to 0, the output of gate 4 to remain at 1, and the output of the flip-flop at Q to go to 1. To change to the reset state, the inputs must be $S = 0$, $R = 1$, and $CP = 1$.

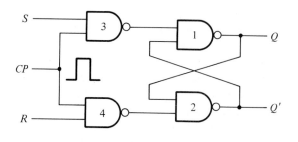

Q	S	R	Q(t + 1)
0	0	0	0
0	0	1	0
0	1	0	1
0	1	1	Indeterminate
1	0	0	1
1	0	1	0
1	1	0	1
1	1	1	Indeterminate

(a) Logic diagram

(b) Characteristic table

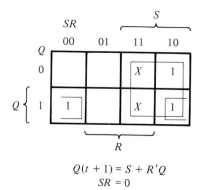

$$Q(t + 1) = S + R'Q$$
$$SR = 0$$

(c) Characteristic equation

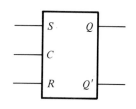

(d) Graphic symbol

FIGURE 6-4
RS flip-flop

In either case, when *CP* returns to 0, the circuit remains in its previous state. When *CP* = 1 and both the *S* and *R* inputs are equal to 0, the state of the circuit does not change.

An indeterminate condition occurs when *CP* = 1 and both *S* and *R* are equal to 1. This condition places 0's in the outputs of gates 3 and 4 and 1's in both outputs *Q* and *Q'*. When the *CP* input goes back to 0 (while *S* and *R* are maintained at 1), it is not possible to determine the next state, as it depends on whether the output of gate 3 or gate 4 goes to 1 first. This indeterminate condition makes the circuit of Fig. 6-4(a) difficult to manage and it is seldom used in practice. Nevertheless, it is an important circuit because all other flip-flops are constructed from it.

The characteristic table of the flip-flop is shown in Fig. 6-4(b). This table shows the operation of the flip-flop in tabular form. *Q* is an abbreviation of *Q*(*t*) and stands for the binary state of the flip-flop before the application of a clock pulse, referred to as the *present state*. The *S* and *R* columns give the possible values of the inputs, and *Q*(*t* + 1) is the state of the flip-flop after the application of a single pulse, referred to as the *next state*. Note that the *CP* input is not included in the characteristic table. The table must be interpreted as follows: Given the present state Q and the inputs S and R, the application of a single pulse in the CP input causes the flip-flop to go to the next state, $Q(t + 1)$.

The characteristic equation of the flip-flop is derived in the map of Fig. 6-4(c). This equation specifies the value of the next state as a function of the present state and the inputs. The characteristic equation is an algebraic expression for the binary information of the characteristic table. The two indeterminate states are marked with dont't-care X's in the map, since they may result in either 1 or 0. However, the relation $SR = 0$ must be included as part of the characteristic equation to specify that both S and R cannot equal to 1 simultaneously.

The graphic symbol of the RS flip-flop is shown in Fig. 6-4(d). It consists of a rectangular-shape block with inputs S, R, and C. The outputs are Q and Q', where Q' is the complement of Q (except in the indeterminate state).

D Flip-Flop

One way to eliminate the undesirable condition of the indeterminate state in the RS flip-flop is to ensure that inputs S and R are never equal to 1 at the same time. This is done in the D flip-flop shown in Fig. 6-5(a). The D flip-flop has only two inputs: D and CP. The D input goes directly to the S input and its complement is applied to the R input. As long as the pulse input is at 0, the outputs of gates 3 and 4 are at the 1 level and the circuit cannot change state regardless of the value of D. The D input is sampled when $CP = 1$. If D is 1, the Q output goes to 1, placing the circuit in the set state. If D is 0, output Q goes to 0 and the circuit switches to the clear state.

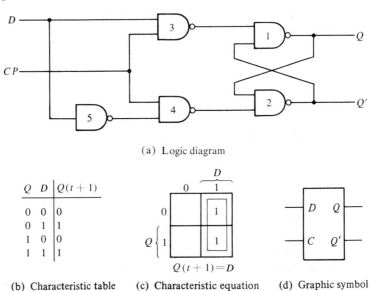

(a) Logic diagram

Q	D	$Q(t+1)$
0	0	0
0	1	1
1	0	0
1	1	1

(b) Characteristic table

$Q(t+1)=D$

(c) Characteristic equation

(d) Graphic symbol

FIGURE 6-5

D flip-flop

The *D* flip-flop receives the designation from its ability to hold *data* into its internal storage. This type of flip-flop is sometimes called a *gated D-latch*. The *CP* input is often given the designation *G* (for *gate*) to indicate that this input enables the gated latch to make possible data entry into the circuit. The binary information present at the data input of the *D* flip-flop is transferred to the *Q* output when the *CP* input is enabled. The output follows the data input as long as the pulse remains in its 1 state. When the pulse goes to 0, the binary information that was present at the data input at the time the pulse transition occurred is retained at the *Q* output until the pulse input is enabled again.

The characteristic table for the D flip-flop is shown in Fig. 6-5(b). It shows that the next state of the flip-flop is independent of the present state since $Q(t + 1)$ is equal to input *D* whether *Q* is equal to 0 or 1. This means that an input pulse will transfer the value of input *D* into the output of the flip-flop independent of the value of the output before the pulse was applied. The characteristic equation shows clearly that $Q(t + 1)$ is equal to *D*.

The graphic symbol for the level sensitive *D* flip-flop is shown in Fig. 6-5(d). The graphic symbol for a transition-sensitive *D* flip-flop is shown later in Fig. 6-14.

JK and T Flip-Flops

A *JK* flip-flop is a refinement of the *RS* flip-flop in that the indeterminate state of the *RS* type is defined in the *JK* type. Inputs *J* and *K* behave like inputs *S* and *R* to set and clear the flip-flop, respectively. The input marked *J* is for *set* and the input marked *K* is for *reset*. When both inputs *J* and *K* are equal to 1, the flip-flop switches to its complement state, that is, if *Q* = 1, it switches to *Q* = 0, and vice versa.

A *JK* flip-flop constructed with two cross-coupled NOR gates and two AND gates is shown in Fig. 6-6(a). Output *Q* is ANDed with *K* and *CP* inputs so that the flip-flop is cleared during a clock pulse only if *Q* was previously 1. Similarly, output *Q′* is ANDed with *J* and *CP* inputs so that the flop-flop is set with a clock pulse only when *Q′* was previously 1. When both *J* and *K* are 1, the input pulse is transmitted through one AND gate only: the one whose input is connected to the flip-flop output that is presently equal to 1. Thus, if *Q* = 1, the output of the upper AND gate becomes 1 upon application of the clock pulse, and the flip-flop is cleared. If *Q′* = 1, the output of the lower AND gate becomes 1 and the flip-flop is set. In either case, the output state of the flip-flop is complemented. The behavior of the *JK* flip-flop is demonstrated in the characteristic table of Fig. 6-6(b).

It is very important to realize that because of the feedback connection in the *JK* flip-flop, a *CP* pulse that remains in the 1 state while both *J* and *K* are equal to 1 will cause the output to complement again and repeat complementing until the pulse goes back to 0. To avoid this undesirable operation, the clock pulse must have a time duration that is shorter than the propagation delay time of the flip-flop. This is a restrictive requirement, since the operation of the circuit depends on the width of the pulse. For this reason, *JK* flip-flops are never constructed as shown in Fig. 6-6(a). The restriction on the pulse width can be eliminated with a master–slave or edge-triggered construction, as discussed in the next section. The same reasoning applies to the *T* flip-flop.

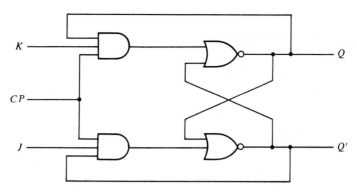

(a) Logic diagram

Q	J	K	Q(t+1)
0	0	0	0
0	0	1	0
0	1	0	1
0	1	1	1
1	0	0	1
1	0	1	0
1	1	0	1
1	1	1	0

(b) Characteristic table

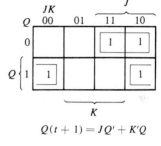

$Q(t+1) = JQ' + K'Q$

(c) Characteristic equation

FIGURE 6-6

JK flip-flop

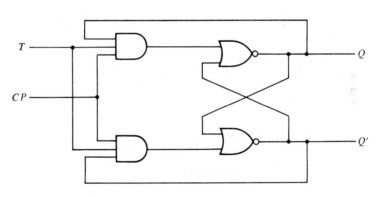

(a) Logic diagram

Q	T	Q(t+1)
0	0	0
0	1	1
1	0	1
1	1	0

(b) Characteristic table

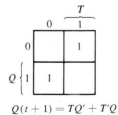

$Q(t+1) = TQ' + T'Q$

(c) Characteristic equation

FIGURE 6-7

T flip-flop

209

The T flip-flop is a single-input version of the JK flip-flop. As shown in Fig. 6-7(a), the T flip-flop is obtained from the JK flip-flop when both inputs are tied together. The designation T comes from the ability of the flip-flop to "toggle," or complement, its state. Regardless of the present state, the flip-flop complements its output when the clock pulse occurs while input T is 1. The characteristic table and characteristic equation show that when $T = 0$, $Q(t + 1) = Q$, that is, the next state is the same as the present state and no change occurs. When $T = 1$, then $Q(t + 1) = Q'$, and the state of the flip-flop is complemented.

6-3 TRIGGERING OF FLIP-FLOPS

The state of a flip-flop is switched by a momentary change in the input signal. This momentary change is called a *trigger* and the transition it causes is said to trigger the flip-flop. Asynchronous flip-flops, such as the basic circuits of Figs. 6-2 and 6-3, require an input trigger defined by a change of signal *level*. This level must be returned to its initial value (0 in the NOR and 1 in the NAND flip-flop) before a second trigger is applied. Clocked flip-flops are triggered by *pulses*. A pulse starts from an initial value of 0, goes momentarily to 1, and after a short time, returns to its initial 0 value. The time interval from the application of the pulse until the output transition occurs is a critical factor that needs further investigation.

As seen from the block diagram of Fig. 6-1, a sequential circuit has a feedback path between the combinational circuit and the memory elements. This path can produce instability if the outputs of memory elements (flip-flops) are changing while the outputs of the combinational circuit that go to flip-flop inputs are being sampled by the clock pulse. This timing problem can be prevented if the outputs of flip-flops do not start changing until the pulse input has returned to 0. To ensure such an operation, a flip-flop must have a signal-propagation delay from input to output in excess of the pulse duration. This delay is usually very difficult to control if the designer depends entirely on the propagation delay of logic gates. One way of ensuring the proper delay is to include within the flip-flop circuit a physical delay unit having a delay equal to or greater than the pulse duration. A better way to solve the feedback timing problem is to make the flip-flop sensitive to the pulse *transition* rather than the pulse duration.

A clock pulse may be either positive or negative. A positive clock source remains at 0 during the interval between pulses and goes to 1 during the occurrence of a pulse. The pulse goes through two signal transitions: from 0 to 1 and the return from 1 to 0. As shown in Fig. 6-8, the positive transition is defined as the *positive edge* and the negative transition as the *negative edge*. This definition applies also to negative pulses.

The clocked flip-flops introduced in Section 6-2 are triggered during the positive edge of the pulse, and the state transition starts as soon as the pulse reaches the logic-1 level. The new state of the flip-flop may appear at the output terminals while the input pulse is still 1. If the other inputs of the flip-flop change while the clock is still 1, the

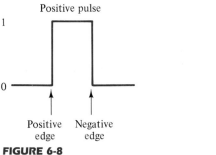 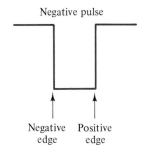

FIGURE 6-8
Definition of clock-pulse transition

flip-flop will start responding to these new values and a new output state may occur. When this happens, the output of one flip-flop cannot be applied to the inputs of another flip-flop when both are triggered by the same clock pulse. However, if we can make the flip-flop respond to the positive- (or negative-) edge transition *only,* instead of the entire pulse duration, then the multiple-transition problem can be eliminated.

One way to make the flip-flop respond only to a pulse transition is to use capacitive coupling. In this configuration, an *RC* (resistor–capacitor) circuit is inserted in the clock input of the flip-flop. This circuit generates a spike in response to a momentary change of input signal. A positive edge emerges from such a circuit with a positive spike, and a negative edge emerges with a negative spike. Edge triggering is achieved by designing the flip-flop to neglect one spike and trigger on the occurrence of the other spike. Another way to achieve edge triggering is to use a master–slave or edge-triggered flip-flop as discussed in what follows.

Master–Slave Flip-Flop

A master–slave flip-flop is constructed from two separate flip-flops. One circuit serves as a master and the other as a slave, and the overall circuit is referred to as a *master–slave flip-flop*. The logic diagram of an *RS* master–slave flip-flop is shown in Fig. 6-9. It consists of a master flip-flop, a slave flip-flop, and an inverter. When clock pulse *CP* is 0, the output of the inverter is 1. Since the clock input of the slave is 1, the flip-flop is enabled and output Q is equal to Y, while Q' is equal to Y'. The master flip-flop is disabled because $CP = 0$. When the pulse becomes 1, the information then at the external R and S inputs is transmitted to the master flip-flop. The slave flip-flop, however, is isolated as long as the pulse is at its 1 level, because the output of the inverter is 0. When the pulse returns to 0, the master flip-flop is isolated, which prevents the external inputs from affecting it. The slave flip-flop then goes to the same state as the master flip-flop.

The timing relationships shown in Fig. 6-10 illustrate the sequence of events that occur in a master–slave flip-flop. Assume that the flip-flop is in the clear state prior to the

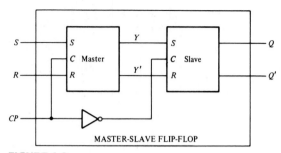

FIGURE 6-9
Logic diagram of a master–slave flip-flop

occurrence of a pulse, so that $Y = 0$ and $Q = 0$. The input conditions are $S = 1$, $R = 0$, and the next clock pulse should change the flip-flop to the set state with $Q = 1$. During the pulse transition from 0 to 1, the master flip-flop is set and changes Y to 1. The slave flip-flop is not affected because its CP input is 0. Since the master flip-flop is an internal circuit, its change of state is not noticeable in the outputs Q and Q'. When the pulse returns to 0, the information from the master is allowed to pass through to the slave, making the external output $Q = 1$. Note that the external S input can be changed at the same time that the pulse goes through its negative-edge transition. This is because, once the CP reaches 0, the master is disabled and its R and S inputs have no influence until the next clock pulse occurs. Thus, in a master–slave flip-flop, it is possible to switch the output of the flip-flop and its input information with the same clock pulse. It must be realized that the S input could come from the output of another master–slave flip-flop that was switched with the same clock pulse.

The behavior of the master–slave flip-flop just described dictates that the state changes in all flip-flops coincide with the negative-edge transition of the pulse. However, some IC master–slave flip-flops change output states in the positive-edge transi-

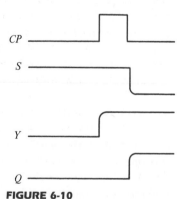

FIGURE 6-10
Timing relationships in a master–slave flip-flop

tion of clock pulses. This happens in flip-flops that have an additional inverter between the *CP* terminal and the input of the master. Such flip-flops are triggered with negative pulses (see Fig. 6-8), so that the negative edge of the pulse affects the master and the positive edge affects the slave and the output terminals.

The master–slave combination can be constructed for any type of flip-flop by adding a clocked *RS* flip-flop with an inverted clock to form the slave. An example of a master–slave *JK* flip-flop constructed with NAND gates is shown in Fig. 6-11. It consists of two flip-flops; gates 1 through 4 form the master flip-flop, and gates 5 through 8 form the slave flip-flop. The information present at the *J* and *K* inputs is transmitted to the master flip-flop on the positive edge of a clock pulse and is held there until the negative edge of the clock pulse occurs, after which it is allowed to pass through to the slave flip-flop. The clock input is normally 0, which keeps the outputs of gates 1 and 2 at the 1 level. This prevents the *J* and *K* inputs from affecting the master flip-flop. The slave flip-flop is a clocked *RS* type, with the master flip-flop supplying the inputs and the clock input being inverted by gate 9. When the clock is 0, the output of gate 9 is 1, so that output *Q* is equal to *Y*, and *Q'* is equal to *Y'*. When the positive edge of a clock pulse occurs, the master flip-flop is affected and may switch states. The slave flip-flop is isolated as long as the clock is at the 1 level, because the output of gate 9 provides a 1 to both inputs of the NAND basic flip-flop of gates 7 and 8. When the clock input returns to 0, the master flip-flop is isolated from the *J* and *K* inputs and the slave flip-flop goes to the same state as the master flip-flop.

Now consider a digital system containing many master–slave flip-flops, with the outputs of some flip-flops going to the inputs of other flip-flops. Assume that clock-pulse inputs to all flip-flops are synchronized (occur at the same time). At the beginning of each clock pulse, some of the master elements change state, but all flip-flop outputs remain at their previous values. After the clock pulse returns to 0, some of the outputs

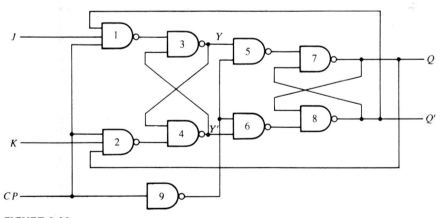

FIGURE 6-11
Clocked master–slave *JK* flip-flop

change state, but none of these new states have an effect on any of the master elements until the next clock pulse. Thus, the states of flip-flops in the system can be changed simultaneously during the same clock pulse, even though outputs of flip-flops are connected to inputs of flip-flops. This is possible because the new state appears at the output terminals only after the clock pulse has returned to 0. Therefore, the binary content of one flip-flop can be transferred to a second flip-flop and the content of the second transferred to the first, and both transfers can occur during the same clock pulse.

Edge-Triggered Flip-Flop

Another type of flip-flop that synchronizes the state changes during a clock-pulse transition is the *edge-triggered* flip-flop. In this type of flip-flop, output transitions occur at a specific level of the clock pulse. When the pulse input level exceeds this threshold level, the inputs are locked out and the flip-flop is therefore unresponsive to further changes in inputs until the clock pulse returns to 0 and another pulse occurs. Some edge-triggered flip-flops cause a transition on the positive edge of the pulse, and others cause a transition on the negative edge of the pulse.

The logic diagram of a D-type positive-edge-triggered flip-flop is shown in Fig. 6-12. It consists of three basic flip-flops of the type shown in Fig. 6-3. NAND gates 1 and 2 make up one basic flip-flop and gates 3 and 4 another. The third basic flip-flop comprising gates 5 and 6 provides the outputs to the circuit. Inputs S and R of the third basic flip-flop must be maintained at logic-1 for the outputs to remain in their steady-state values. When $S = 0$ and $R = 1$, the output goes to the set state with $Q = 1$. When $S = 1$ and $R = 0$, the output goes to the clear state with $Q = 0$. Inputs S and R

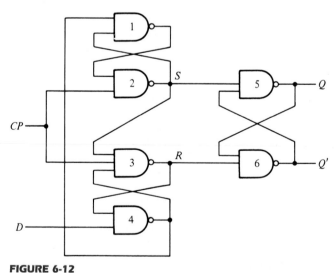

FIGURE 6-12
D-type positive-edge-triggered flip-flop

are determined from the states of the other two basic flip-flops. These two basic flip-flops respond to the external inputs D (data) and CP (clock pulse).

The operation of the circuit is explained in Fig. 6-13, where gates 1–4 are redrawn to show all possible transitions. Outputs S and R from gates 2 and 3 go to gates 5 and 6, as shown in Fig. 6-12, to provide the actual outputs of the flip-flop. Figure 6-13(a) shows the binary values at the outputs of the four gates when $CP = 0$. Input D may be equal to 0 or 1. In either case, a CP of 0 causes the outputs of gates 2 and 3 to go to 1, thus making $S = R = 1$, which is the condition for a steady-state output. When

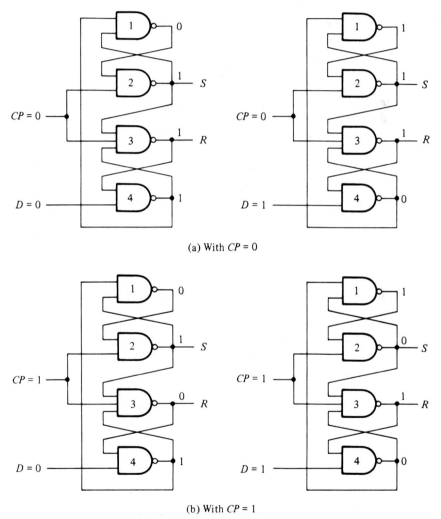

(a) With $CP = 0$

(b) With $CP = 1$

FIGURE 6-13
Operation of the D-type edged-triggered flip-flop

$D = 0$, gate 4 has a 1 output, which causes the output of gate 1 to go to 0. When $D = 1$, gate 4 goes to 0, which causes the output of gate 1 to go to 1. These are the two possible conditions when the CP terminal, being 0, disables any changes at the outputs of the flip-flop, no matter what the value of D happens to be.

There is a definite time, called the *setup* time, in which the D input must be maintained at a constant value prior to the application of the pulse. The setup time is equal to the propagation delay through gates 4 and 1 since a change in D causes a change in the outputs of these two gates. Assume now that D does not change during the setup time and that input CP becomes 1. This situation is depicted in Fig. 6-13(b). If $D = 0$ when CP becomes 1, then S remains 1 but R changes to 0. This causes the output of the flip-flop Q to go to 0 (in Fig. 6-12). If now, while $CP = 1$, there is a change in the D input, the output of gate 4 will remain at 1 (even if D goes to 1), since one of the gate inputs comes from R, which is maintained at 0. Only when CP returns to 0 can the output of gate 4 change; but then both R and S become 1, disabling any changes in the output of the flip-flop. However, there is a definite time, called the *hold time*, that the D input must not change after the application of the positive-going transition of the pulse. The hold time is equal to the propagation delay of gate 3, since it must be ensured that R becomes 0 in order to maintain the output of gate 4 at 1, regardless of the value of D.

If $D = 1$ when $CP = 1$, then S changes to 0, but R remains at 1, which causes the output of the flip-flop Q to go to 1. A change in D while $CP = 1$ does not alter S and R, because gate 1 is maintained at 1 by the 0 signal from S. When CP goes to zero, both S and R go to 1 to prevent the output from undergoing any changes.

In summary, when the input clock pulse makes a positive-going transition, the value of D is transferred to Q. Changes in D when CP is maintained at a steady 1 value do not affect Q. Moreover, a negative pulse transition does not affect the output, nor does it when $CP = 0$. Hence, the edge-triggered flip-flop eliminates any feedback problems in sequential circuits just as a master–slave flip-flop does. The setup time and hold time must be taken into consideration when using this type of flip-flop.

When using different types of flip-flops in the same sequential circuit, one must ensure that all flip-flop outputs make their transitions at the same time, i.e., during either the negative edge or the positive edge of the pulse. Those flip-flops that behave opposite from the adopted polarity transition can be changed easily by the addition of inverters in their clock inputs. An alternate procedure is to provide both positive and negative pulses (by means of an inverter), and then apply the positive pulses to flip-flops that trigger during the negative edge and negative pulses to flip-flops that trigger during the positive edge, or vice versa.

Graphic Symbols

The graphic symbols for four flip-flops are shown in Fig. 6-14. The input letter symbols in each diagram designate the type of flip-flop such as RS, JK, D, and T. The clock-pulse input is recognized in the diagram from the arrowhead-shape symbol. This is a symbol of a *dynamic indicator* and denotes that the flip-flop responds to a positive-edge transition of the clock. The presence of a small circle outside the block along the dy-

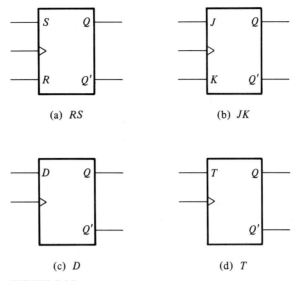

FIGURE 6-14

Graphic symbols for flip-flops

namic indicator designates a negative-edge transition for triggering the flip-flop. The letter symbol C is used for the clock input when the flip-flop responds to a pulse level rather than a pulse transition. This was shown in Fig. 6-5(d) for the level-sensitive D flip-flop.

The outputs of the flip-flop are marked with the letter symbol Q and Q' within the block. The flip-flop may be assigned a different variable name even though Q is written inside the block. In that case, the letter symbol for the flip-flop output is marked outside the block along the output line. The state of the flip-flop is determined from the value of its normal output Q. If one wishes to obtain the complement output, it is not necessary to use an inverter because the complement value is available directly from Q'.

Direct Inputs

Flip-flops available in IC packages sometimes provide special inputs for setting or clearing the flip-flop asynchronously. These inputs are usually called *direct preset* and *direct clear*. They affect the flip-flop on a positive (or negative) value of the input signal without the need for a clock pulse. These inputs are useful for bringing all flip-flops to an initial state prior to their clocked operation. For example, after power is turned on in a digital system, the states of its flip-flops are indeterminate. A *clear* switch clears all the flip-flops to an initial cleared state and a *start* switch begins the system's clocked operation. The clear switch must clear all flip-flops asynchronously without the need for a pulse.

The graphic symbol of a negative-edge-triggered JK flip-flop with direct clear is shown in Fig. 6-15. The clock-pulse input CP has a small circle under the dynamic

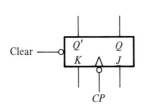

Function table

	Inputs			Outputs	
Clear	Clock	J	K	Q	Q'
0	X	X	X	0	1
1	↓	0	0	No change	
1	↓	0	1	0	1
1	↓	1	0	1	0
1	↓	1	1	Toggle	

FIGURE 6-15
JK flip-flop with direct clear

symbol to indicate that the outputs change in response to a negative transition of the clock. The direct-clear input also has a small circle to indicate that, normally, this input must be maintained at 1. If the clear input is maintained at 0, the flip-flop remains cleared, regardless of the other inputs or the clock pulse. The function table specifies the circuit operation. The X's are don't-care conditions, which indicate that a 0 in the direct-clear input disables all other inputs. Only when the clear input is 1 would a negative transition of the clock have an effect on the outputs. The outputs do not change if $J = K = 0$. The flip-flop toggles, or complements, when $J = K = 1$. Some flip-flops may also have a direct-preset input, which sets the output Q to 1 (and Q' to 0) asynchronously.

6-4 ANALYSIS OF CLOCKED SEQUENTIAL CIRCUITS

The behavior of a sequential circuit is determined from the inputs, the outputs, and the state of its flip-flops. The outputs and the next state are both a function of the inputs and the present state. The analysis of a sequential circuit consists of obtaining a table or a diagram for the time sequence of inputs, outputs, and internal states. It is also possible to write Boolean expressions that describe the behavior of the sequential circuit. However, these expressions must include the necessary time sequence, either directly or indirectly.

A logic diagram is recognized as a clocked sequential circuit if it includes flip-flops. The flip-flops may be of any type and the logic diagram may or may not include combinational circuit gates. In this section, we first introduce a specific example of a clocked sequential circuit with D flip-flops and use it to present the basic methods for describing the behavior of sequential circuits. Additional examples are used throughout the discussion to illustrate other procedures.

Sequential-Circuit Example

An example of a clocked sequential circuit is shown in Fig. 6-16. The circuit consists of two D flip-flops A and B, an input x, and an output y. Since the D inputs determine

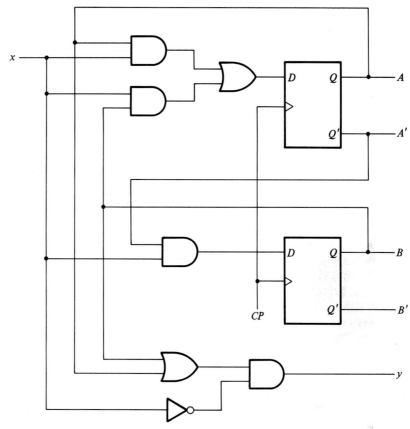

FIGURE 6-16
Example of a sequential circuit

the flip-flops' next state, it is possible to write a set of next-state equations for the circuit:

$$A(t + 1) = A(t)x(t) + B(t)x(t)$$

$$B(t + 1) = A'(t)x(t)$$

A state equation is an algebraic expression that specifies the condition for a flip-flop state transition. The left side of the equation denotes the next state of the flip-flop and the right side of the equation is a Boolean expression that specifies the present state and input conditions that make the next state equal to 1. Since all the variables in the Boolean expressions are a function of the present state, we can omit the designation (t) after each variable for convenience. The previous equations can be expressed in more compact form as follows:

$$A(t + 1) = Ax + Bx$$

$$B(t + 1) = A'x$$

The Boolean expressions for the next state can be derived directly from the gates that form the combinational-circuit part of the sequential circuit. The outputs of the combinational circuit are applied to the D inputs of the flip-flops. The D input values determine the next state.

Similarly, the present-state value of the output can be expressed algebraically as follows:

$$y(t) = [A(t) + B(t)]x'(t)$$

Removing the symbol (t) for the present state, we obtain the output Boolean function:

$$y = (A + B)x'$$

State Table

The time sequence of inputs, outputs, and flip-flop states can be enumerated in a *state table*. The state table for the circuit of Fig. 6-16 is shown in Table 6-1. The table consists of four sections labeled *present state*, *input*, *next state*, and *output*. The present-state section shows the states of flip-flops A and B at any given time t. The input section gives a value of x for each possible present state. The next-state section shows the states of the flip-flops one clock period later at time $t + 1$. The output section gives the value of y for each present state.

The derivation of a state table consists of first listing all possible binary combinations of present state and inputs. In this case, we have eight binary combinations from 000 to 111. The next-state values are then determined from the logic diagram or from the state equations. The next state of flip-flop A must satisfy the state equation

$$A(t + 1) = Ax + Bx$$

The next-state section in the state table under column A has three 1's where the present

TABLE 6-1
State Table for the Circuit of Fig. 6-16

Present State	Input	Next State	Output
A B	x	A B	y
0 0	0	0 0	0
0 0	1	0 1	0
0 1	0	0 0	1
0 1	1	1 1	0
1 0	0	0 0	1
1 0	1	1 0	0
1 1	0	0 0	1
1 1	1	1 0	0

TABLE 6-2
Second Form of the State Table

Present State	Next State		Output	
	$x = 0$	$x = 1$	$x = 0$	$x = 1$
AB	AB	AB	y	y
00	00	01	0	0
01	00	11	1	0
10	00	10	1	0
11	00	10	1	0

state and input value satisfy the conditions that the present state of A and input x are both equal to 1 or the present state of B and input x are both equal to 1. Similarly, the next state of flip-flop B is derived from the state equation

$$B(t + 1) = A'x$$

It is equal to 1 when the present state of A is 0 and input x is equal to 1. The output column is derived from the output equation

$$y = Ax' + Bx$$

The state table of any sequential circuit with D-type flip-flops is obtained by the same procedure outlined in the previous example. In general, a sequential circuit with m flip-flops and n inputs needs 2^{m+n} rows in the state table. The binary numbers from 0 through $2^{m+n} - 1$ are listed under the present-state and input columns. The next-state section has m columns, one for each flip-flop. The binary values for the next state are derived directly from the state equations. The output section has as many columns as there are output variables. Its binary value is derived from the circuit or from the Boolean function in the same manner as in a truth table. Note that the examples in this chapter use only one input and one output variable, but, in general, a sequential circuit may have two or more inputs or outputs.

It is sometimes convenient to express the state table in a slightly different form. In the other configuration, the state table has only three sections: present state, next state, and output. The input conditions are enumerated under the next-state and output sections. The state table of Table 6-1 is repeated in Table 6-2 using the second form. For each present state, there are two possible next states and outputs, depending on the value of the input. We will use both forms of the state table. One form may be preferable over the other, depending on the application.

State Diagram

The information available in a state table can be represented graphically in a state diagram. In this type of diagram, a state is represented by a circle, and the transition between states is indicated by directed lines connecting the circles. The state diagram of

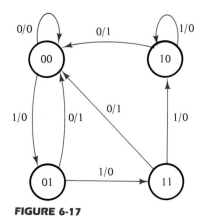

FIGURE 6-17

State diagram of the circuit of Fig. 6-16

the sequential circuit of Fig. 6-16 is shown in Fig. 6-17. The state diagram provides the same information as the state table and is obtained directly from Table 6-2. The binary number inside each circle identifies the state of the flip-flops. The directed lines are labeled with two binary numbers separated by a slash. The input value during the present state is labeled first and the number after the slash gives the output during the present state. For example, the directed line from state 00 to 01 is labeled 1/0, meaning that when the sequential circuit is in the present state 00 and the input is 1, the output is 0. After a clock transition, the circuit goes to the next state, 01. The same clock transition may change the input value. If the input changes to 0, then the output becomes 1, but if the input remains at 1, the output stays at 0. This information is obtained from the state diagram along the two directed lines emanating from the circle representing state 01. A directed line connecting a circle with itself indicates that no change of state occurs.

There is no difference between a state table and a state diagram except in the manner of representation. The state table is easier to derive from a given logic diagram and the state diagram follows directly from the state table. The state diagram gives a pictorial view of state transitions and is the form suitable for human interpretation of the circuit operation. For example, the state diagram of Fig. 6-17 clearly shows that, starting from state 00, the output is 0 as long as the input stays at 1. The first 0 input after a string of 1's gives an output of 1 and transfers the circuit back to the initial state 00.

Flip-Flop Input Functions

The logic diagram of a sequential circuit consists of flip-flops and gates. The interconnections among the gates form a combinational circuit and may be specified algebraically with Boolean functions. Thus, knowledge of the type of flip-flops and a list of the Boolean functions of the combinational circuit provide all the information needed to draw the logic diagram of a sequential circuit. The part of the combinational circuit that generates external outputs is described algebraically by the *circuit output functions*.

The part of the circuit that generates the inputs to flip-flops are described algebraically by a set of Boolean functions called *flip-flop input functions,* or sometimes *input equations.*

We shall adopt the convention of using two letters to designate a flip-flop input function; the first to designate the name of the input and the second the name of the flip-flop. As an example, consider the following flip-flop input functions:

$$JA = BC'x + B'Cx'$$

$$KA = B + y$$

JA and KA designate two Boolean variables. The first letter in each denotes the J and K input, respectively, of a JK flip-flop. The second letter, A, is the symbol name of the flip-flop. The right side of each equation is a Boolean function for the corresponding flip-flop input variable. The implementation of the two input functions is shown in the logic diagram of Fig. 6-18. The JK flip-flop has an output symbol A and two inputs labeled J and K. The combinational circuit drawn in the diagram is the implementation of the algebraic expression given by the input functions. The outputs of the combinational circuit are denoted by JA and KA in the input functions and go to the J and K inputs, respectively, of flip-flop A.

From this example, we see that a flip-flop input function is an algebraic expression for a combinational circuit. The two-letter designation is a variable name for an *output* of the combinational circuit. This output is always connected to the *input* (designated by the first letter) of a flip-flop (designated by the second letter).

The sequential circuit of Fig. 6-16 has one input x, one output y, and two D flip-flops A and B. The logic diagram can be expressed algebraically with two flip-flop input functions and one output-circuit function:

$$DA = Ax + Bx$$

$$DB = A'x$$

$$y = (A + B)x'$$

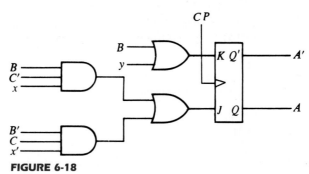

FIGURE 6-18

Implementation of the flip-flop input functions
$JA = BC'x + B'Cx'$ and $KA = B + y$

This set of Boolean functions provides all the necessary information for drawing the logic diagram of the sequential circuit. The symbol *DA* specifies a *D* flip-flop labeled *A*. *DB* specifies a second *D* flip-flop labeled *B*. The Boolean expressions associated with these two variables and the expression for output *y* specify the combinational-circuit part of the sequential circuit.

The flip-flop input functions constitute a convenient algebraic form for specifying a logic diagram of a sequential circuit. They imply the type of flip-flop from the first letter of the input variable and they fully specify the combinational circuit that drives the flip-flop. Time is not included explicitly in these equations, but is implied from the clock-pulse operation. It is sometimes convenient to specify a sequential circuit algebraically with circuit output functions and flip-flop input functions instead of drawing the logic diagram.

Characteristic Tables

The analysis of a sequential circuit with flip-flops other than the *D* type is complicated because the relationship between the inputs of the flip-flop and the next state is not straightforward. This relationship is best described by means of a characteristic table rather than a state equation. The characteristic tables of four flip-flops were presented in Section 6-2. When analyzing sequential circuits, it is more convenient to present the characteristic table in a somewhat different form. The modified form of the characteristic tables of four types of flip-flops are shown in Table 6-3. They define the next state as a function of the inputs and present state. $Q(t)$ refers to the present state prior to the application of a pulse. $Q(t + 1)$ is the next state one clock period later. Note that the clock-pulse input is not listed in the characteristic table, but is implied to occur between time t and $t + 1$.

The characteristic table for the *JK* flip-flop shows that the next state is equal to the

TABLE 6-3
Flip-Flop Characteristic Tables

	JK Flip-Flop				RS Flip-Flop	
J K	$Q(t + 1)$			*S R*	$Q(t + 1)$	
0 0	$Q(t)$	No change		0 0	$Q(t)$	No change
0 1	0	Reset		0 1	0	Reset
1 0	1	Set		1 0	1	Set
1 1	$Q'(t)$	Complement		1 1	?	Unpredictable

	D Flip-Flop				T Flip-Flop	
D	$Q(t + 1)$			*T*	$Q(t + 1)$	
0	0	Reset		0	$Q(t)$	No change
1	1	Set		1	$Q'(t)$	Complement

present state when inputs J and K are both equal to 0. This can be expressed as $Q(t + 1) = Q(t)$, indicating that the clock pulse produces no change of state. When $K = 1$ and $J = 0$, the clock pulse resets the flip-flop and $Q(t + 1) = 0$. With $J = 1$ and $K = 0$, the flip-flop sets and $Q(t + 1) = 1$. When both J and K are equal to 1, the next state changes to the complement of the present state, which can be expressed as $Q(t + 1) = Q'(t)$.

The RS flip-flop is similar to the JK when S is replaced by J and R by K except for the indeterminate case. The question mark for the next state when S and R are both equal to 1 indicates an unpredictable next state.

The next state of a D flip-flop is dependent only on the D input and independent of the present state, which can be expressed as $Q(t + 1) = D$. This means that the next-state value can be obtained directly from the binary logic value of the D input. Note that the D flip-flop does not have a "no-change" condition. This condition can be accomplished either by disabling the clock pulses or by leaving the clock pulses and connecting the output back into the D input when the state of the flip-flop must remain the same.

The T flip-flop is obtained from a JK flip-flop when inputs J and K are tied together. The characteristic table has only two conditions. When $T = 0$ ($J = K = 0$), a clock pulse does not change the state. When $T = 1$ ($J = K = 1$), a clock pulse complements the state of the flip-flop.

Analysis with *JK* and Other Flip-Flops

It was shown previously that the next-state values of a sequential circuit with D flip-flops can be derived directly from the next-state equations. When other types of flip-flops are used, it is necessary to refer to the characteristic table. The next-state values of a sequential circuit that uses any other type of flip-flop such as JK, RS, or T can be derived by following a two-step procedure:

1. Obtain the binary values of each flip-flop input function in terms of the present-state and input variables.
2. Use the corresponding flip-flop characteristic table to determine the next state.

To illustrate this procedure, consider the sequential circuit with two JK flip-flops A and B and one input x, as shown in Fig. 6-19. The circuit has no outputs and, therefore, the state table does not need an output column. (The outputs of the flip-flops may be considered as the outputs in this case.) The circuit can be specified by the following flip-flop input functions:

$$JA = B \qquad\qquad JB = x'$$
$$KA = Bx' \qquad\quad KB = A'x + Ax' = A \oplus x$$

The state table of the sequential circuit is shown in Table 6-4. First, we derive the binary values listed under the columns labeled *flip-flop inputs*. These columns are not part of the state table, but they are needed for the purpose of evaluating the next state

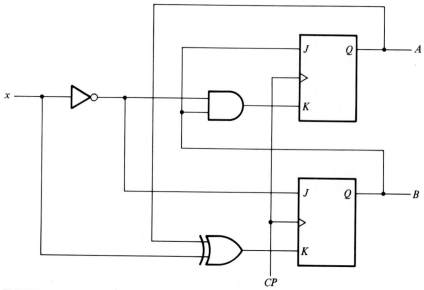

FIGURE 6-19
Sequential circuit with *JK* flip-flops

as specified in step 1 of the procedure. These binary values are obtained directly from the four input flip-flop functions in a manner similar to that for obtaining a truth table from an algebraic expression. The next state of each flip-flop is evaluated from the cor-responding *J* and *K* inputs and the characteristic table of the *JK* flip-flop listed in Table 6-3. There are four cases to consider. When $J = 1$ and $K = 0$, the next state is 1. When $J = 0$ and $K = 1$, the next state is 0. When $J = K = 0$, there is no change of state and the next-state value is the same as the present state. When $J = K = 1$, the

TABLE 6-4
State Table for Sequential Circuit with *JK* flip-Flops

Present state A B	Input x	Next state A B	Flip-flop inputs			
			JA	KA	JB	KB
0 0	0	0 1	0	0	1	0
0 0	1	0 0	0	0	0	1
0 1	0	1 1	1	1	1	0
0 1	1	1 0	1	0	0	1
1 0	0	1 1	0	0	1	1
1 0	1	1 0	0	0	0	0
1 1	0	0 0	1	1	1	1
1 1	1	1 1	1	0	0	0

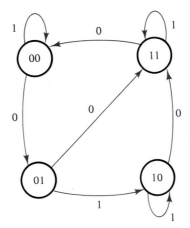

FIGURE 6-20
State diagram of the circuit of Fig. 6-19

next-state bit is the complement of the present-state bit. Examples of the last two cases occur in the table when the present state AB is 10 and input x is 0. JA and KA are both equal to 0 and the present state of A is 1. Therefore, the next state of A remains the same and is equal to 1. In the same row of the table, JB and KB are both equal to 1. Since the present state of B is 0, the next state of B is complemented and changes to 1.

The state diagram of the sequential circuit is shown in Fig. 6-20. Note that since the circuit has no outputs, the directed lines out of the circles are marked with one binary number only to designate the value of input x.

Mealy and Moore Models

The most general model of a sequential circuit has inputs, outputs, and internal states. It is customary to distinguish between two models of sequential circuits: the Mealy model and the Moore model. In the Mealy model, the outputs are functions of both the present state and inputs. In the Moore model, the outputs are a function of the present state only. An example of a Mealy model is shown in Fig. 6-16. Output y is a function of both input x and the present state of A and B. The corresponding state diagram shown in Fig. 6-17 has both the input and output values included along the directed lines between the circles. An example of a Moore model is shown in Fig. 6-19. Here the outputs are taken from the flip-flops and are a function of the present state only. The corresponding state diagram in Fig. 6-20 has only the inputs marked along the directed lines. The outputs are the flip-flop states marked inside the circles. The outputs of a Moore model can be a combination of flip-flop variables such as $A \oplus B$. This output is a function of the present state only even though it requires an additional exclusive-OR gate to generate it.

The state table of a Mealy model sequential circuit must include an output section that is a function of both the present state and inputs. When the outputs are taken directly from the flip-flops, the state table can exclude the output section because the outputs are already listed in the present-state columns of the state table. In a general Moore model sequential circuit, there may be an output section, but it will be a function of the present state only.

In a Moore model, the outputs of the sequential circuit are synchronized with the clock because they depend on only flip-flop outputs that are synchronized with the clock. In a Mealy model, the outputs may change if the inputs change during the clock-pulse period. Moreover, the outputs may have momentary false values because of the delay encountered from the time that the inputs change and the time that the flip-flop outputs change. In order to synchronize a Mealy type circuit, the inputs of the sequential circuit must be synchronized with the clock and the outputs must be sampled only during the clock-pulse transition.

6-5   STATE REDUCTION AND ASSIGNMENT

The analysis of sequential circuits starts from a circuit diagram and culminates in a state table or diagram. The design of a sequential circuit starts from a set of specifications and culminates in a logic diagram. Design procedures are presented starting from Section 6-7. This section discusses certain properties of sequential circuits that may be used to reduce the number of gates and flip-flops during the design.

State Reduction

Any design process must consider the problem of minimizing the cost of the final circuit. The two most obvious cost reductions are reductions in the number of flip-flops and the number of gates. Because these two items seem the most obvious, they have been extensively studied and investigated. In fact, a large portion of the subject of switching theory is concerned with finding algorithms for minimizing the number of flip-flops and gates in sequential circuits.

The reduction of the number of flip-flops in a sequential circuit is referred to as the *state-reduction* problem. State-reduction algorithms are concerned with procedures for reducing the number of states in a state table while keeping the external input–output requirements unchanged. Since m flip-flops produce 2^m states, a reduction in the number of states may (or may not) result in a reduction in the number of flip-flops. An unpredictable effect in reducing the number of flip-flops is that sometimes the equivalent circuit (with less flip-flops) may require more combinational gates.

We shall illustrate the need for state reduction with an example. We start with a sequential circuit whose specification is given in the state diagram of Fig. 6-21. In this example, only the input–output sequences are important; the internal states are used merely to provide the required sequences. For this reason, the states marked inside the circles are denoted by letter symbols instead of by their binary values. This is in contrast to a binary counter, where the binary-value sequence of the states themselves are taken as the outputs.

There are an infinite number of input sequences that may be applied to the circuit;

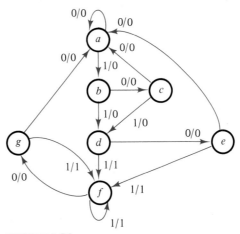

FIGURE 6-21
State diagram

each results in a unique output sequence. As an example, consider the input sequence 01010110100 starting from the initial state *a*. Each input of 0 or 1 produces an output of 0 or 1 and causes the circuit to go to the next state. From the state diagram, we obtain the output and state sequence for the given input sequence as follows: With the circuit in initial state *a*, an input of 0 produces an output of 0 and the circuit remains in state *a*. With present state *a* and input of 1, the output is 0 and the next state is *b*. With present state *b* and input of 0, the output is 0 and next state is *c*. Continuing this process, we find the complete sequence to be as follows:

state	*a*	*a*	*b*	*c*	*d*	*e*	*f*	*f*	*g*	*f*	*g*	*a*
input	0	1	0	1	0	1	1	0	1	0	0	
output	0	0	0	0	0	1	1	0	1	0	0	

In each column, we have the present state, input value, and output value. The next state is written on top of the next column. It is important to realize that in this circuit, the states themselves are of secondary importance because we are interested only in output sequences caused by input sequences.

Now let us assume that we have found a sequential circuit whose state diagram has less than seven states and we wish to compare it with the circuit whose state diagram is given by Fig. 6-21. If identical input sequences are applied to the two circuits and identical outputs occur for all input sequences, then the two circuits are said to be equivalent (as far as the input–output is concerned) and one may be replaced by the other. The problem of state reduction is to find ways of reducing the number of states in a sequential circuit without altering the input–output relationships.

We shall now proceed to reduce the number of states for this example. First, we need the state table; it is more convenient to apply procedures for state reduction here than in state diagrams. The state table of the circuit is listed in Table 6-5 and is obtained directly from the state diagram of Fig. 6-21.

TABLE 6-5
State Table

Present State	Next State x = 0	Next State x = 1	Output x = 0	Output x = 1
a	*a*	*b*	0	0
b	*c*	*d*	0	0
c	*a*	*d*	0	0
d	*e*	*f*	0	1
e	*a*	*f*	0	1
f	*g*	*f*	0	1
g	*a*	*f*	0	1

An algorithm for the state reduction of a completely specified state table is given here without proof: "Two states are said to be equivalent if, for each member of the set of inputs, they give exactly the same output and send the circuit either to the same state or to an equivalent state. When two states are equivalent, one of them can be removed without altering the input–output relationships."

We shall apply this algorithm to Table 6-5. Going through the state table, we look for two present states that go to the same next state and have the same output for both input combinations. States *g* and *e* are two such states: they both go to states *a* and *f* and have outputs of 0 and 1 for $x = 0$ and $x = 1$, respectively. Therefore, states *g* and *e* are equivalent; one can be removed. The procedure of removing a state and replacing it by its equivalent is demonstrated in Table 6-6. The row with present state *g* is crossed out and state *g* is replaced by state *e* each time it occurs in the next-state columns.

Present state *f* now has next states *e* and *f* and outputs 0 and 1 for $x = 0$ and $x = 1$, respectively. The same next states and outputs appear in the row with present state *d*. Therefore, states *f* and *d* are equivalent; state *f* can be removed and replaced by *d*. The

TABLE 6-6
Reducing the State Table

Present State	Next State x = 0	Next State x = 1	Output x = 0	Output x = 1
a	*a*	*b*	0	0
b	*c*	*d*	0	0
c	*a*	*d*	0	0
d	*e*	f̶*d*	0	1
e	*a*	f̶*d*	0	1
f̶	g̶*e*	*f*	0	1
g̶	*a*	*f*	0	1

TABLE 6-7
Reduced State Table

Present State	Next state		Output	
	$x = 0$	$x = 1$	$x = 0$	$x = 1$
a	a	b	0	0
b	c	d	0	0
c	a	d	0	0
d	e	d	0	1
e	a	d	0	1

final reduced table is shown in Table 6-7. The state diagram for the reduced table consists of only five states and is shown in Fig. 6-22. This state diagram satisfies the original input–output specifications and will produce the required output sequence for any given input sequence. The following list derived from the state diagram of Fig. 6-22 is for the input sequence used previously. We note that the same output sequence results although the state sequence is different:

state	a	a	b	c	d	e	d	d	e	d	e	a
input	0	1	0	1	0	1	1	0	1	0	0	
output	0	0	0	0	0	1	1	0	1	0	0	

In fact, this sequence is exactly the same as that obtained for Fig. 6-21, if we replace g by e and f by d.

   The checking of each pair of states for possible equivalence can be done systematically by means of a procedure that employs an implication table. The implication table consists of squares, one for every suspected pair of possible equivalent states. By judicious use of the table, it is possible to determine all pairs of equivalent states in a state table. The use of the implication table for reducing the number of states in a state table is demonstrated in Section 9-5.

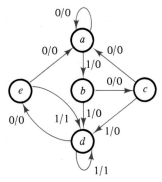

FIGURE 6-22
Reduced state diagram

It is worth noting that the reduction in the number of states of a sequential circuit is possible if one is interested only in external input–output relationships. When external outputs are taken directly from flip-flops, the outputs must be independent of the number of states before state-reduction algorithms are applied.

The sequential circuit of this example was reduced from seven to five states. In either case, the representation of the states with physical components requires that we use three flip-flops, because m flip-flops can represent up to 2^m distinct states. With three flip-flops, we can formulate up to eight binary states denoted by binary numbers 000 through 111, with each bit designating the state of one flip-flop. If the state table of Table 6-5 is used, we must assign binary values to seven states; the remaining state is unused. If the state table of Table 6-7 is used, only five states need binary assignment, and we are left with three unused states. Unused states are treated as don't-care conditions during the design of the circuit. Since don't-care conditions usually help in obtaining a simpler Boolean function, it is more likely that the circuit with five states will require fewer combinational gates than the one with seven states. In any case, the reduction from seven to five states does not reduce the number of flip-flops. In general, reducing the number of states in a state table is likely to result in a circuit with less equipment. However, the fact that a state table has been reduced to fewer states does not guarantee a saving in the number of flip-flops or the number of gates.

State Assignment

The cost of the combinational-circuit part of a sequential circuit can be reduced by using the known simplification methods for combinational circuits. However, there is another factor, known as the *state-assignment* problem, that comes into play in minimizing the combinational gates. State-assignment procedures are concerned with methods for assigning binary values to states in such a way as to reduce the cost of the combinational circuit that drives the flip-flops. This is particularly helpful when a sequential circuit is viewed from its external input–output terminals. Such a circuit may follow a sequence of internal states, but the binary values of the individual states may be of no consequence as long as the circuit produces the required sequence of outputs for any given sequence of inputs. This does not apply to circuits whose external outputs are taken directly from flip-flops with binary sequences fully specified.

TABLE 6-8
Three Possible Binary State Assignments

State	Assignment 1	Assignment 2	Assignment 3
a	001	000	000
b	010	010	100
c	011	011	010
d	100	101	101
e	101	111	011

TABLE 6-9
Reduced State Table with Binary Assignment 1

Present state	Next State		Output	
	$x = 0$	$x = 1$	$x = 0$	$x = 1$
001	001	010	0	0
010	011	100	0	0
011	001	100	0	0
100	101	100	0	1
101	001	100	0	1

The binary state-assignment alternatives available can be demonstrated in conjunction with the sequential circuit specified in Table 6-7. Remember that, in this example, the binary values of the states are immaterial as long as their sequence maintains the proper input–output relationships. For this reason, any binary number assignment is satisfactory as long as each state is assigned a unique number. Three examples of possible binary assignments are shown in Table 6-8 for the five states of the reduced table. Assignment 1 is a straight binary assignment for the sequence of states from *a* through *e*. The other two assignments are chosen arbitrarily. In fact, there are 140 different distinct assignments for this circuit.

Table 6-9 is the reduced state table with binary assignment 1 substituted for the letter symbols of the five states. It is obvious that a different binary assignment will result in a state table with different binary values for the states, whereas the input–output relationships remain the same. The binary form of the state table is used to derive the combinational-circuit part of the sequential circuit. The complexity of the combinational circuit obtained depends on the binary state assignment chosen.

Various procedures have been suggested that lead to a particular binary assignment from the many available. The most common criterion is that the chosen assignment should result in a simple combinational circuit for the flip-flop inputs. However, to date, there are no state-assignment procedures that guarantee a minimal-cost combinational circuit. State assignment is one of the challenging problems of switching theory. The interested reader will find a rich and growing literature on this topic. Techniques for dealing with the state-assignment problem are beyond the scope of this book.

6-6 FLIP-FLOP EXCITATION TABLES

The characteristic table is useful for analysis and for defining the operation of the flip-flop. It specifies the next state when the inputs and present state are known. During the design process, we usually know the transition from present state to next state and wish to find the flip-flop input conditions that will cause the required transition. For this reason, we need a table that lists the required inputs for a given change of state. Such a list is called an *excitation table*.

TABLE 6-10
Flip-Flop Excitation Tables

$Q(t)$	$Q(t+1)$	S	R
0	0	0	X
0	1	1	0
1	0	0	1
1	1	X	0

(a) *RS*

$Q(t)$	$Q(t+1)$	J	K
0	0	0	X
0	1	1	X
1	0	X	1
1	1	X	0

(b) *JK*

$Q(t)$	$Q(t+1)$	D
0	0	0
0	1	1
1	0	0
1	1	1

(c) *D*

$Q(t)$	$Q(t+1)$	T
0	0	0
0	1	1
1	0	1
1	1	0

(b) *T*

Table 6-10 presents the excitation tables for the four flip-flops. Each table consists of two columns, $Q(t)$ and $Q(t+1)$, and a column for each input to show how the required transition is achieved. There are four possible transitions from present state to next state. The required input conditions for each of the four transitions are derived from the information available in the characteristic table. The symbol X in the tables represents a don't-care condition, i.e., it does not matter whether the input is 1 or 0.

RS Flip-Flop

The excitation table for the *RS* flip-flop is shown in Table 6-10(a). The first row shows the flip-flop in the 0-state at time t. It is desired to leave it in the 0-state after the occurrence of the pulse. From the characteristic table, Table 6-3, we find that if S and R are both 0, the flip-flop will not change state. Therefore, both S and R inputs should be 0. However, it really doesn't matter if R is made a 1 when the pulse occurs, since it results in leaving the flip-flop in the 0-state. Thus, R can be 1 or 0 and the flip-flop will remain in the 0-state at $t + 1$. Therefore, the entry under R is marked by the don't-care condition X.

If the flip-flop is in the 0-state and it is desired to have it go to the 1-state, then from the characteristic table, we find that the only way to make $Q(t+1)$ equal to 1 is to make $S = 1$ and $R = 0$. If the flip-flop is to have a transition from the 1-state to the 0-state, we must have $S = 0$ and $R = 1$.

The last condition that may occur is for the flip-flop to be in the 1-state and remain in the 1-state. Certainly, R must be 0; we do not want to clear the flip-flop. However, S may be either a 0 or a 1. If it is 0, the flip-flop does not change and remains in the 1-

where A and B are the present-state values of flip-flops A and B, x is the input, and DA and DB are the input functions. The minterms for output y are obtained from the output column in the state table.

The Boolean functions are simplified by means of the maps plotted in Fig. 6-27. The simplified functions are

$$DA = AB' + Bx'$$

$$DB = A'x + B'x + ABx'$$

$$y = B'x$$

The logic diagram of the sequential circuit is shown in Fig. 6-28.

Design with Unused States

A circuit with m flip-flops would have 2^m states. There are occasions when a sequential circuit may use less than this maximum number of states. States that are not used in specifying the sequential circuit are not listed in the state table. When simplifying the input functions to flip-flops, the unused states can be treated as don't-care conditions.

Consider the state table shown in Table 6-14. There are five states listed in the table: 001, 010, 011, 100, and 101. The other three states, 000, 110, and 111, are not used. When an input of 0 or 1 is included with these unused states, we obtain six minterms: 0, 1, 12, 13, 14, and 15. These six binary combinations are not listed in the table under present state and input and are treated as don't-care conditions.

The state table is extended into an excitation table with RS flip-flops. The flip-flop input conditions are derived from the present-state and next-state values of the state table. Since RS flip-flops are used, we need to refer to Table 6-10(a) for the excitation

TABLE 6-14
State Table with Unused States

Present State			Input	Next State			Flip-Flop Inputs						Output
A	B	C	x	A	B	C	SA	RA	SB	RB	SC	RC	y
0	0	1	0	0	0	1	0	X	0	X	X	0	0
0	0	1	1	0	1	0	0	X	1	0	0	1	0
0	1	0	0	0	1	1	0	X	X	0	1	0	0
0	1	0	1	1	0	0	1	0	0	1	0	X	0
0	1	1	0	0	0	1	0	X	0	1	X	0	0
0	1	1	1	1	0	0	1	0	0	1	0	1	0
1	0	0	0	1	0	1	X	0	0	X	1	0	0
1	0	0	1	1	0	0	X	0	0	X	0	X	1
1	0	1	0	0	0	1	0	1	0	X	X	0	0
1	0	1	1	1	0	0	X	0	0	X	0	1	1

conditions of this type of flip-flop. The three flip-flops are given variable names A, B, and C. The input variable is x and the output variable is y. The excitation table of the circuit provides all the information needed for the design of the sequential circuit.

The combinational-circuit part of the sequential circuit is simplified in the maps of Fig. 6-29. There are seven maps in the diagram. Six maps are for simplifying the input functions for the three RS flip-flops. The seventh map is for simplifying the output y. Each map has six X's in the squares of the don't-care minterms 0, 1, 2, 13, 14, and 15.

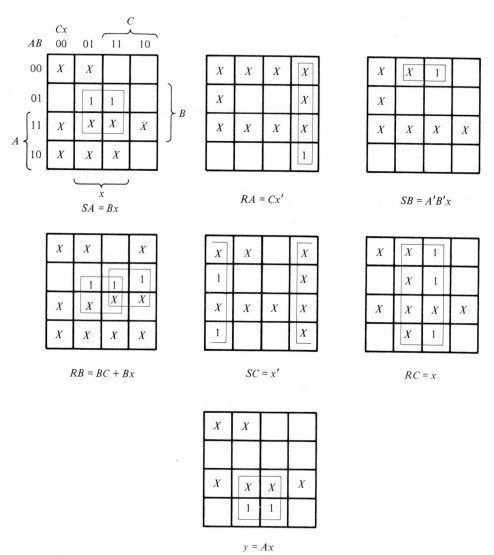

FIGURE 6-29

Maps for simplifying the sequential circuit

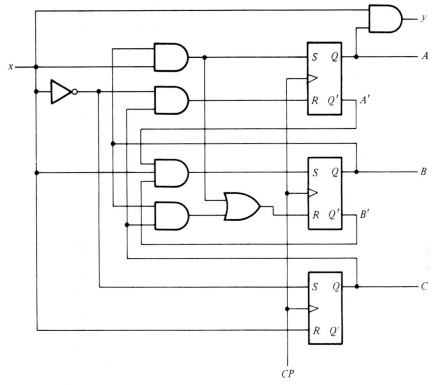

FIGURE 6-30
Logic diagram with *RS* flip-flops

The other don't-care terms in the maps come from the X's in the flip-flop input columns of the table. The simplified functions are listed under each map. The logic diagram obtained from these Boolean functions is shown in Fig. 6-30.

One factor neglected up to this point in the design is the initial state of a sequential circuit. When power is first turned on in a digital system, one does not know in what state the flip-flops will settle. It is customary to provide a *master-reset* input whose purpose is to initialize the states of all flip-flops in the system. Typically, the master reset is a signal applied to all flip-flops asynchronously before the clocked operations start. In most cases, flip-flops are cleared to 0 by the master-reset signal, but some may be set to 1. For example, the circuit of Fig. 6.30 may initially be reset to a state $ABC = 001$, since state 000 is not a valid state for this circuit.

But what if a circuit is not reset to an initial valid state? Or worse, what if, because of a noise signal or any other unforeseen reason, the circuit finds itself in one of its invalid states? In that case, it is necessary to ensure that the circuit eventually goes into one of the valid states so it can resume normal operation. Otherwise, if the sequential circuit circulates among invalid states, there will be no way to bring it back to its intended sequence of state transitions. Although one can assume that this undesirable condition is not supposed to occur, a careful designer must ensure that this situation never occurs.

It was stated previously that unused states in a sequential circuit can be treated as don't-care conditions. Once the circuit is designed, the m flip-flops in the system can be in any one of 2^m possible states. If some of these states were taken as don't-care conditions, the circuit must be investigated to determine the effect of these unused states. The next state from invalid states can be determined from the analysis of the circuit. In any case, it is always wise to analyze a circuit obtained from a design to ensure that no mistakes were made during the design process.

Analysis of Previously Designed Circuit

We wish to analyze the sequential circuit of Fig. 6-30 to determine whether it operates according to the original state table and also determine the effect of the unused states on the circuit operation. The unused states are 000, 110, and 111. The analysis of the circuit can be done by the method outlined in Section 6-4. The maps of Fig. 6-29 may also help in the analysis. What is needed here is to start with the circuit diagram of Fig. 6-30 and derive the state table or diagram. If the derived state table is identical to the state-table part of Table 6-14, then we know that the design is correct. In addition, we must determine the next states from the unused states 000, 110, and 111.

The maps of Fig. 6-29 can help in finding the next state from each of the unused states. Take, for instance, the unused state 000. If the circuit, for some reason, happens to be in the present state 000, an input $x = 0$ will transfer the circuit to some next state and an input $x = 1$ will transfer it to another (or the same) next state. We first investigate minterm $ABCx = 0000$. From the maps, we see that this minterm is not included in any function except for SC, i.e., the set input of flip-flop C. Therefore, flip-flops A and B will not change, but flip-flop C will be set to 1. Since the present state is $ABC = 000$, the next state will be $ABC = 001$. The maps also show that minterm $ABCx = 0001$ is included in the functions for SB and RC. Therefore, B will be set and C will be cleared. Starting with $ABC = 000$ and setting B, we obtain the next state $ABC = 010$ (C is already cleared). Investigation of the map for output y shows that y will be 0 for these two minterms.

The result of the analysis procedure is shown in the state diagram of Fig. 6-31. The circuit operates as intended, as long as it stays within the states 001, 010, 011, 100, and 101. If it ever finds itself in one of the invalid states, 000, 110, or 111, it goes to one of the valid states within one or two clock pulses. Thus, the circuit is self-correcting, since it eventually goes to a valid state from which it continues to operate as required.

An undesirable situation would have occurred if the next state of 110 for $x = 1$ happened to be 111 and the next state of 111 for $x = 0$ or 1 happened to be 110. Then, if the circuit starts from 110 or 111, it will circulate and stay between these two states forever. Unused states that cause such undesirable behavior should be avoided; if they

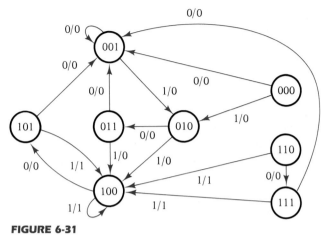

FIGURE 6-31
State diagram for the circuit of Fig. 6-30

are found to exist, the circuit should be redesigned. This can be done most easily by specifying a valid next state for any unused state that is found to circulate among invalid states.

6-8   DESIGN OF COUNTERS

A sequential circuit that goes through a prescribed sequence of states upon the application of input pulses is called a *counter*. The input pulses, called *count pulses,* may be clock pulses or they may originate from an external source and may occur at prescribed intervals of time or at random. In a counter, the sequence of states may follow a binary count or any other sequence of states. Counters are found in almost all equipment containing digital logic. They are used for counting the number of occurrences of an event and are useful for generating timing sequences to control operations in a digital system.

Of the various sequences a counter may follow, the straight binary sequence is the simplest and most straightforward. A counter that follows the binary sequence is called a *binary counter*. An n-bit binary counter consists of n flip-flops and can count in binary from 0 to $2^n - 1$. As an example, the state diagram of a 3-bit counter is shown in Fig. 6-32. As seen from the binary states indicated inside the circles, the flip-flop outputs repeat the binary count sequence with a return to 000 after 111. The directed lines between circles are not marked with input–output values as in other state diagrams. Remember that state transitions in clocked sequential circuits occur during a clock pulse; the flip-flops remain in their present states if no pulse occurs. For this reason, the clock-pulse variable CP does not appear explicitly as an input variable in a state diagram or state table. From this point of view, the state diagram of a counter does not have to show input–output values along the directed lines. The only input to the circuit is the count pulse, and the outputs are directly specified by the present states of the flip-

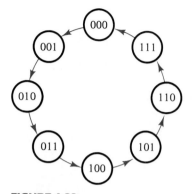

FIGURE 6-32
State diagram of a 3-bit binary counter

flops. The next state of a counter depends entirely on its present state, and the state transition occurs every time the pulse occurs.

Table 6-15 is the excitation table for the 3-bit binary counter. The three flip-flops are given variable designations A_2, A_1, and A_0. Binary counters are most efficiently constructed with T flip-flops (or JK flip-flops with J and K tied together). The flip-flop excitation for the T inputs is derived from the excitation table of the T flip-flop and from inspection of the state transition of the present state to the next state. As an illustration, consider the flip-flop input entries for row 001. The present state here is 001 and the next state is 010, which is the next count in the sequence. Comparing these two counts, we note that A_2 goes from 0 to 0; so TA_2 is marked with a 0 because flip-flop A_2 must remain unchanged when a clock pulse occurs. A_1 goes from 0 to 1; so TA_1 is marked with a 1 because this flip-flop must be complemented in the next clock pulse. Similarly, A_0 goes from 1 to 0, indicating that it must be complemented; so TA_0 is marked with a 1. The last row with present state 111 is compared with the first count 000, which is its

TABLE 6-15
Excitation Table for 3-Bit Counter

Present State			Next State			Flip-Flop Inputs		
A_2	A_1	A_0	A_2	A_1	A_0	TA_2	TA_2	TA_0
0	0	0	0	0	1	0	0	1
0	0	1	0	1	0	0	1	1
0	1	0	0	1	1	0	0	1
0	1	1	1	0	0	1	1	1
1	0	0	1	0	1	0	0	1
1	0	1	1	1	0	0	1	1
1	1	0	1	1	1	0	1	1
1	1	1	0	0	0	1	1	1

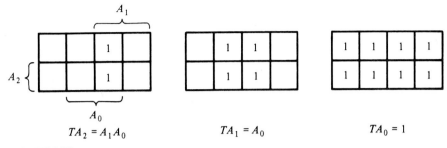

$$TA_2 = A_1 A_0 \qquad\qquad TA_1 = A_0 \qquad\qquad TA_0 = 1$$

FIGURE 6-33
Maps for a 3-bit binary counter

next state. Going from all 1's to all 0's requires that all three flip-flops be complemented.

The flip-flop input functions from the excitation tables are simplified in the maps of Fig. 6-33. The Boolean functions listed under each map specify the combinational-circuit part of the counter. Including these functions with the three flip-flops, we obtain the logic diagram of the counter, as shown in Fig. 6-34.

Counter with Nonbinary Sequence

A counter with n flip-flops may have a binary sequence of less than 2^n states. A BCD counter counts the binary states from 0000 to 1001 and returns to 0000 to repeat the sequence. Other counters may follow an arbitrary sequence that may not be the straight binary sequence. In any case, the design procedure is the same. The state table is obtained from the count sequence and the counter is designed using sequential-circuit design techniques. As an example, consider the counter specified in Table 6-16. The count has a repeated sequence of six states, with flip-flops B and C repeating the binary count 00, 01, 10, while flip-flop A alternates between 0 and 1 every three counts. The count sequence is not straight binary and two states, 011 and 111, are not included in the count. The choice of JK flip-flops results in the flip-flop input conditions listed in

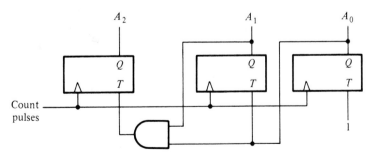

FIGURE 6-34
Logic diagram of a 3-bit binary counter

TABLE 6-16
Excitation Table for Counter

Present State			Next State			Flip-Flop Inputs					
A	B	C	A	B	C	JA	KA	JB	KB	JC	KC
0	0	0	0	0	1	0	X	0	X	1	X
0	0	1	0	1	0	0	X	1	X	X	1
0	1	0	1	0	0	1	X	X	1	0	X
1	0	0	1	0	1	X	0	0	X	1	X
1	0	1	1	1	0	X	0	1	X	X	1
1	1	0	0	0	0	X	1	X	1	0	X

the table. Inputs KB and KC have only 1's and X's in their columns, so these inputs are always equal to 1. The other flip-flop input functions can be simplified using minterms 3 and 7 as don't-care conditions. The simplified functions are

$$JA = B \qquad KA = B$$

$$JB = C \qquad KB = 1$$

$$JC = B' \qquad KC = 1$$

The logic diagram of the counter is shown in Fig. 6-35(a). Since there are two unused states, we analyze the circuit to determine their effect. If the circuit happens to be in state 011 because of an error signal, the circuit goes to state 100 after the application

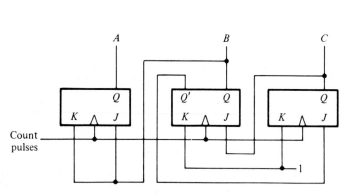

(a) Logic diagram of counter

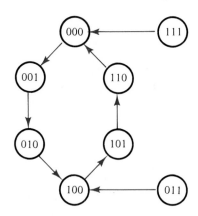

(b) State diagram of counter

FIGURE 6-35
Logic and state diagrams

of a clock pulse. This is obtained by noting that while the circuit is in present state 011, the outputs of the flip-flops are $A = 0$, $B = 1$, and $C = 1$. From the flip-flop input functions, we obtain $JA = KA = 1$, $JB = KB = 1$, $JC = 0$, and $KC = 1$. Therefore, flip-flop A is complemented and goes to 1. Flip-flop B is also complemented and goes to 0. Flip-flop C is reset to 0 because $KC = 1$. This results in next state 100. In a similar manner, we can evaluate the next state from present state 111 to be 000.

The state diagram including the unused states is shown in Fig. 6-35(b). If the circuit ever goes to one of the unused states because of an error, the next count pulse transfers it to one of the valid states and the circuit continues to count correctly. Thus, the counter is self-correcting. A self-correcting counter is one that if it happens to be in one of the unused states, it eventually reaches the normal count sequence after one or more clock pulses.

REFERENCES

1. MANO, M. M., *Computer Engineering: Hardware Design*. Englewood Cliffs, NJ: Prentice-Hall, 1988.

2. KOHAVI, Z., *Switching and Automata Theory*, 2nd Ed. New York: McGraw-Hill, 1978.

3. HILL, F. J., and G. R. PETERSON, *Introduction to Switching Theory and Logical Design*, 3rd Ed. New York: John Wiley, 1981.

4. ROTH, C. H., *Fundamentals of Logic Design*, 3rd Ed. New York: West, 1985.

5. SHIVA, S. G., *Introduction to Logic Design*. Glenview, IL: Scott, Foresman, 1988.

6. MCCLUSKEY, E. J., *Logic Design Principles*. Englewood Cliffs, NJ: Prentice-Hall, 1986.

7. BREEDING, K. J., *Digital Design Fundamentals*. Englewood Cliffs, NJ: Prentice-Hall, 1989.

8. ERCEGOVAC, M. D., and T. Lang, *Digital Systems and Hardware/Firmware Algorithms*. New York: John Wiley, 1985.

9. MANGE, D., *Analysis and Synthesis of Logic Systems*. Norwood, MA: Artech House, 1986.

10. DIETMEYER, D. L., *Logic Design of Digital Systems*. Boston: Allyn and Bacon, 1988.

PROBLEMS

6-1 Construct a D flip-flop that has the same characteristics as the one shown in Fig. 6-5, but instead of using NAND gates, use NOR and AND gates. (Remember that a one-input NOR gate is equivalent to an inverter.)

6-2 Construct a D flip-flop that has the same characteristics as the one shown in Fig. 6-5, but instead of using NAND gates, use NOR gates.

6-3   The D flip-flop shown in Fig. 6-5 can be constructed with only four NAND gates. This can be done by removing gate number 5 from the circuit and, instead, connecting the output of gate number 3 to the input of gate number 4. Draw the modified circuit and show that it operates the same way as the original circuit.

6-4   Draw the logic diagram of a master–slave D flip-flop. Use NAND gates.

6-5   The D-type positive-edge-triggered flip-flop of Fig. 6-12 is modified by including an asynchronous-clear input in the circuit. The asynchronous-clear input is connected to a third input in gate 2 and also to a third input in gate 6.
(a) Draw the logic diagram of the flip-flop, including the asynchronous-clear input.
(b) Analyze the circuit and show that when the asynchronous-clear input is logic-0, the Q output is cleared to 0 regardless of the values of the other two inputs, D and CP.
(c) Show that when the asynchronous-clear input is logic-1, it has no effect on the normal operation of the circuit.

6-6   A sequential circuit with two D flip-flops, A and B; two inputs, x and y; and one output, z, is specified by the following next-state and output equations:

$$A(t + 1) = x'y + xA$$

$$B(t + 1) = x'B + xA$$

$$z = B$$

(a) Draw the logic diagram of the circuit.
(b) Derive the state table.
(c) Derive the state diagram.

6-7   A sequential circuit has three D flip-flops, A, B, and C, and one input, x. It is described by the following flip-flop input functions:

$$DA = (BC' + B'C)x + (BC + B'C')x'$$

$$DB = A$$

$$DC = B$$

(a) Derive the state table for the circuit.
(b) Draw two state diagrams: one for $x = 0$ and the other for $x = 1$.

6-8   A sequential circuit has one flip-flop, Q; two inputs, x and y; and one output, S. It consists of a full-adder circuit connected to a D flip-flop, as shown in Fig. P6-8. Derive the state table and state diagram of the sequential circuit.

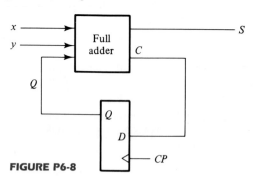

FIGURE P6-8

6-9 Derive the state table and the state diagram of the sequential circuit shown in Fig. P6-9. Explain the function that the circuit performs.

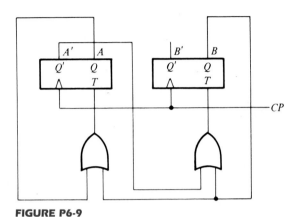

FIGURE P6-9

6-10 A *JN* flip-flop has two inputs, *J* and *N*. Input *J* behaves like the *J* input of a *JK* flip-flop and input *N* behaves like the complement of the *K* input of a *JK* flip-flop (that is, $N = K'$).
(a) Tabulate the characteristic table of the flip-flop (as in Table 6-3).
(b) Tabulate the excitation table of the flip-flop (as in Table 6-10).
(c) Show that by connecting the two inputs together, one obtains a *D* Flip-flop.

6-11 A sequential circuit has two *JK* flip-flops, one input *x*, and one output *y*. The logic diagram of the circuit is shown in Fig. P6-11. Derive the state table and state diagram of the circuit.

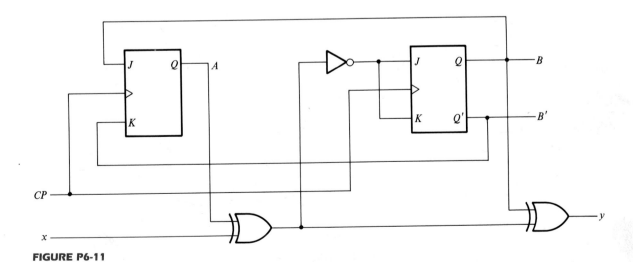

FIGURE P6-11

6-12 A sequential circuit has two *JK* flip-flops, *A* and *B*; two inputs, *x* and *y*; and one output, *z*. The flip-flop input functions and the circuit output function are as follows:

$$JA = Bx + B'y' \qquad KA = B'xy'$$

$$JB = A'x \qquad\qquad KB = A + xy'$$

$$z = Axy + Bx'y'$$

(a) Draw the logic diagram of the circuit.
(b) Tabulate the state table.
(c) Derive the next-state equations for *A* and *B*.

6-13 Starting from state 00 in the state diagram of Fig. 6-17, determine the state transitions and output sequence that will be generated when an input sequence of 010110111011110 is applied.

6-14 Reduce the number of states in the following state table and tabulate the reduced state table.

Present State	Next state		Output	
	$x = 0$	$x = 1$	$x = 0$	$x = 1$
a	*f*	*b*	0	0
b	*d*	*c*	0	0
c	*f*	*e*	0	0
d	*g*	*a*	1	0
e	*d*	*c*	0	0
f	*f*	*b*	1	1
g	*g*	*h*	0	1
h	*g*	*a*	1	0

6-15 Starting from state *a* of the state table in problem 6-14, find the output sequence generated with an input sequence 01110010011.

6-16 Repeat Problem 6-15 using the reduced table of Problem 6-14. Show that the same output sequence is obtained.

6-17 Substitute binary assignment 2 of Table 6-8 to the states in Table 6-7 and obtain the binary state table. Repeat with binary assignment 3.

6-18 Analyze the circuit of Fig. P6-18 and prove that it is equivalent to a *T* flip-flop.

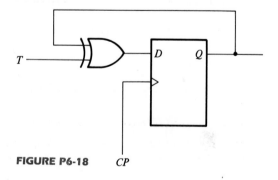

FIGURE P6-18   CP

6-19 Convert a D flip-flop to a JK flip-flop by including input gates to the D flip-flop. The gates needed for the input of the D flip-flop can be determined by means of sequential-circuit design procedures. The sequential circuit to be considered will have one D flip-flop and two inputs, J and K.

6-20 Design a sequential circuit with two D flip-flops, A and B, and one input, x. When $x = 0$, the state of the circuit remains the same. When $x = 1$, the circuit goes through the state transitions from 00 to 01 to 11 to 10 back to 00, and repeats.

6-21 Design a sequential circuit with two JK flip-flops, A and B, and two inputs, E and x. If $E = 0$, the circuit remains in the same state regardless of the value of x. When $E = 1$ and $x = 1$, the circuit goes through the state transitions from 00 to 01 to 10 to 11 back to 00, and repeats. When $E = 1$ and $x = 0$, the circuit goes through the state transitions from 00 to 11 to 10 to 01 back to 00, and repeats.

6-22 A sequential circuit has three flip-flops, A, B, C; one input, x; and one output, y. The state diagram is shown in Fig. P6-22. The circuit is to be designed by treating the unused states as don't-care conditions. The final circuit must be analyzed to ensure that it is self-correcting.
(a) Use D flip-flops in the design.
(b) Use JK flip-flops in the design.

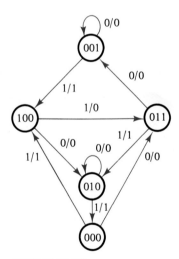

FIGURE P6-22

6-23 Design the sequential circuit specified by the state diagram of Fig. 6-23 using RS flip-flops.

6-24 Design the sequential circuit specified by the state diagram of Fig. 6-23 using T flip-flops.

6-25 Design the following nonbinary sequence counters as specified in each case. Treat the unused states as don't-care conditions. Analyze the final circuit to ensure that it is self-correcting. If your design produces a nonself-correcting counter, you must modify the circuit to make it self-correcting.

(a) Design a counter with the following repeated binary sequence: 0, 1, 2, 3, 4, 5, 6. Use *JK* flip-flops.

(b) Design a counter with the following repeated binary sequence: 0, 1, 2, 4, 6. Use *D* flip-flops.

(c) Design a counter with the following repeated binary sequence: 0, 1, 3, 5, 7. Use *T* flip-flops.

(d) Design a counter with the following repeated binary sequence: 0, 1, 3, 7, 6, 4. Use *T* flip-flops.

7
Registers, Counters, and the Memory Unit

7-1 INTRODUCTION

A clocked sequential circuit consists of a group of flip-flops and combinational gates connected to form a feedback path. The flip-flops are essential because, in their absence, the circuit reduces to a purely combinational circuit (provided there is no feedback path). A circuit with only flip-flops is considered a sequential circuit even in the absence of combinational gates. Certain MSI circuits that include flip-flops are classified by the operation that they perform rather than the name sequential circuit. Two such MSI components are registers and counters.

A register is a group of binary cells suitable for holding binary information. A group of flip-flops constitutes a register, since each flip-flop is a binary cell capable of storing one bit of information. An n-bit register has a group of n flip-flops and is capable of storing any binary information containing n bits. In addition to the flip-flops, a register may have combinational gates that perform certain data-processing tasks. In its broadest definition, a register consists of a group of flip-flops and gates that affect their transition. The flip-flops hold binary information and the gates control when and how new information is transferred into the register.

Counters were introduced in Section 6-8. A counter is essentially a register that goes through a predetermined sequence of states upon the application of input pulses. The gates in a counter are connected in such a way as to produce a prescribed sequence of binary states in the register. Although counters are a special type of register, it is common to differentiate them by giving them a special name.

A memory unit is a collection of storage cells together with associated circuits needed to transfer information in and out of storage. A random-access memory (RAM) differs from a read-only memory (ROM) in that a RAM can transfer the stored information out (read) and is also capable of receiving new information in for storage (write). A more appropriate name for such a memory would be *read–write memory*.

Registers, counters, and memories are extensively used in the design of digital systems in general and digital computers in particular. Registers can also be used to facilitate the design of sequential circuits. Counters are useful for generating timing variables to sequence and control the operations in a digital system. Memories are essential for storage of programs and data in a digital computer. Knowledge of the operation of these components is indispensable for the understanding of the organization and design of digital systems.

7-2 REGISTERS

Various types of registers are available in MSI circuits. The simplest possible register is one that consists of only flip-flops without any external gates. Figure 7-1 shows such a register constructed with four D-type flip-flops and a common clock-pulse input. The clock pulse input, CP, enables all flip-flops, so that the information presently available at the four inputs can be transferred into the 4-bit register. The four outputs can be sampled to obtain the information presently stored in the register.

The way that the flip-flops in a register are triggered is of primary importance. If the flip-flops are constructed with gated D-type latches, as in Fig. 6-5, then information present at a data (D) input is transferred to the Q output when the enable (CP) is 1, and the Q output follows the input data as long as the CP signal remains 1. When CP goes to 0, the information that was present at the data input just before the transition is retained at the Q output. In other words, the flip-flops are sensitive to the pulse duration, and the register is enabled for as long as $CP = 1$. A register that responds to the pulse duration is commonly called a *gated latch,* and the CP input is frequently labeled with the variable G (instead of CP). Latches are suitable for use as temporary storage of binary information that is to be transferred to an external destination.

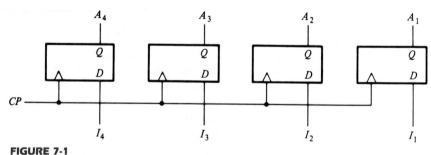

FIGURE 7-1
4-bit register

As explained in Section 6-3, a flip-flop can be used in the design of clocked sequential circuits provided that its clock input responds to the pulse transition rather than the pulse duration. This means that the flip-flops in the register must be of the edge-triggered or master–slave type. A group of flip-flops sensitive to pulse duration is usually called a latch, whereas a group of flip-flops sensitive to pulse transition is called a register. In subsequent discussions, we will assume that any group of flip-flops drawn constitutes a register and that all flip-flops are of the edge-triggered or master–slave type.

Register with Parallel Load

The transfer of new information into a register is referred to as *loading* the register. If all the bits of the register are loaded simultaneously with a single clock pulse, we say that the loading is done in parallel. A pulse applied to the *CP* input of the register of Fig. 7-1 will load all four inputs in parallel. In this configuration, the clock pulse must be inhibited from the *CP* terminal if the content of the register must be left unchanged. In other words, the *CP* input acts as an enable signal that controls the loading of new information into the register. When *CP* goes to 1, the input information is loaded into the register. If *CP* remains at 0, the content of the register is not changed. Note that the change of state in the outputs occurs at the positive edge of the pulse. If a flip-flop changes state at the negative edge, there will be a small circle under the triangle symbol in the *CP* input of the flip-flop.

Most digital systems have a master-clock generator that supplies a continuous train of clock pulses. All clock pulses are applied to all flip-flops and registers in the system. The master-clock generator acts like a pump that supplies a constant beat to all parts of the system. A separate control signal then decides what specific clock pulses will have an effect on a particular register. In such a system, the clock pulses must be ANDed with the control signal, and the output of the AND gate is then applied to the *CP* terminal of the register shown in Fig. 7-1. When the control signal is 0, the output of the AND gate is 0, and the stored information in the register remains unchanged. Only when the control signal is a 1 does the clock pulse pass through the AND gate and into the *CP* terminal for new information to be loaded into the register. Such a control variable is called a *load* control input.

Inserting an AND gate in the path of clock pulses means that logic is performed with clock pulses. The insertion of logic gates produces propagation delays between the master-clock generator and the clock inputs of flip-flops. To fully synchronize the system, we must ensure that all clock pulses arrive at the same time to all inputs of all flip-flops so that they can all change simultaneously. Performing logic with clock pulses inserts variable delays and may throw the system out of synchronism. For this reason, it is advisable (but not necessary, as long as the delays are taken into consideration) to apply clock pulses directly to all flip-flops and control the operation of the register with other inputs, such as the *R* and *S* inputs of an *RS* flip-flop.

A 4-bit register with a load control input using *RS* flip-flops is shown in Fig. 7-2. The *CP* input of the register receives continuous synchronized pulses, which are applied

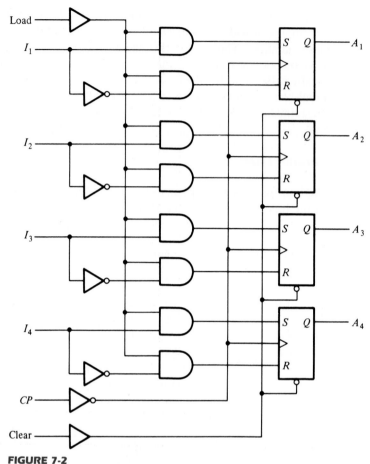

FIGURE 7-2
4-bit register with parallel load

to all flip-flops. The inverter in the *CP* path causes all flip-flops to be triggered by the negative edge of the incoming pulses. The purpose of the inverter is to reduce the loading of the master-clock generator. This is because the *CP* input is connected to only one gate (the inverter) instead of the four gate inputs that would have been required if the connections were made directly into the flip-flop clock inputs (marked with small triangles).

The *clear* input goes to a special terminal in each flip-flop through a noninverting buffer gate. When this terminal goes to 0, the flip-flop is cleared asynchronously. The clear input is useful for clearing the register to all 0's prior to its clocked operation. The clear input must be maintained at 1 during normal clocked operations (see Fig. 6-15).

The *load* input goes through a buffer gate (to reduce loading) and through a series of AND gates to the *R* and *S* inputs of each flip-flop. Although clock pulses are continuously present, it is the load input that controls the operation of the register. The two AND gates and the inverter associated with each input I determine the values of R and S. If the load input is 0, both R and S are 0, and no change of state occurs with any clock pulse. Thus, the load input is a control variable that can prevent any information change in the register as long as its input is 0. When the load control goes to 1, inputs I_1 through I_4 specify what binary information is loaded into the register on the next clock pulse. For each I that is equal to 1, the corresponding flip-flop inputs are $S = 1$, $R = 0$. For each I that is equal to 0, the corresponding flip-flop inputs are $S = 0$, $R = 1$. Thus, the input value is transferred into the register provided the load input is

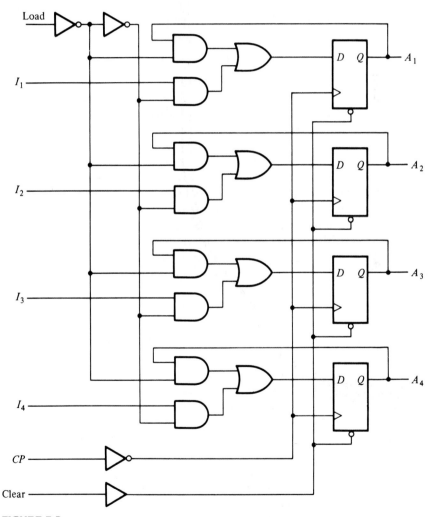

FIGURE 7-3
Register with parallel load using D flip-flops

1, the clear input is 1, and a clock pulse goes from 1 to 0. This type of transfer is called a *parallel-load* transfer because all bits of the register are loaded simultaneously. If the buffer gate associated with the load input is changed to an inverter gate, then the register is loaded when the load input is 0 and inhibited when the load input is 1.

A register with parallel load can be constructed with D flip-flops, as shown in Fig. 7-3. The clock and clear inputs are the same as before. When the load input is 1, the I inputs are transferred into the register on the next clock pulse. When the load input is 0, the circuit inputs are inhibited and the D flip-flops are reloaded with their present value, thus maintaining the content of the register. The feedback connection in each flip-flop is necessary when a D type is used because a D flip-flop does not have a "no-change" input condition. With each clock pulse, the D input determines the next state of the output. To leave the output unchanged, it is necessary to make the D input equal to the present Q output in each flip-flop.

Sequential-Logic Implementation

We saw in Chapter 6 that a clocked sequential circuit consists of a group of flip-flops and combinational gates. Since registers are readily available as MSI circuits, it becomes convenient at times to employ a register as part of the sequential circuit. A block diagram of a sequential circuit that uses a register is shown in Fig. 7-4. The present state of the register and the external inputs determine the next state of the register and the values of external outputs. Part of the combinational circuit determines the next state and the other part generates the outputs. The next state value from the combinational circuit is loaded into the register with a clock pulse. If the register has a load input, it must be set to 1; otherwise, if the register has no load input (as in Fig. 7-1), the next state value will be transferred automatically every clock pulse.

The combinational-circuit part of a sequential circuit can be implemented by any of the methods discussed in Chapter 5. It can be constructed with SSI gates, with ROM, or with a programmable logic array (PLA). By using a register, it is possible to reduce the design of a sequential circuit to that of a combinational circuit connected to a register.

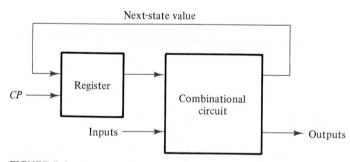

FIGURE 7-4

Block diagram of a sequential circuit

**Example
7-1**

Design the sequential circuit whose state table is listed in Fig. 7-5(a).

The state table specifies two flip-flops, A_1 and A_2; one input, x; and one output, y. The next-state and output information is obtained directly from the table:

$$A_1(t + 1) = \Sigma \ (4, 6)$$
$$A_2(t + 1) = \Sigma \ (1, 2, 5, 6)$$
$$y(A_1, A_2, x) = \Sigma \ (3, 7)$$

Present state		Input	Next state		Output
A_1	A_2	x	A_1	A_2	y
0	0	0	0	0	0
0	0	1	0	1	0
0	1	0	0	1	0
0	1	1	0	0	1
1	0	0	1	0	0
1	0	1	0	1	0
1	1	0	1	1	0
1	1	1	0	0	1

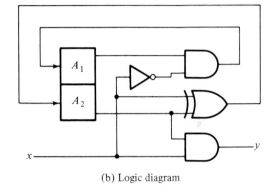

(a) State table   (b) Logic diagram

FIGURE 7-5

Example of sequential-circuit implementation

The minterm values are for variables A_1, A_2, and x, which are the present-state and input variables. The functions for the next state and output can be simplified by means of maps to give

$$A_1(t + 1) = A_1 x'$$
$$A_2(t + 1) = A_2 \oplus x$$
$$y = A_2 x$$

The logic diagram is shown in Fig. 7-5(b).   ■

**Example
7-2**

Repeat Example 7-1, but now use a ROM and a register.

The ROM can be used to implement the combinational circuit and the register will provide the flip-flops. The number of inputs to the ROM is equal to the number of flip-flops plus the number of external inputs. The number of outputs of the ROM is equal to the number of flip-flops plus the number of external outputs. In this case, we have three inputs and three outputs for the ROM; so its size must be 8 × 3. The implementation is shown in Fig. 7-6. The ROM truth table is identical to the state table with "present state" and "inputs" specifying the address of ROM and "next state" and "outputs" specifying the ROM outputs. The next-state values must be connected from the ROM outputs to the register inputs.   ■

ROM truth table

Address			Outputs		
1	2	3	1	2	3
0	0	0	0	0	0
0	0	1	0	1	0
0	1	0	0	1	0
0	1	1	0	0	1
1	0	0	1	0	0
1	0	1	0	1	0
1	1	0	1	1	0
1	1	1	0	0	1

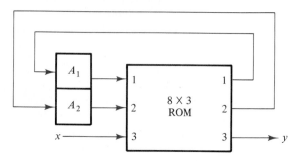

FIGURE 7-6

Sequential circuit using a register and a ROM

7-3 SHIFT REGISTERS

A register capable of shifting its binary information either to the right or to the left is called a *shift register*. The logical configuration of a shift register consists of a chain of flip-flops connected in cascade, with the output of one flip-flop connected to the input of the next flip-flop. All flip-flops receive a common clock pulse that causes the shift from one stage to the next.

The simplest possible shift register is one that uses only flip-flops, as shown in Fig. 7-7. The Q output of a given flip-flop is connected to the D input of the flip-flop at its right. Each clock pulse shifts the contents of the register one bit position to the right. The *serial input* determines what goes into the leftmost flip-flop during the shift. The *serial output* is taken from the output of the rightmost flip-flop prior to the application of a pulse. Although this register shifts its contents to the right, if we turn the page upside down, we find that the register shifts its contents to the left. Thus, a unidirectional shift register can function either as a shift-right or as a shift-left register.

The register in Fig. 7-7 shifts its contents with every clock pulse during the negative edge of the pulse transition. (This is indicated by the small circle associated with the clock input in all flip-flops.) If we want to control the shift so that it occurs only with certain pulses but not with others, we must control the CP input of the register. It will be shown later that the shift operations can be controlled through the D inputs of the flip-flops rather than through the CP input. If, however, the shift register in Fig. 7-7 is

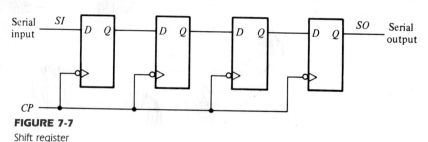

FIGURE 7-7

Shift register

used, the shift can easily be controlled by means of an external AND gate, as shown in what follows.

Serial Transfer

A digital system is said to operate in a serial mode when information is transferred and manipulated one bit at a time. The content of one register is transferred to another by shifting the bits from one register to the other. The information is transferred one bit at a time by shifting the bits out of the source register into the destination register.

The serial transfer of information from register A to register B is done with shift registers, as shown in the block diagram of Fig. 7-8(a). The serial output (SO) of register A goes to the serial input (SI) of register B. To prevent the loss of information stored in the source register, the A register is made to circulate its information by connecting the serial output to its serial input terminal. The initial content of register B is shifted out through its serial output and is lost unless it is transferred to a third shift register. The shift-control input determines when and by how many times the registers are shifted. This is done by the AND gate that allows clock pulses to pass into the CP terminals only when the shift control is 1.

Suppose the shift registers have four bits each. The control unit that supervises the transfer must be designed in such a way that it enables the shift registers, through the

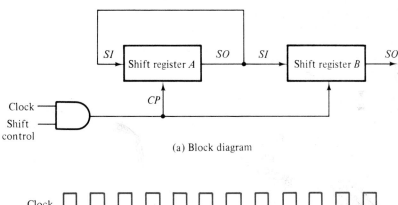

(a) Block diagram

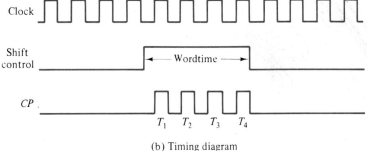

(b) Timing diagram

FIGURE 7-8
Serial transfer from register A to register B

shift-control signal, for a fixed time duration equal to four clock pulses. This is shown in the timing diagram of Fig. 7-8(b). The shift-control signal is synchronized with the clock and changes value just after the negative edge of a clock pulse. The next four clock pulses find the shift-control signal in the 1 state, so the output of the AND gate connected to the CP terminals produces four pulses, T_1, T_2, T_3, and T_4. The fourth pulse changes the shift control to 0 and the shift registers are disabled.

Assume that the binary content of A before the shift is 1011 and that of B, 0010. The serial transfer from A to B will occur in four steps, as shown in Table 7-1. After the first pulse, T_1, the rightmost bit of A is shifted into the leftmost bit of B and, at the same time, this bit is circulated into the leftmost position of A. The other bits of A and B are shifted once to the right. The previous serial output from B is lost and its value changes from 0 to 1. The next three pulses perform identical operations, shifting the bits of A into B, one at a time. After the fourth shift, the shift control goes to 0 and both registers A and B have the value 1011. Thus, the content of A is transferred into B, while the content of A remains unchanged.

The difference between serial and parallel modes of operation should be apparent from this example. In the parallel mode, information is available from all bits of a register and all bits can be transferred simultaneously during one clock pulse. In the serial mode, the registers have a single serial input and a single serial output. The information is transferred one bit at a time while the registers are shifted in the same direction.

Computers may operate in a serial mode, a parallel mode, or in a combination of both. Serial operations are slower because of the time it takes to transfer information in and out of shift registers. Serial computers, however, require less hardware to perform operations because one common circuit can be used over and over again to manipulate the bits coming out of shift registers in a sequential manner. The time interval between clock pulses is called the *bit time,* and the time required to shift the entire contents of a shift register is called the *word time.* These timing sequences are generated by the control section of the system. In a parallel computer, control signals are enabled during one clock-pulse interval. Transfers into registers are in parallel, and they occur upon application of a single clock pulse. In a serial computer, control signals must be maintained for a period equal to one word time. The pulse applied every bit time transfers the result of the operation, one at a time, into a shift register. Most computers operate in a parallel mode because this is a faster mode of operation.

TABLE 7-1
Serial-Transfer Example

Timing Pulse	Shift Register A				Shift Register B				Serial Output of B
Initial value	1	0	1	1	0	0	1	0	0
After T_1	1	1	0	1	1	0	0	1	1
After T_2	1	1	1	0	1	1	0	0	0
After T_3	0	1	1	1	0	1	1	0	0
After T_4	1	0	1	1	1	0	1	1	1

Bidirectional Shift Register with Parallel Load

Shift registers can be used for converting serial data to parallel data, and vice versa. If we have access to all the flip-flop outputs of a shift register, then information entered serially by shifting can be taken out in parallel from the outputs of the flip-flops. If a parallel-load capability is added to a shift register, then data entered in parallel can be taken out in serial fashion by shifting the data stored in the register.

Some shift registers provide the necessary input and output terminals for parallel transfer. They may also have both shift-right and shift-left capabilities. The most general shift register has all the capabilities listed below. Others may have only some of these functions, with at least one shift operation.

1. A *clear* control to clear the register to 0.
2. A *CP* input for clock pulses to synchronize all operations.
3. A *shift-right* control to enable the shift-right operation and the *serial input* and *output* lines associated with the shift right.
4. A *shift-left* control to enable the shift-left operation and the *serial input* and *output* lines associated with the shift left.
5. A *parallel-load* control to enable a parallel transfer and the *n* input lines associated with the parallel transfer.
6. *n* parallel output lines.
7. A control state that leaves the information in the register unchanged even though clock pulses are continuously applied.

A register capable of shifting both right and left is called a *bidirectional shift register*. One that can shift in only one direction is called a *unidirectional shift register*. If the register has both shift and parallel-load capabilities, it is called a *shift register with parallel load*.

The diagram of a shift register that has all the capabilities listed above is shown in Fig. 7-9. It consists of four *D* flip-flops, although *RS* flip-flops could be used provided an inverter is inserted between the *S* and *R* terminals. The four multiplexers (MUX) are part of the register and are drawn here in block diagram form. (See Fig. 5-16 for the logic diagram of the multiplexer.) The four multiplexers have two common selection variables, s_1 and s_0. Input 0 in each MUX is selected when $s_1 s_0 = 00$, input 1 is selected when $s_1 s_0 = 01$, and similarly for the other two inputs to the multiplexers.

The s_1 and s_0 inputs control the mode of operation of the register as specified in the function entries of Table 7-2. When $s_1 s_0 = 00$, the present value of the register is applied to the *D* inputs of the flip-flops. This condition forms a path from the output of each flip-flop into the input of the same flip-flop. The next clock pulse transfers into each flip-flop the binary value it held previously, and no change of state occurs. When $s_1 s_0 = 01$, terminals 1 of the multiplexer inputs have a path to the *D* inputs of the flip-flops. This causes a shift-right operation, with the serial input transferred into flip-flop A_4. When $s_1 s_0 = 10$, a shift-left operation results, with the other serial input going into

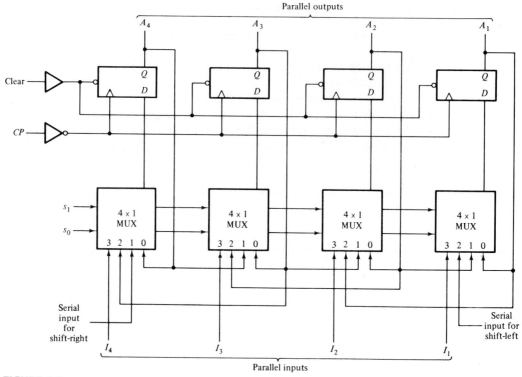

FIGURE 7-9
4-bit bidirectional shift register with parallel load

flip-flop A_1. Finally, when $s_1 s_0 = 11$, the binary information on the parallel input lines is transferred into the register simultaneously during the next clock pulse.

A bidirectional shift register with parallel load is a general-purpose register capable of performing three operations: shift left, shift right, and parallel load. Not all shift registers available in MSI circuits have all these capabilities. The particular application dictates the choice of one MSI shift register over another.

TABLE 7-2
Function Table for the Register of Fig. 7-9

Mode Control		Register Operation
s_1	s_0	
0	0	No change
0	1	Shift right
1	0	Shift left
1	1	Parallel load

Serial Addition

Operations in digital computers are mostly done in parallel because this is a faster mode of operation. Serial operations are slower but require less equipment. To demonstrate the serial mode of operation, we present here the design of a serial adder. The parallel counterpart was discussed in Section 5-2.

The two binary numbers to be added serially are stored in two shift registers. Bits are added one pair at a time, sequentially, through a single full-adder (FA) circuit, as shown in Fig. 7-10. The carry out of the full-adder is transferred to a D flip-flop. The output of this flip-flop is then used as an input carry for the next pair of significant bits. The two shift registers are shifted to the right for one word-time period. The sum bits from the S output of the full-adder could be transferred into a third shift register. By shifting the sum into A while the bits of A are shifted out, it is possible to use one register for storing both the augend and the sum bits. The serial input (SI) of register B is able to receive a new binary number while the addend bits are shifted out during the addition.

The operation of the serial adder is as follows. Initially, the A register holds the augend, the B register holds the addend, and the carry flip-flop is cleared to 0. The serial outputs (SO) of A and B provide a pair of significant bits for the full-adder at x and y. Output Q of the flip-flop gives the input carry at z. The shift-right control enables both registers and the carry flip-flop; so at the next clock pulse, both registers are shifted

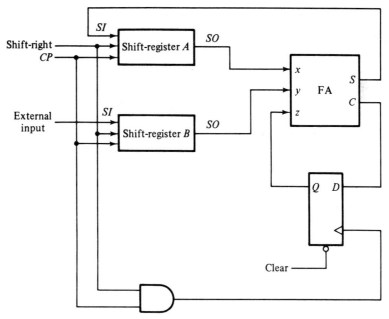

FIGURE 7-10
Serial adder

once to the right, the sum bit from S enters the leftmost flip-flop of A, and the output carry is transferred into flip-flop Q. The shift-right control enables the registers for a number of clock pulses equal to the number of bits in the registers. For each succeeding clock pulse, a new sum bit is transferred to A, a new carry is transferred to Q, and both registers are shifted once to the right. This process continues until the shift-right control is disabled. Thus, the addition is accomplished by passing each pair of bits together with the previous carry through a single full-adder circuit and transferring the sum, one bit at a time, into register A.

If a new number has to be added to the contents of register A, this number must be first transferred serially into register B. Repeating the process once more will add the second number to the previous number in A.

Comparing the serial adder with the parallel adder described in Section 5-2, we note the following differences. The parallel adder must use registers with parallel-load capability, whereas the serial adder uses shift registers. The number of full-adder circuits in the parallel adder is equal to the number of bits in the binary numbers, whereas the serial adder requires only one full-adder circuit and a carry flip-flop. Excluding the registers, the parallel adder is a purely combinational circuit, whereas the serial adder is a sequential circuit. The sequential circuit in the serial adder consists of a full-adder circuit and a flip-flop that stores the output carry. This is typical in serial operations because the result of a bit-time operation may depend not only on the present inputs but also on previous inputs.

To show that bit-time operations in serial computers may require a sequential circuit, we will redesign the serial adder by considering it a sequential circuit.

Example 7-3

Design a serial adder using a sequential-logic procedure.

First, we must stipulate that two shift registers are available to store the binary numbers to be added serially. The serial outputs from the registers are designated by variables x and y. The sequential circuit to be designed will not include the shift registers; they will be inserted later to show the complete unit. The sequential circuit proper has two inputs, x and y, that provide a pair of significant bits, an output S that generates the sum bit, and flip-flop Q for storing the carry. The present state of Q provides the present value of the carry. The clock pulse that shift the registers enables flip-flop Q to load the next carry. This carry is then used with the next pair of bits in x and y. The state table that specifies the sequential circuit is given in Table 7-3.

The present state of Q is the present value of the carry. The present carry in Q is added together with inputs x and y to produce the sum bit in output S. The next state of Q is equivalent to the output carry. Note that the state-table entries are identical to the entries in a full-adder truth table, except that the input carry is now the present state of Q and the output carry is now the next state of Q.

If we use a D flip-flop for Q, we obtain the same circuit as in Fig. 7-10 because the input requirements of the D input are the same as the next-state values. If we use a JK flip-flop for Q, we obtain the input excitation requirements listed in Table 7-3. The three Boolean functions of interest are the flip-flop input functions for JQ and KQ and

TABLE 7-3
Excitation Table for a Serial Adder

Present State	Inputs		Next State	Output	Flip-Flop Inputs	
Q	x	y	Q	S	JQ	KQ
0	0	0	0	0	0	X
0	0	1	0	1	0	X
0	1	0	0	1	0	X
0	1	1	1	0	1	X
1	0	0	0	1	X	1
1	0	1	1	0	X	0
1	1	0	1	0	X	0
1	1	1	1	1	X	0

output S. These functions are specified in the excitation table and can be simplified by means of maps:

$$JQ = xy$$
$$KQ = x'y' = (x + y)'$$
$$S = x \oplus y \oplus Q$$

As shown in Fig. 7-11, the circuit consists of three gates and a JK flip-flop. The two shift registers are also included in the diagram to show the complete serial adder. Note that output S is a function not only of x and y, but also of the present state of Q. The next state of Q is a function of the present values of x and y that come out of the serial outputs of the shift registers.

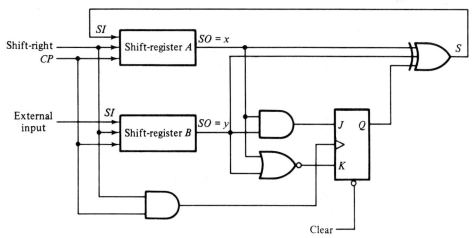

FIGURE 7-11

Second form of serial adder

7-4   RIPPLE COUNTERS

MSI counters come in two categories: ripple counters and synchronous counters. In a ripple counter, the flip-flop output transition serves as a source for triggering other flip-flops. In other words, the *CP* inputs of all flip-flops (except the first) are triggered not by the incoming pulses, but rather by the transition that occurs in other flip-flops. In a

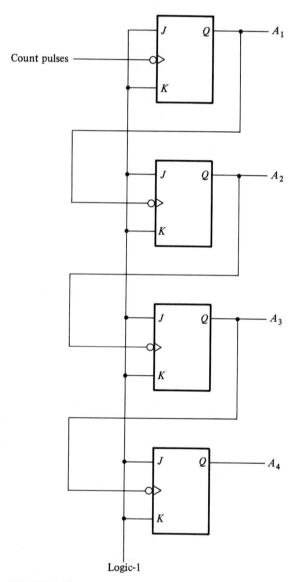

FIGURE 7-12
4-bit binary ripple counter

synchronous counter, the input pulses are applied to all CP inputs of all flip-flops. The change of state of a particular flip-flop is dependent on the present state of other flip-flops. Synchronous MSI counters are discussed in the next section. Here we present some common MSI ripple counters and explain their operation.

Binary Ripple Counter

A binary ripple counter consists of a series connection of complementing flip-flops (T or JK type), with the output of each flip-flop connected to the CP input of the next higher-order flip-flop. The flip-flop holding the least significant bit receives the incoming count pulses. The diagram of a 4-bit binary ripple counter is shown in Fig. 7-12. All J and K inputs are equal to 1. The small circle in the CP input indicates that the flip-flop complements during a negative-going transition or when the output to which it is connected goes from 1 to 0. To understand the operation of the binary counter, refer to its count sequence given in Table 7-4. It is obvious that the lowest-order bit A_1 must be complemented with each count pulse. Every time A_1 goes from 1 to 0, it complements A_2. Every time A_2 goes from 1 to 0, it complements A_3, and so on. For example, take the transition from count 0111 to 1000. The arrows in the table emphasize the transitions in this case. A_1 is complemented with the count pulse. Since A_1 goes from 1 to 0, it triggers A_2 and complements it. As a result, A_2 goes from 1 to 0, which in turn complements A_3. A_3 now goes from 1 to 0, which complements A_4. The output transition of A_4, if connected to a next stage, will not trigger the next flip-flop since it goes from 0 to 1. The flip-flops change one at a time in rapid succession, and the signal propagates through the counter in a *ripple* fashion. Ripple counters are sometimes called *asynchronous counters*.

TABLE 7-4
Count Sequence for a Binary Ripple Counter

Count Sequence				Conditions for Complementing Flip-Flops
A_4	A_3	A_2	A_1	
0	0	0	0	Complement A_1
0	0	0	1	Complement A_1 A_1 will go from 1 to 0 and complement A_2
0	0	1	0	Complement A_1
0	0	1	1	Complement A_1 A_1 will go from 1 to 0 and complement A_2; A_2 will go from 1 to 0 and complement A_3
0	1	0	0	Complement A_1
0	1	0	1	Complement A_1 A_1 will go from 1 to 0 and complement A_2
0	1	1	0	Complement A_1
0	1	1	1	Complement A_1 A_1 will go from 1 to 0 and complement A_2; A_2 will go from 1 to 0 and complement A_3; A_3 will go from 1 to 0 and complement A_4
1	0	0	0	and so on . . .

A binary counter with a reverse count is called a binary *down-counter*. In a down-counter, the binary count is decremented by 1 with every input count pulse. The count of a 4-bit down-counter starts from binary 15 and continues to binary counts 14, 13, 12, . . . , 0 and then back to 15. The circuit of Fig. 7-12 will function as a binary down-counter if the outputs are taken from the complement terminals Q' of all flip-flops. If only the normal outputs of flip-flops are available, the circuit must be modified slightly as described next.

A list of the count sequence of a count-down binary counter shows that the lowest-order bit must be complemented with every count pulse. Any other bit in the sequence is complemented if its previous lower-order bit goes from 0 to 1. Therefore, the diagram of a binary down-counter looks the same as in Fig. 7-12, provided all flip-flops trigger on the positive edge of the pulse. (The small circles in the CP inputs must be absent.) If negative-edge-triggered flip-flops are used, then the CP input of each flip-flop must be connected to the Q' output of the previous flip-flop. Then when Q goes from 0 to 1, Q' will go from 1 to 0 and complement the next flip-flop as required.

BCD Ripple Counter

A decimal counter follows a sequence of ten states and returns to 0 after the count of 9. Such a counter must have at least four flip-flops to represent each decimal digit, since a decimal digit is represented by a binary code with at least four bits. The sequence of states in a decimal counter is dictated by the binary code used to represent a decimal digit. If BCD is used, the sequence of states is as shown in the state diagram of Fig. 7-13. This is similar to a binary counter, except that the state after 1001 (code for decimal digit 9) is 0000 (code for decimal digit 0).

The design of a decimal ripple counter or of any ripple counter not following the binary sequence is not a straightforward procedure. The formal tools of logic design can serve only as a guide. A satisfactory end product requires the ingenuity and imagination of the designer.

The logic diagram of a BCD ripple counter is shown in Fig. 7-14. The four outputs are designated by the letter symbol Q with a numeric subscript equal to the binary weight of the corresponding bit in the BCD code. The flip-flops trigger on the negative

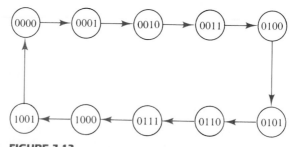

FIGURE 7-13
State diagram of a decimal BCD counter

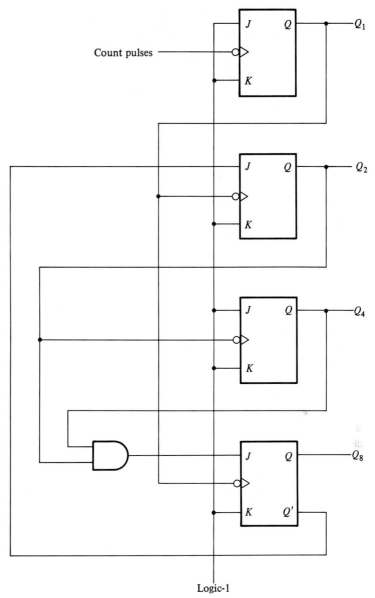

FIGURE 7-14

BCD ripple counter

edge, i.e., when the CP signal goes from 1 to 0. Note that the output of Q_1 is applied to the CP inputs of both Q_2 and Q_8 and the output of Q_2 is applied to the CP input of Q_4. The J and K inputs are connected either to a permanent 1 signal or to outputs of flip-flops, as shown in the diagram.

A ripple counter is an asynchronous sequential circuit and cannot be described by Boolean equations developed for describing clocked sequential circuits. Signals that affect the flip-flop transition depend on the order in which they change from 1 to 0. The operation of the counter can be explained by a list of conditions for flip-flop transitions. These conditions are derived from the logic diagram and from knowledge of how a JK flip-flop operates. Remember that when the CP input goes from 1 to 0, the flip-flop is set if $J = 1$, is cleared if $K = 1$, is complemented if $J = K = 1$, and is left unchanged if $J = K = 0$. The following are the conditions for each flip-flop state transition:

1. Q_1 is complemented on the negative edge of every count pulse.
2. Q_2 is complemented if $Q_8 = 0$ and Q_1 goes from 1 to 0. Q_2 is cleared if $Q_8 = 1$ and Q_1 goes from 1 to 0.
3. Q_4 is complemented when Q_2 goes from 1 to 0.
4. Q_8 is complemented when $Q_4 Q_2 = 11$ and Q_1 goes from 1 to 0. Q_8 is cleared if either Q_4 or Q_2 is 0 and Q_1 goes from 1 to 0.

To verify that these conditions result in the sequence required by a BCD ripple counter, it is necessary to verify that the flip-flop transitions indeed follow a sequence of states as specified by the state diagram of Fig. 7-13. Another way to verify the operation of the counter is to derive the timing diagram for each flip-flop from the conditions just listed. This diagram is shown in Fig. 7-15 with the binary states listed after each clock pulse. Q_1 changes state after each clock pulse. Q_2 complements every time Q_1 goes from 1 to 0 as long as $Q_8 = 0$. When Q_8 becomes 1, Q_2 remains cleared at 0. Q_4 complements every time Q_2 goes from 1 to 0. Q_8 remains cleared as long as Q_2 or Q_4 is 0. When both Q_2 and Q_4 become 1's, Q_8 complements when Q_1 goes from 1 to 0. Q_8 is cleared on the next transition of Q_1.

The BCD counter of Fig. 7-14 is a *decade* counter, since it counts from 0 to 9. To count in decimal from 0 to 99, we need a two-decade counter. To count from 0 to 999,

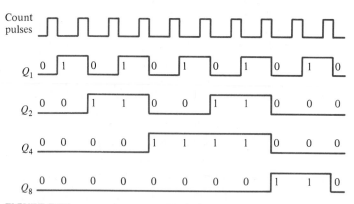

FIGURE 7-15

Timing diagram for the decimal counter of Fig. 7-14

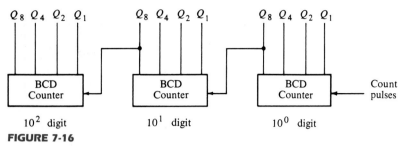

FIGURE 7-16

Block diagram of a three-decade decimal BCD counter

we need a three-decade counter. Multiple-decade counters can be constructed by connecting BCD counters in cascade, one for each decade. A three-decade counter is shown in Fig. 7-16. The inputs to the second and third decades come from Q_8 of the previous decade. When Q_8 in one decade goes from 1 to 0, it triggers the count for the next higher-order decade while its own decade goes from 9 to 0. For instance, the count after 399 will be 400.

7-5 SYNCHRONOUS COUNTERS

Synchronous counters are distinguished from ripple counters in that clock pulses are applied to the *CP* inputs of *all* flip-flops. The common pulse triggers all the flip-flops simultaneously, rather than one at a time in succession as in a ripple counter. The decision whether a flip-flop is to be complemented or not is determined from the values of the *J* and *K* inputs at the time of the pulse. If $J = K = 0$, the flip-flop remains unchanged. If $J = K = 1$, the flip-flop complements.

A design procedure for any type of synchronous counter was presented in Section 6-8. The design of a 3-bit binary counter was carried out in detail and is illustrated in Fig. 6-34. In this section, we present some typical MSI synchronous counters and explain their operation. It must be realized that there is no need to design a counter if it is already available commercially in IC form.

Binary Counter

The design of synchronous binary counters is so simple that there is no need to go through a rigorous sequential-logic design process. In a synchronous binary counter, the flip-flop in the lowest-order position is complemented with every pulse. This means that its *J* and *K* inputs must be maintained at logic-1. A flip-flop in any other position is complemented with a pulse provided all the bits in the lower-order positions are equal to 1, because the lower-order bits (when all 1's) will change to 0's on the next count pulse. The binary count dictates that the next higher-order bit be complemented. For example, if the present state of a 4-bit counter is $A_4 A_3 A_2 A_1 = 0011$, the next count will be 0100. A_1 is always complemented. A_2 is complemented because the present state

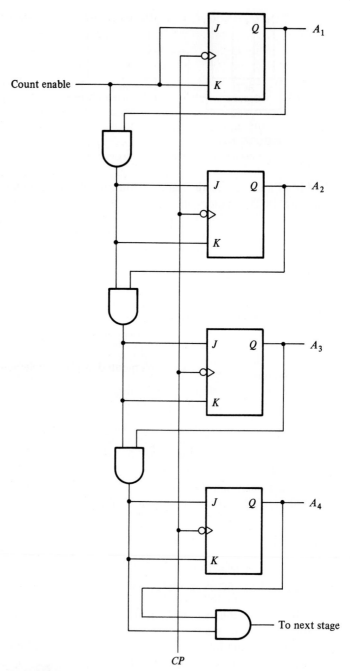

FIGURE 7-17
4-bit synchronous binary counter

of $A_1 = 1$. A_3 is complemented because the present state of $A_2 A_1 = 11$. But A_4 is not complemented because the present state of $A_3 A_2 A_1 = 011$, which does not give an all-1's condition.

Synchronous binary counters have a regular pattern and can easily be constructed with complementing flip-flops and gates. The regular pattern can be clearly seen from the 4-bit counter depicted in Fig. 7-17. The CP terminals of all flip-flops are connected to a common clock-pulse source. The first stage A_1 has its J and K equal to 1 if the counter is enabled. The other J and K inputs are equal to 1 if all previous low-order bits are equal to 1 and the count is enabled. The chain of AND gates generates the required logic for the J and K inputs in each stage. The counter can be extended to any number of stages, with each stage having an additional flip-flop and an AND gate that gives an output of 1 if all previous flip-flop outputs are 1's.

Note that the flip-flops trigger on the negative edge of the pulse. This is not essential here as it was with the ripple counter. The counter could also be triggered on the positive edge of the pulse.

Binary Up–Down Counter

In a synchronous count-down binary counter, the flip-flop in the lowest-order position is complemented with every pulse. A flip-flop in any other position is complemented with a pulse provided all the lower-order bits are equal to 0. For example, if the present state of a 4-bit count-down binary counter is $A_4 A_3 A_2 A_1 = 1100$, the next count will be 1011. A_1 is always complemented. A_2 is complemented because the present state of $A_1 = 0$. A_3 is complemented because the present state of $A_2 A_1 = 00$. But A_4 is not complemented because the present state of $A_3 A_2 A_1 = 100$, which is not an all-0's condition.

A count-down binary counter can be constructed as shown in Fig. 7-17, except that the inputs to the AND gates must come from the complement outputs Q' and not from the normal outputs Q of the previous flip-flops. The two operations can be combined in one circuit. A binary counter capable of counting either up or down is shown in Fig. 7-18. The T flip-flops employed in this circuit may be considered as JK flip-flops with the J and K terminals tied together. When the up input control is 1, the circuit counts up, since the T inputs receive their signals from the values of the previous normal outputs of the flip-flops. When the down input control is 1 and the up input is 0, the circuit counts down, since the complemented outputs of the previous flip-flops are applied to the T inputs. When the up and down inputs are both 0, the circuit does not change state but remains in the same count. When the up and down inputs are both 1, the circuit counts up. This ensures that only one operation is performed at any given time.

BCD Counter

A BCD counter counts in binary-coded decimal from 0000 to 1001 and back to 0000. Because of the return to 0 after a count of 9, a BCD counter does not have a regular

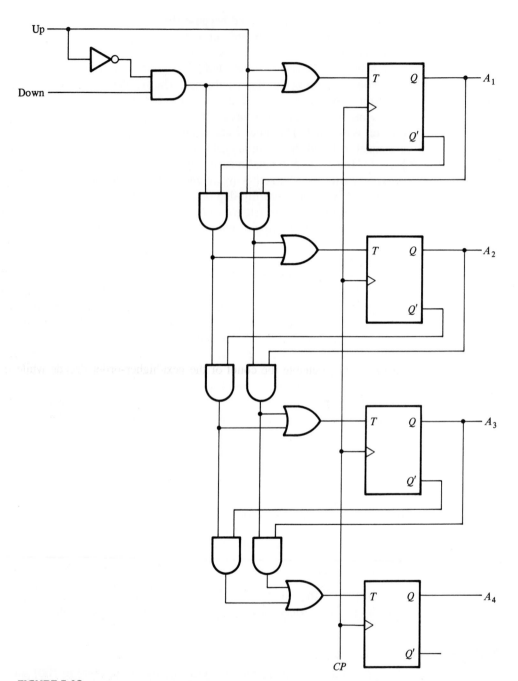

FIGURE 7-18
4-bit up–down counter

TABLE 7-5
Excitation Table for BCD Counter

Present State				Next State				Output	Flip-Flop Inputs			
Q_8	Q_4	Q_2	Q_1	Q_8	Q_4	Q_2	Q_1	y	TQ_8	TQ_4	TQ_2	TQ_1
0	0	0	0	0	0	0	1	0	0	0	0	1
0	0	0	1	0	0	1	0	0	0	0	1	1
0	0	1	0	0	0	1	1	0	0	0	0	1
0	0	1	1	0	1	0	0	0	0	1	1	1
0	1	0	0	0	1	0	1	0	0	0	0	1
0	1	0	1	0	1	1	0	0	0	0	1	1
0	1	1	0	0	1	1	1	0	0	0	0	1
0	1	1	1	1	0	0	0	0	1	1	1	1
1	0	0	0	1	0	0	1	0	0	0	0	1
1	0	0	1	0	0	0	0	1	1	0	0	1

pattern as in a straight binary count. To derive the circuit of a BCD synchronous counter, it is necessary to go through a design procedure as discussed in Section 6-8.

The excitation table of a BCD counter is given in Table 7-5. The excitation for the T flip-flops is obtained from the present and next state conditions. An output y is also shown in the table. This output is equal to 1 when the counter present state is 1001. In this way, y can enable the count of the next-higher-order decade while the same pulse switches the present decade from 1001 to 0000.

The flip-flop input functions from the excitation table can be simplified by means of maps. The unused states for minterms 10 to 15 are taken as don't-care terms. The simplified functions are

$$TQ_1 = 1$$

$$TQ_2 = Q_8' Q_1$$

$$TQ_4 = Q_2 Q_1$$

$$TQ_8 = Q_8 Q_1 + Q_4 Q_2 Q_1$$

$$y = Q_8 Q_1$$

The circuit can be easily drawn with four T flip-flops, five AND gates, and one OR gate.

Synchronous BCD counters can be cascaded to form a counter for decimal numbers of any length. The cascading is done as in Fig. 7-16, except that output y must be connected to the count input of the next-higher-order decade.

Binary Counter with Parallel Load

Counters employed in digital systems quite often require a parallel-load capability for transferring an initial binary number prior to the count operation. Figure 7-19 shows the logic diagram of a register that has a parallel-load capability and can also operate as

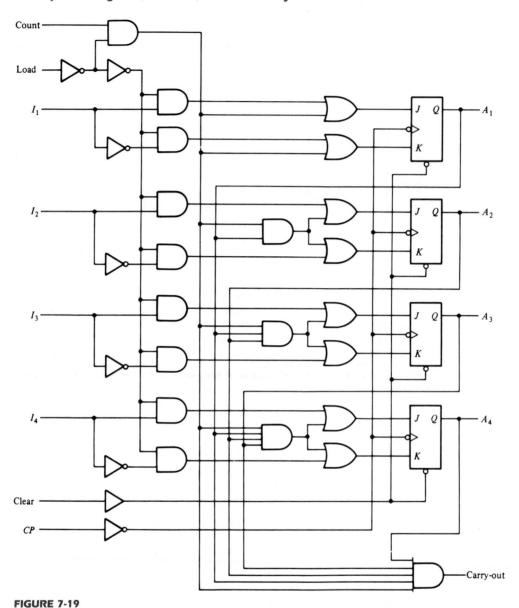

FIGURE 7-19
4-bit binary counter with parallel load

a counter. The input load control when equal to 1 disables the count sequence and causes a transfer of data from inputs I_1 through I_4 into flip-flops A_1 through A_4, respectively. If the load input is 0 and the count input control is 1, the circuit operates as a counter. The clock pulses then cause the state of the flip-flops to change according to the binary count sequence. If both control inputs are 0, clock pulses do not change the state of the register.

The carry-out terminal becomes a 1 if all flip-flops are equal to 1 while the count input is enabled. This is the condition for complementing the flip-flop holding the next-higher-order bit. This output is useful for expanding the counter to more than four bits. The speed of the counter is increased if this carry is generated directly from the outputs of all four flip-flops instead of going through a chain of AND gates. Similarly, each flip-flop is associated with an AND gate that receives all previous flip-flop outputs directly to determine when the flip-flop should be complemented.

The operation of the counter is summarized in Table 7-6. The four control inputs: clear, CP, load, and count determine the next output state. The clear input is asynchronous and, when equal to 0, causes the counter to be cleared to all 0's, regardless of the presence of clock pulses or other inputs. This is indicated in the table by the X entries, which symbolize don't-care conditions for the other inputs, so their value can be either 0 or 1. The clear input must go to the 1 state for the clocked operations listed in the next three entries in the table. With the load and count inputs both at 0, the outputs do not change, whether a pulse is applied in the CP terminal or not. A load input of 1 causes a transfer from inputs I_1–I_4 into the register during the positive edge of an input pulse. The input information is loaded into the register regardless of the value of the count input, because the count input is inhibited when the load input is 1. If the load input is maintained at 0, the count input controls the operation of the counter. The outputs change to the next binary count on the positive-edge transition of every clock pulse, but no change of state occurs if the count input is 0.

The 4-bit counter shown in Fig. 7-19 can be enclosed in one IC package. Two ICs are necessary for the construction of an 8-bit counter; four ICs for a 16-bit counter; and so on. The carry output of one IC must be connected to the count input of the IC holding the four next-higher-order bits of the counter.

Counters with parallel-load capability having a specified number of bits are very useful in the design of digital systems. Later, we will refer to them as registers with load and increment capabilities. The *increment* function is an operation that adds 1 to the present content of a register. By enabling the count control during one clock pulse period, the content of the register can be incremented by 1.

A counter with parallel load can be used to generate any desired number of count sequences. A modulo-N (abbreviated mod-N) counter is a counter that goes through a repeated sequence of N counts. For example, a 4-bit binary counter is a mod-16 counter. A BCD counter is a mod-10 counter. In some applications, one may not be concerned with the particular N states that a mod-N counter uses. If this is the case, then a coun-

TABLE 7-6
Function Table for the Counter of Fig. 7-19

Clear	CP	Load	Count	Function
0	X	X	X	Clear to 0
1	X	0	0	No change
1	↑	1	X	Load inputs
1	↑	0	1	Count next binary state

ter with parallel load can be used to construct any mod-N counter, with N being any value desired. This is shown in the following example.

Example 7-4

Construct a mod-6 counter using the MSI circuit specified in Fig. 7-19.

Figure 7-20 shows four ways in which a counter with parallel load can be used to generate a sequence of six counts. In each case, the count control is set to 1 to enable the count through the pulses in the CP input. We also use the facts that the load control inhibits the count and that the clear operation is independent of other control inputs.

The AND gate in Fig. 7-20(a) detects the occurrence of state 0101 in the output. When the counter is in this state, the load input is enabled and an all-0's input is loaded into the register. Thus, the counter goes through binary states 0, 1, 2, 3, 4, and 5 and then returns to 0. This produces a sequence of six counts.

The clear input of the register is asynchronous, i.e., it does not depend on the clock. In Fig. 7-20(b), the NAND gate detects the count of 0110, but as soon as this count occurs, the register is cleared. The count 0110 has no chance of staying on for any ap-

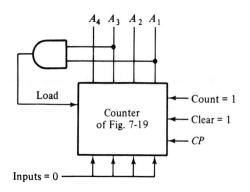

(a) Binary states 0, 1, 2, 3, 4, 5.

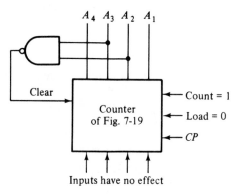

(b) Binary states 0, 1, 2, 3, 4, 5.

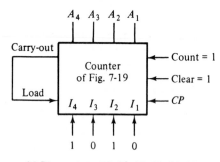

(c) Binary states 10, 11, 12, 13, 14, 15.

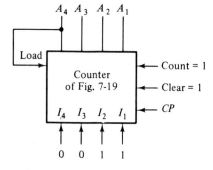

(d) Binary states 3, 4, 5, 6, 7, 8.

FIGURE 7-20

Four ways to achieve a mod-6 counter using a counter with parallel load

preciable time because the register goes immediately to 0. A momentary spike occurs in output A_2 as the count goes from 0101 to 0110 and immediately to 0000. This momentary spike may be undesirable, and for this reason, this configuration is not recommended. If the counter has a synchronous clear input, it would be possible to clear the counter with the clock after an occurrence of the 0101 count.

Instead of using the first six counts, we may want to choose the last six counts from 10 to 15. In this case, it is possible to take advantage of the output carry to load a number in the register. In Fig. 7-20(c), the counter starts with count 1010 and continues to 1111. The output carry generated during the last state enables the load control, which then loads the input, which is set at 1010.

It is also possible to choose any intermediate count of six states. The mod-6 counter of Fig. 7-20(d) goes through the count sequence 3, 4, 5, 6, 7, and 8. When the last count 1000 is reached, output A_4 goes to 1 and the load control is enabled. This loads the value of 0011 into the register, and the binary count continues from this state. ∎

7-6 TIMING SEQUENCES

The sequence of operations in a digital system are specified by a control unit. The control unit that supervises the operations in a digital system would normally consist of timing signals that determine the time sequence in which the operations are executed. The timing sequences in the control unit can be easily generated by means of counters or shift registers. This section demonstrates the use of these MSI functions in the generation of timing signals for a control unit.

Word-Time Generation

First, we demonstrate a circuit that generates the required timing signal for serial mode of operation. Serial transfer of information was discussed in Section 7-3, with an example depicted in Fig. 7-8. The control unit in a serial computer must generate a *word-time* signal that stays on for a number of pulses equal to the number of bits in the shift registers. The word-time signal can be generated by means of a counter that counts the required number of pulses.

Assume that the word-time signal to be generated must stay on for a period of eight clock pulses. Figure 7-21(a) shows a counter circuit that accomplishes this task. Initially, the 3-bit counter is cleared to 0. A start signal will set flip-flop Q. The output of this flip-flop supplies the word-time control and also enables the counter. After the count of eight pulses, the flip-flop is reset and Q goes to 0. The timing diagram of Fig. 7-21(b) demonstrates the operation of the circuit. The start signal is synchronized with the clock and stays on for one clock-pulse period. After Q is set to 1, the counter starts counting the clock pulses. When the counter reaches the count of 7 (binary 111), it sends a stop signal to the reset input of the flip-flop. The stop signal becomes a 1 after the negative-edge transition of pulse 7. The next clock pulse switches the counter to the 000 state and also clears Q. Now the counter is disabled and the word-time signal stays

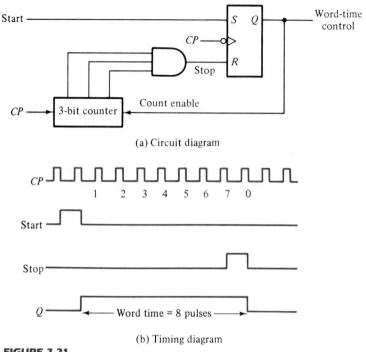

(a) Circuit diagram

(b) Timing diagram

FIGURE 7-21

Generation of a word-time control for serial operations

at 0. Note that the word-time control stays on for a period of eight pulses. Note also that the stop signal in this circuit can be used to start another word-count control in another circuit just as the start signal is used in this circuit.

Timing Signals

In a parallel mode of operation, a single clock pulse can specify the time at which an operation should be executed. The control unit in a digital system that operates in the parallel mode must generate timing signals that stay on for only one clock pulse period, but these timing signals must be distinguished from each other.

Timing signals that control the sequence of operations in a digital system can be generated with a shift register or a counter with a decoder. A *ring counter* is a circular shift register with only one flip-flop being set at any particular time; all others are cleared. The single bit is shifted from one flip-flop to the other to produce the sequence of timing signals. Figure 7-22(a) shows a 4-bit shift register connected as a ring counter. The initial value of the register is 1000, which produces the variable T_0. The single bit is shifted right with every clock pulse and circulates back from T_3 to T_0. Each flip-flop is in the 1 state once every four clock pulses and produces one of the four timing signals shown in Fig. 7-22(c). Each output becomes a 1 after the negative-edge transition of a clock pulse and remains 1 during the next clock pulse.

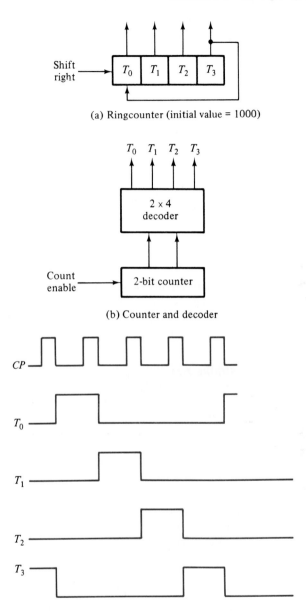

(a) Ringcounter (initial value = 1000)

(b) Counter and decoder

(c) Sequence of four timing signals

FIGURE 7-22
Generation of timing signals

The timing signals can be generated also by continuously enabling a 2-bit counter that goes through four distinct states. The decoder shown in Fig. 7-22(b) decodes the four states of the counter and generates the required sequence of timing signals.

The timing signals, when enabled by the clock pulses, will provide multiple-phase clock pulses. For example, if T_0 is ANDed with CP, the output of the AND gate will generate clock pulses at one-fourth the frequency of the master-clock pulses. Multiple-phase clock pulses can be used for controlling different registers with different time scales.

To generate 2^n timing signals, we need either a shift register with 2^n flip-flops or an n-bit counter together with an n-to-2^n-line decoder. For example, 16 timing signals can be generated with a 16-bit shift register connected as a ring counter or with a 4-bit counter and a 4-to-16-line decoder. In the first case, we need 16 flip-flops. In the second case, we need four flip-flops and 16 4-input AND gates for the decoder. It is also possible to generate the timing signals with a combination of a shift register and a decoder. In this way, the number of flip-flops is less than a ring counter, and the decoder requires only 2-input gates. This combination is sometimes called a *Johnson counter*.

Johnson Counter

A k-bit ring counter circulates a single bit among the flip-flops to provide k distinguishable states. The number of states can be doubled if the shift register is connected as a *switch-tail* ring counter. A switch-tail ring counter is a circular shift register with the complement output of the last flip-flop connected to the input of the first flip-flop. Figure 7-23(a) shows such a shift register. The circular connection is made from the complement output of the rightmost flip-flop to the input of the leftmost flip-flop. The register shifts its contents once to the right with every clock pulse, and at the same time, the complement value of the E flip-flop is transferred into the A flip-flop. Starting from a cleared state, the switch-tail ring counter goes through a sequence of eight states, as listed in Fig. 7-23(b). In general, a k-bit switch-tail ring counter will go through a sequence of $2k$ states. Starting from all 0's, each shift operation inserts 1's from the left until the register is filled with all 1's. In the following sequences, 0's are inserted from the left until the register is again filled with all 0's.

A Johnson counter is a k-bit switch-tail ring counter with $2k$ decoding gates to provide outputs for $2k$ timing signals. The decoding gates are not shown in Fig. 7-23, but are specified in the last column of the table. The eight AND gates listed in the table, when connected to the circuit, will complete the construction of the Johnson counter. Since each gate is enabled during one particular state sequence, the outputs of the gates generate eight timing sequences in succession.

The decoding of a k-bit switch-tail ring counter to obtain $2k$ timing sequences follows a regular pattern. The all-0's state is decoded by taking the complement of the two extreme flip-flop outputs. The all-1's state is decoded by taking the normal outputs of the two extreme flip-flops. All other states are decoded from an adjacent 1, 0 or 0, 1 pattern in the sequence. For example, sequence 7 has an adjacent 0, 1 pattern in flip-flops B and C. The decoded output is then obtained by taking the complement of B and the normal output of C, or $B'C$.

One disadvantage of the circuit in Fig. 7-23(a) is that if it finds itself in an unused state, it will persist in moving from one invalid state to another and never find its way

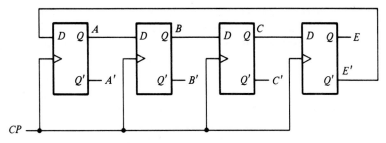

(a) Four-stage switch-tail ring counter

Sequence number	Flip-flop outputs				AND gate required for output
	A	*B*	*C*	*E*	
1	0	0	0	0	$A'E'$
2	1	0	0	0	AB'
3	1	1	0	0	BC'
4	1	1	1	0	CE'
5	1	1	1	1	AE
6	0	1	1	1	$A'B$
7	0	0	1	1	$B'C$
8	0	0	0	1	$C'E$

(b) Count sequence and required decoding.

FIGURE 7-23
Construction of a Johnson counter

to a valid state. This difficulty can be corrected by modifying the circuit to avoid this undesirable condition. One correcting procedure is to disconnect the output from flip-flop *B* that goes to the *D* input of flip-flop *C*, and instead enable the input of flip-flop *C* by the function:

$$DC = (A + C)B$$

where *DC* is the flip-flop input function for the *D* input of flip-flop *C*.

Johnson counters can be constructed for any number of timing sequences. The number of flip-flops needed is one-half the number of timing signals. The number of decoding gates is equal to the number of timing signals and only 2-input gates are employed.

7-7 RANDOM-ACCESS MEMORY (RAM)

A memory unit is a collection of storage cells together with associated circuits needed to transfer information in and out of the device. Memory cells can be accessed for information transfer to or from any desired random location and hence the name *random-access memory*, abbreviated RAM.

A memory unit stores binary information in groups of bits called *words*. A word in memory is an entity of bits that move in and out of storage as a unit. A memory word is a group of 1's and 0's and may represent a number, an instruction, one or more alphanumeric characters, or any other binary-coded information. A group of eight bits is called a *byte*. Most computer memories use words that are multiples of 8 bits in length. Thus, a 16-bit word contains two bytes, and a 32-bit word is made up of four bytes. The capacity of a memory unit is usually stated as the total number of bytes that it can store.

The communication between a memory and its environment is achieved through data input and output lines, address selection lines, and control lines that specify the direction of transfer. A block diagram of the memory unit is shown in Fig. 7-24. The n data input lines provide the information to be stored in memory and the n data output lines supply the information coming out of memory. The k address lines specify the particular word chosen among the many available. The two control inputs specify the direction of transfer desired: The write input causes binary data to be transferred into the memory, and the read input causes binary data to be transferred out of memory.

The memory unit is specified by the number of words it contains and the number of bits in each word. The address lines select one particular word. Each word in memory is assigned an identification number, called an address, starting from 0 and continuing with 1, 2, 3, up to $2^k - 1$, where k is the number of address lines. The selection of a specific word inside the memory is done by applying the k-bit binary address to the address lines. A decoder inside the memory accepts this address and opens the paths needed to select the word specified. Computer memories may range from 1024 words, requiring an address of 10 bits, to 2^{32} words, requiring 32 address bits. It is customary to refer to the number of words (or bytes) in a memory with one of the letters K (kilo), M (mega), or G (giga). K is equal to 2^{10}, M is equal to 2^{20}, and G is equal to 2^{30}. Thus, $64K = 2^{16}$, $2M = 2^{21}$, and $4G = 2^{32}$.

Consider, for example, the memory unit with a capacity of 1K words of 16 bits each. Since $1K = 1024 = 2^{10}$ and 16 bits constitute two bytes, we can say that the memory can accommodate $2048 = 2K$ bytes. Figure 7-25 shows the possible content of the first three and the last three words of this memory. Each word contains 16 bits,

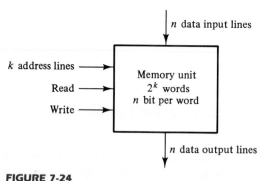

FIGURE 7-24
Block diagram of a memory unit

Memory address

Binary	decimal	Memory content
0000000000	0	1011010101011101
0000000001	1	1010101110001001
0000000010	2	0000110101000110
⋮	⋮	⋮
1111111101	1021	1001110100010100
1111111110	1022	0000110100011110
1111111111	1023	1101111000100101

FIGURE 7-25

Content of a 1024 × 16 memory

which can be divided into two bytes. The words are recognized by their decimal address from 0 to 1023. The equivalent binary address consists of 10 bits. The first address is specified with ten 0's, and the last address is specified with ten 1's. This is because 1023 in binary is equal to 1111111111. A word in memory is selected by its binary address. When a word is read or written, the memory operates on all 16 bits as a single unit.

The 1K × 16 memory of Fig. 7-25 has 10 bits in the address and 16 bits in each word. As another example, a 64K × 10 memory will have 16 bits in the address (since $64K = 2^{16}$) and each word will consist of 10 bits. The number of address bits needed in a memory is dependent on the total number of words that can be stored in the memory and is independent of the number of bits in each word. The number of bits in the address is determined from the relationship $2^k = m$, where m is the total number of words, and k is the number of address bits.

Write and Read Operations

The two operations that a random-access memory can perform are the write and read operations. The write signal specifies a transfer-in operation and the read signal specifies a transfer-out operation. On accepting one of these control signals, the internal circuits inside the memory provide the desired function. The steps that must be taken for the purpose of transferring a new word to be stored into memory are as follows:

1. Transfer the binary address of the desired word to the address lines.
2. Transfer the data bits that must be stored in memory to the data input lines.
3. Activate the *write* input.

TABLE 7-7
Control Inputs to Memory Chip

Memory Enable	Read/Write	Memory Operation
0	X	None
1	0	Write to selected word
1	1	Read from selected word

The memory unit will then take the bits from the input data lines and store them in the word specified by the address lines.

The steps that must be taken for the purpose of transferring a stored word out of memory are as follows:

1. Transfer the binary address of the desired word to the address lines.

2. Activate the *read* input.

The memory unit will then take the bits from the word that has been selected by the address and apply them to the output data lines. The content of the selected word does not change after reading.

Commercial memory components available in integrated-circuit chips sometimes provide the two control inputs for reading and writing in a somewhat different configuration. Instead of having separate read and write inputs to control the two operations, some integrated circuits provide two other control inputs: one input selects the unit and the other determines the operation. The memory operations that result from these control inputs are specified in Table 7-7.

The memory enable (sometimes called the chip select) is used to enable the particular memory chip in a multichip implementation of a large memory. When the memory enable is inactive, the memory chip is not selected and no operation is performed. When the memory enable input is active, the read/write input determines the operation to be performed.

Types of Memories

The mode of access of a memory system is determined by the type of components used. In a random-access memory, the word locations may be thought of as being separated in space, with each word occupying one particular location. In a sequential-access memory, the information stored in some medium is not immediately accessible, but is available only at certain intervals of time. A magnetic-tape unit is of this type. Each memory location passes the read and write heads in turn, but information is read out only when the requested word has been reached. The *access time* of a memory is the time required to select a word and either read or write it. In a random-access memory, the access time is always the same regardless of the particular location of the word. In a sequential-access memory, the time it takes to access a word depends on the position of the word with respect to the reading-head position and therefore, the access time is variable.

Integrated-circuit RAM units are available in two possible operating modes, *static* and *dynamic*. The static RAM consists essentially of internal flip-flops that store the binary information. The stored information remains valid as long as power is applied to the unit. The dynamic RAM stores the binary information in the form of electric charges that are applied to capacitors. The capacitors are provided inside the chip by MOS transistors. The stored charge on the capacitors tends to discharge with time and the capacitors must be periodically recharged by *refreshing* the dynamic memory. Refreshing is done by cycling through the words every few milliseconds to restore the decaying charge. Dynamic RAM offers reduced power consumption and larger storage capacity in a single memory chip, but static RAM is easier to use and has shorter read and write cycles.

Memory units that lose the stored information when power is turned off are said to be *volatile*. Integrated-circuit RAMs, both static and dynamic, are of this category since the binary cells need external power to maintain the stored information. In contrast, a nonvolatile memory, such as magnetic disk, retains its stored information after removal of power. This is because the data stored on magnetic components is manifested by the direction of magnetization, which is retained after power is turned off. Another nonvolatile memory is the read-only memory (ROM) discussed in Section 5-7. A nonvolatile property is desirable in digital computers to store programs that are needed while the computer is in operation. Programs and data that cannot be altered are stored in ROM. Other large programs are maintained on magnetic disks. When power is turned on, the computer can use the programs from ROM. The other programs residing on disks can be transferred into the computer RAM as needed. Before turning the power off, the user transfers the binary information from the computer RAM into a disk if this information must be retained.

7-8 MEMORY DECODING

In addition to the storage components in a memory unit, there is a need for decoding circuits to select the memory word specified by the input address. In this section, we present the internal construction of a random-access memory and demonstrate the operation of the decoder. To be able to include the entire memory in one diagram, the memory unit presented here has a small capacity of 12 bits arranged in 4 words of 3 bits each. In addition to internal decoders, a memory unit may also need external decoders. This happens when integrated-circuit RAM chips are connected in a multichip memory configuration. The use of an external decoder to provide a large capacity memory will be demonstrated by means of an example.

Internal Construction

The internal construction of a random-access memory of m words with n bits per word consists of $m \times n$ binary storage cells and associated decoding circuits for selecting individual words. The binary storage cell is the basic building block of a memory unit.

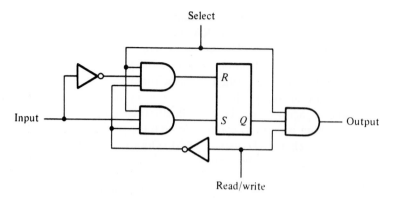

(a) Logic diagram

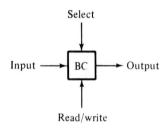

(b) Block diagram

FIGURE 7-26

Memory cell

The equivalent logic of a binary cell that stores one bit of information is shown in Fig. 7-26. Although the cell is shown to include gates and a flip-flop, internally, it is constructed with two transistors having multiple inputs. A binary storage cell must be very small in order to be able to pack as many cells as possible in the area available in the integrated-circuit chip. The binary cell stores one bit in its internal flip-flop. It has three inputs and one output. The select input enables the cell for reading or writing and the read/write input determines the cell operation when it is selected. A 1 in the read/ write input provides the read operation by forming a path from the flip-flop to the output terminal. A 0 in the read/write input provides the write operation by forming a path from the input terminal to the flip-flop. Note that the flip-flop operates without a clock and is similar to an *SR* latch (see Fig. 6-2).

The logical construction of a small RAM is shown in Fig. 7-27. It consists of 4 words of 3 bits each and has a total of 12 binary cells. Each block labeled *BC* represents the binary cell with its three inputs and one output, as specified in Fig. 7-26(b). A memory with four words needs two address lines. The two address inputs go through a 2 × 4 decoder to select one of the four words. The decoder is enabled with the memory-enable input. When the memory enable is 0, all outputs of the decoder are 0

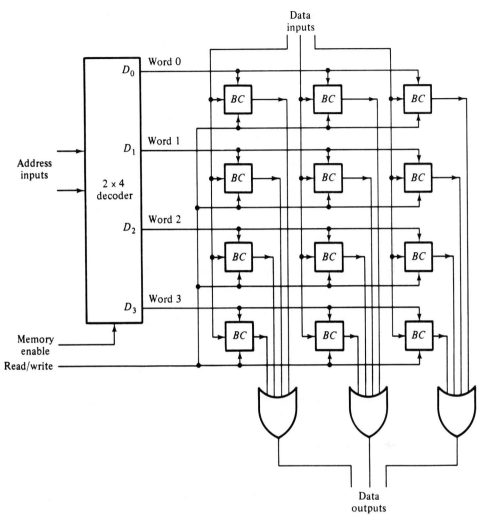

FIGURE 7-27

Logical construction of a 4 × 3 RAM

and none of the memory words are selected. With the memory enable at 1, one of the four words is selected, dictated by the value in the two address lines. Once a word has been selected, the read/write input determines the operation. During the read operation, the four bits of the selected word go through OR gates to the output terminals. During the write operation, the data available in the input lines are transferred into the four binary cells of the selected word. The binary cells that are not selected are disabled and their previous binary values remain unchanged. When the memory-enable input that goes into the decoder is equal to 0, none of the words are selected and the contents of all cells remain unchanged regardless of the value of the read/write input.

Commercial random-access memories may have a capacity of thousands of words and each word may range from 1 to 64 bits. The logical construction of a large capacity memory would be a direct extension of the configuration shown here. A memory with 2^k words of n bits per word requires k address lines that go into a $k \times 2^k$ decoder. Each one of the decoder outputs selects one word of n bits for reading or writing.

Array of RAM Chips

Integrated-circuit RAM chips are available in a variety of sizes. If the memory unit needed for an application is larger than the capacity of one chip, it is necessary to combine a number of chips in an array to form the required memory size. The capacity of the memory depends on two parameters: the number of words and the number of bits per word. An increase in the number of words requires that we increase the address length. Every bit added to the length of the address doubles the number of words in memory. The increase in the number of bits per word requires that we increase the length of the data input and output lines, but the address length remains the same.

To demonstrate with an example, let us first introduce a typical RAM chip, as shown in Fig. 7-28. The capacity of the RAM is 1024 words of 8 bits each. It requires a 10-bit address and 8 input and output lines. These are shown in the block diagram by a single line and a number indicating the total number of inputs or outputs. The chip-select (*CS*) input selects the particular RAM chip and the read/write (RW) input specifies the read or write operation when the chip is selected.

Suppose that we want to increase the number of words in the memory by using two or more RAM chips. Since every bit added to the address doubles the binary number that can be formed, it is natural to increase the number of words in factors of 2. For example, two RAM chips will double the number of words and add one bit to the composite address. Four RAM chips multiply the number of words by 4 and add two bits to the composite address.

Consider the possiblity of constructing a 4K $\times$ 8 RAM with four 1K $\times$ 8 RAM chips. This is shown in Fig. 7-29. The 8 input data lines go to all the chips. The outputs must be ORed together to form the common 8 output data lines. (The OR gates are not shown in the diagram.) The 4K word memory requires a 12-bit address. The 10 least significant bits of the address are applied to the address inputs of all four chips.

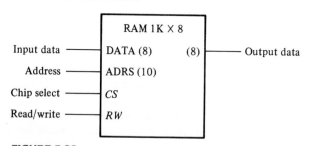

FIGURE 7-28
Block diagram of a 1K $\times$ 8 RAM chip.

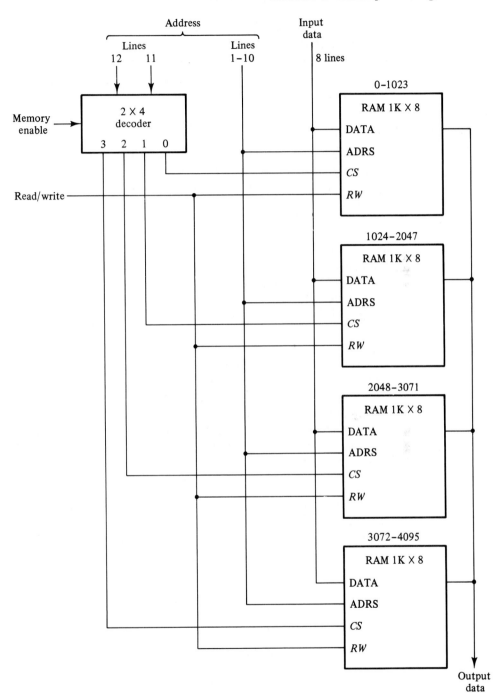

FIGURE 7-29

Block diagram of a 4K × 8 RAM.

The other two most significant bits are applied to a 2×4 decoder. The four outputs of the decoder are applied to the CS inputs of each chip. The memory is disabled when the memory-enable input of the decoder is equal to 0. This causes all four outputs of the decoder to be in the 0 state and none of the chips are selected. When the decoder is enabled, address bits 12 and 11 determine the particular chip that is selected. If bits 12 and 11 are equal to 00, the first RAM chip is selected. The remaining ten address bits select a word within the chip in the range from 0 to 1023. The next 1024 words are selected from the second RAM chip with a 12-bit address that starts with 01 and follows by the ten bits from the common address lines. The address range for each chip is listed in decimal over its block diagram in Fig. 7-29.

It is also possible to combine two chips to form a composite memory containing the same number of words but with twice as many bits in each word. Figure 7-30 shows the interconnection of two $1K \times 8$ chips to form a $1K \times 16$ memory. The 16 input and output data lines are split between the two chips. Both receive the same 10-bit address and the common CS and RW control inputs.

The two techniques just described may be combined to assemble an array of identical chips into a large-capacity memory. The composite memory will have a number of bits per word that is a multiple of that for one chip. The total number of words will increase in factors of 2 times the word capacity of one chip. An external decoder is needed to

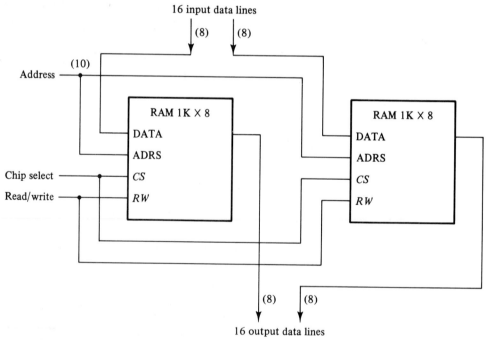

FIGURE 7-30
Block diagram of a 1K × 16 RAM.

select the individual chips from the additional address bits of the composite memory.

To reduce the number of pins in the package, many RAM integrated circuits provide common terminals for the input data and output data. The common terminals are said to be *bidirectional,* which means that for the read operation, they act as outputs, and for the write operation, they act as inputs.

7-9 ERROR-CORRECTING CODE

The complexity level of a memory array may cause occasional errors in storing and retrieving the binary information. The reliability of a memory unit may be improved by employing error-detecting and correcting codes. The most common error-detection scheme is the parity bit. (See Section 4-9.) A parity bit is generated and stored along with the data word in memory. The parity of the word is checked after reading it from memory. The data word is accepted if the parity sense is correct. If the parity checked results in an inversion, an error is detected, but it cannot be corrected.

An error-correcting code generates multiple check bits that are stored with the data word in memory. Each check bit is a parity over a group of bits in the data word. When the word is read from memory, the associated parity bits are also read from memory and compared with a new set of check bits generated from the read data. If the check bits compare, it signifies that no errror has occurred. If the check bits do not compare with the stored parity, they generate a unique pattern, called a *syndrome,* that can be used to identify the bit in error. A single error occurs when a bit changes in value from 1 to 0 or from 0 to 1 during the write or read operation. If the specific bit in error is identified, then the error can be corrected by complementing the erroneous bit.

Hamming Code

One of the most common error-correcting codes used in random-access memories was devised by R. W. Hamming. In the Hamming code, k parity bits are added to an n-bit data word, forming a new word of $n + k$ bits. The bit positions are numbered in sequence from 1 to $n + k$. Those positions numbered as a power of 2 are reserved for the parity bits. The remaining bits are the data bits. The code can be used with words of any length. Before giving the general characteristics of the code, we will illustrate its operation with a data word of eight bits.

Consider, for example, the 8-bit data word 11000100. We include four parity bits with the 8-bit word and arrange the 12 bits as follows:

Bit position:	1	2	3	4	5	6	7	8	9	10	11	12
	P_1	P_2	1	P_4	1	0	0	P_8	0	1	0	0

The four parity bits, P_1, P_2, P_4, and P_8, are in positions 1, 2, 4, and 8, respectively. The eight bits of the data word are in the remaining positions. Each parity bit is calculated as follows:

$$P_1 = \text{XOR of bits } (3, 5, 7, 9, 11) = 1 \oplus 1 \oplus 0 \oplus 0 \oplus 0 = 0$$

$$P_2 = \text{XOR of bits } (3, 6, 7, 10, 11) = 1 \oplus 0 \oplus 0 \oplus 1 \oplus 0 = 0$$

$$P_4 = \text{XOR of bits } (5, 6, 7, 12) = 1 \oplus 0 \oplus 0 \oplus 0 = 1$$

$$P_8 = \text{XOR of bits } (9, 10, 11, 12) = 0 \oplus 1 \oplus 0 \oplus 0 = 1$$

Remember that the exclusive-OR operation performs the odd function. It is equal to 1 for an odd number of 1's in the variables and to 0 for an even number of 1's. Thus, each parity bit is set so that the total number of 1's in the checked positions, including the parity bit, is always even.

The 8-bit data word is stored in memory together with the 4 parity bits as a 12-bit composite word. Substituting the four P bits in their proper positions, we obtain the 12-bit composite word stored in memory:

Bit position: 1 2 3 4 5 6 7 8 9 10 11 12
                0 0 1 1 1 0 0 1 0 1 0 0

When the 12 bits are read from memory, they are checked again for possible errors. The parity is checked over the same combination of bits including the parity bit. The four check bits are evaluated as follows:

$$C_1 = \text{XOR of bits } (1, 3, 5, 7, 9, 11)$$

$$C_2 = \text{XOR of bits } (2, 3, 6, 7, 10, 11)$$

$$C_4 = \text{XOR of bits } (4, 5, 6, 7, 12)$$

$$C_8 = \text{XOR of bits } (8, 9, 10, 11, 12)$$

A 0 check bit designates an even parity over the checked bits and a 1 designates an odd parity. Since the bits were stored with even parity, the result, $C = C_8C_4C_2C_1 = 0000$, indicates that no error has occurred. However, if $C \neq 0$, then the 4-bit binary number formed by the check bits gives the position of the erroneous bit. For example, consider the following three cases:

Bit position: 1 2 3 4 5 6 7 8 9 10 11 12
                0 0 1 1 1 0 0 1 0 1 0 0   No error
                1 0 1 1 1 0 0 1 0 1 0 0   Error in bit 1
                0 0 1 1 0 0 0 1 0 1 0 0   Error in bit 5

In the first case, there is no error in the 12-bit word. In the second case, there is an error in bit position number 1 because it changed from 0 to 1. The third case shows an error in bit position 5 with a change from 1 to 0. Evaluating the XOR of the corresponding bits, we determine the four check bits to be as follows:

	C_8	C_4	C_2	C_1
For no error:	0	0	0	0
With error in bit 1:	0	0	0	1
With error in bit 5:	0	1	0	1

Thus, for no error, we have $C = 0000$; with an error in bit 1, we obtain $C = 0001$; and with an error in bit 5, we get $C = 0101$. The binary number of C, when it is not equal to 0000, gives the position of the bit in error. The error can be corrected by complementing the corresponding bit. Note that an error can occur in the data word or in one of the parity bits.

The Hamming code can be used for data words of any length. In general, the Hamming code consists of k check bits and n data bits for a total of $n + k$ bits. The syndrome value C consists of k bits and has a range of 2^k values between 0 and $2^k - 1$. One of these values, usually zero, is used to indicate that no error was detected, leaving $2^k - 1$ values to indicate which of the $n + k$ bits was in error. Each of these $2^k - 1$ values can be used to uniquely describe a bit in error. Therefore, the range of k must be equal to or greater than $n + k$, giving the relationship

$$2^k - 1 \geq n + k$$

Solving for n in terms of k, we obtain

$$2^k - 1 - k \geq n$$

This relationship gives a formula for evaluating the number of data bits that can be used in conjunction with k check bits. For example, when $k = 3$, the number of data bits that can be used is $n \leq (2^3 - 1 - 3) = 4$. For $k = 4$, we have $2^4 - 1 - 4 = 11$, giving $n \leq 11$. The data word may be less than 11 bits, but must have at least 5 bits, otherwise, only 3 check bits will be needed. This justifies the use of 4 check bits for the 8 data bits in the previous example. Ranges of n for various values of k are listed in Table 7-8.

The grouping of bits for parity generation and checking can be determined from a list of the binary numbers from 0 through $2^k - 1$. (Table 1-1 gives such a list.) The least significant bit is a 1 in the binary numbers 1, 3, 5, 7, and so on. The second significant bit is a 1 in the binary numbers 2, 3, 6, 7, and so on. Comparing these numbers with the bit positions used in generating and checking parity bits in the Hamming code, we note the relationship between the bit groupings in the code and the position of the 1 bits in the binary count sequence. Note that each group of bits starts with a number that is a power of 2 such as 1, 2, 4, 8, 16, etc. These numbers are also the position numbers for the parity bits.

TABLE 7-8
Range of Data Bits for k Check Bits

Number of Check Bits, k	Range of Data Bits, n
3	2–4
4	5–11
5	12–26
6	27–57
7	58–120

Single-Error Correction, Double-Error Detection

The Hamming code can detect and correct only a single error. Multiple errors are not detected. By adding another parity bit to the coded word, the Hamming code can be used to correct a single error and detect double errors. If we include this additional parity bit, then the previous 12-bit coded word becomes $001110010100P_{13}$, where P_{13} is evaluated from the exclusive-OR of the other 12 bits. This produces the 13-bit word 0011100101001 (even parity). When the 13-bit word is read from memory, the check bits are evaluated and also the parity P over the entire 13 bits. If $P = 0$, the parity is correct (even parity), but if $P = 1$, then the parity over the 13 bits is incorrect (odd parity). The following four cases can occur:

If $C = 0$ and $P = 0$	No error occurred
If $C \neq 0$ and $P = 1$	A single error occurred, which can be corrected
If $C \neq 0$ and $P = 0$	A double error occurred, which is detected but cannot be corrected
If $C = 0$ and $P = 1$	An error occurred in the P_{13} bit

Note that this scheme cannot detect more than two errors.

Integrated circuits that use a modified Hamming code to generate and check parity bits for a single-error correction, double-error detection scheme are available commercially. One that uses an 8-bit data word and a 5-bit check word is IC type 74637. Other integrated circuits are available for data words of 16 and 32 bits. These circuits can be used in conjunction with a memory unit to correct a single error or detect double errors during the write and read operations.

REFERENCES

1. MANO. M. M., *Computer System Architecture,* 2nd Ed. Englewood Cliffs, NJ: Prentice-Hall, 1982.

2. PEATMAN, J. B., *Digital Hardware Design.* New York: McGraw-Hill, 1980.

3. BLAKESLEE, T. R., *Digital Design With Standard MSI And LSI,* 2nd Ed. New York: John Wiley, 1979.

4. FLETCHER, W. I., *An Engineering Approach to Digital Design.* Englewood Cliffs, NJ: Prentice-Hall, 1979.

5. SANDIGE, R. S., *Digital Concepts Using Standard Integrated Circuits.* New York: McGraw-Hill, 1978.

6. SHIVA, S. G., *Introduction to Logic Design.* Glenview, IL: Scott, Foresman, 1988.

7. ROTH, C. H., *Fundamentals of Logic Design,* 3rd Ed. New York: West, 1985.

8. BOOTH, T. L., *Introduction to Computer Engineering,* 3rd Ed. New York: John Wiley, 1984

9. *The TTL Logic Data Book.* Dallas: Texas Instruments, 1988.

10. *LSI Logic Data Book.* Dallas: Texas Instruments, 1986.

11. *Memory Components Handbook.* Santa Clara, CA: Intel, 1986.

12. HAMMING, R. W., "Error Detecting and Error Correcting Codes." *Bell Syst. Tech. J.*, **29** (1950) 147–160.

13. LIN, S., and D. J. COSTELLO, JR., *Error Control Coding*. Englewood Cliffs, NJ: Prentice-Hall, 1983.

PROBLEMS

7-1 Include a 2-input NAND gate with the register of Fig. 7-1 and connect the gate output to the *CP* inputs of all the flip-flops. One input of the NAND gate receives the clock pulses from the clock-pulse generator. The other input of the NAND gate provides a parallel-load control. Explain the operation of the modified register.

7-2 Change the asynchronous-clear circuit to a synchronous-clear circuit in the register of Fig. 7-2. The modified register will have a parallel-load capability and a synchronous-clear capability, but no asynchronous-clear circuit. The register is cleared synchronously when the clock pulse in the *CP* input goes through a negative transition provided $R = 1$ and $S = 0$ in all the flip-flops.

7-3 Repeat Problem 7-2 for the register of Fig. 7-3. Here the circuit will be cleared synchronously when the *CP* input goes through a negative transition while the *D* inputs of all flip-flops are equal to 0.

7-4 Design a sequential circuit whose state diagram is given in Fig. 6-31 using a 3-bit register and a 16×4 ROM.

7-5 The content of a 4-bit register is initially 1101. The register is shifted six times to the right with the serial input being 101101. What is the content of the register after each shift?

7-6 What is the difference between a serial and parallel transfer? Explain how to convert serial data to parallel and parallel data to serial. What type of register is needed?

7-7 The 4-bit bidirectional shift register with parallel load shown in Fig. 7-9 is enclosed within one IC package.
(a) Draw a block diagram of the IC showing all inputs and outputs. Include two pins for the power supply.
(b) Draw a block diagram using two ICs to produce an 8-bit bidirectional shift register with parallel load.

7-8 Design a shift register with parallel load that operates according to the following function table:

Shift	Load	Register Operation
0	0	No change
0	1	Load parallel data
1	X	Shift right

7-9 Draw the logic diagram of a 4-bit register with four *D* flip-flops and four 4×1 multiplexers with mode-selection inputs s_1 and s_0. The register operates according to the following function table:

S_1	S_0	Register Operation
0	0	No change
0	1	Complement the four outputs
1	0	Clear register to 0 (synchronous with the clock)
1	1	Load parallel data

7-10 The serial adder of Fig. 7-10 uses two 4-bit registers. Register A holds the binary number 0101 and register B holds 0111. The carry flip-flop is initially reset to 0. List the binary values in register A and the carry flip-flop after each shift.

7-11 What changes are needed in Fig. 7-11 to convert it to a serial subtractor that subtracts the content of register B from the content of register A?

7-12 It was stated in Section 1-5 that the 2's complement of a binary number can be formed by leaving all least significant 0's and the first 1 unchanged and complementing all other higher significant bits. Design a serial 2's complementer using this procedure. The circuit needs a shift register to store the binary number and an RS flip-flop to be set when the first least significant 1 occurs. An exclusive-OR gate can be used to transfer the unchanged bits ($x \oplus 0 = x$) or complement the bits ($x \oplus 1 = x'$).

7-13 Draw the logic diagram of a 4-bit binary ripple counter using flip-flops that trigger on the positive-edge transition.

7-14 Draw the logic diagram of a 4-bit binary ripple down-counter using the following:
(a) Flip-flops that trigger on the positive-edge transition of the clock.
(b) Flip-flops that trigger on the negative-edge transition of the clock.

7-15 Construct a BCD ripple counter using a 4-bit binary ripple counter that can be cleared asynchronously (similar to the clear input in Fig. 7-2) and an external NAND gate.

7-16 A flip-flop has a 10-nanosecond delay from the time its CP input goes from 1 to 0 to the time the output is complemented. What is the maximum delay in a 10-bit binary ripple counter that uses these flip-flops? What is the maximum frequency the counter can operate reliably?

7-17 How many flip-flops will be complemented in a 10-bit binary ripple counter to reach the next count after the following count:
(a) 1001100111;
(b) 0011111111.

7-18 Determine the next state for each of the six unused states in the BCD ripple counter shown in Fig. 7-14. Determine whether the counter is self-correcting.

7-19 Design a 4-bit binary ripple counter with D flip-flops.

7-20 Design a 4-bit binary synchronous counter with D flip-flops.

7-21 Modify the counter of Fig. 7-18 so that when both the up and down control inputs are equal to 1, the counter does not change state, but remains in the same count.

7-22 Verify the flip-flop input functions of the synchronous BCD counter specified in Table 7-5. Draw the logic diagram of the BCD counter and include a count-enable control input.

7-23 Design a synchronous BCD counter with JK flip-flops.

7-24 Show the connections between four IC binary counters with parallel load (Fig. 7-19) to produce a 16-bit binary counter with parallel load. Use a block diagram for each IC.

7-25 Construct a BCD counter using the circuit specified in Fig. 7-19 and an AND gate.

7-26 Construct a mod-12 counter using the circuit of Fig. 7-19. Give four alternatives.

7-27 Using two circuits of the type shown in Fig. 7-19, construct a binary counter that counts from 0 through binary 64.

7-28 Using a start signal as in Fig. 7-21, construct a word-time control that stays on for a period of 16 clock pulses.

7-29 Add four 2-input AND gates to the circuit of Fig. 7-22(b). One input in each gate is connected to one output of the decoder. The other input in each gate is connected to the clock. Label the outputs of the AND gate as P_0, P_1, P_2, and P_3. Show the timing diagram of the four P outputs.

7-30 Show the circuit and the timing diagram for generating six repeated timing signals, T_0 through T_5.

7-31 Complete the design of the Johnson counter of Fig. 7-23 showing the outputs of the eight timing signals using eight AND gates.

7-32 Construct a Johnson counter for ten timing signals.

7-33 The following memory units are specified by the number of words times the number of bits per word. How many address lines and input–output data lines are needed in each case?
(a) 2K × 16;
(b) 64K × 8;
(c) 16M × 32;
(d) 96K × 12.

7-34 Word number 535 in the memory shown in Fig. 7-25 contains the binary equivalent of 2209. List the 10-bit address and the 16-bit memory content of the word.

7-35 (a) How many 128 × 8 RAM chips are needed to provide a memory capacity of 2048 bytes?
(b) How many lines of the address must be used to access 2048 bytes? How many of these lines are connected to the address inputs of all chips?
(c) How many lines must be decoded for the chip-select inputs? Specify the size of the decoder.

7-36 A computer uses RAM chips of 1024 × 1 capacity.
(a) How many chips are needed and how should their address lines be connected to provide a memory capacity of 1024 bytes?
(b) How many chips are needed to provide a memory capacity of 16K bytes? Explain in words how the chips are to be connected.

7-37 An integrated-circuit RAM chip has a capacity of 1024 words of 8 bits each (1K × 8).
(a) How many address and data lines are there in the chip?
(b) How many chips are needed to construct a 16K × 16 RAM?
(c) How many address and data lines are there in the 16K × 16 RAM?
(d) What size decoder is needed to construct the 16K × 16 memory from the 1K × 8 chips? What are the inputs to the decoder and where are its outputs connected?

7-38 Given the 8-bit data word 01011011, generate the 13-bit composite word for the Hamming code that corrects single errors and detects double errors.

7-39 Given the 11-bit data word 11001001010, generate the 15-bit Hamming-code word.

7-40 A 12-bit Hamming-code word containing 8 bits of data and 4 parity bits is read from memory. What was the original 8-bit data word that was written into memory if the 12-bit word read out is as follows:
(a) 000011101010
(b) 101110000110
(c) 101111110100

7-41 How many parity check bits must be included with the data word to achieve single error-correction and double-error detection when the data word contains: (a) 16 bits; (b) 32 bits; (c) 48 bits.

7-42 It is necessary to formulate the Hamming code for four data bits, D_3, D_5, D_6, and D_7, together with three parity bits, P_1, P_2, and P_4.
(a) Evaluate the 7-bit composite code word for the data word 0010.
(b) Evaluate three check bits, C_4, C_2, and C_1, assuming no error.
(c) Assume an error in bit D_5 during writing into memory. Show how the error in the bit is detected and corrected.
(d) Add parity bit P_8 to include a double-error detection in the code. Assume that errors occurred in bits P_2 and D_5. Show how the double error is detected.

Algorithmic State Machines (ASM)

8-1 INTRODUCTION

The binary information stored in a digital system can be classified as either data or control information. Data are discrete elements of information that are manipulated to perform arithmetic, logic, shift, and other similar data-processing tasks. These operations are implemented with digital components such as adders, decoders, multiplexers, counters, and shift registers. Control information provides command signals that supervise the various operations in the data section in order to accomplish the desired data-processing tasks. The logic design of a digital system can be divided into two distinct parts. One part is concerned with the design of the digital circuits that perform the data-processing operations. The other part is concerned with the design of the control circuit that supervises the operations and their sequence.

The relationship between the control logic and the data processor in a digital system is shown in Fig. 8-1. The data processor subsystem manipulates data in registers according to the system's requirements. The control logic initiates properly sequenced commands to the data processor. The control logic uses status conditions from the data processor to serve as decision variables for determining the sequence of control signals.

The control logic that generates the signals for sequencing the operations in the data processor is a sequential circuit whose internal states dictate the control commands for the system. At any given time, the state of the sequential control initiates a prescribed set of commands. Depending on status conditions and other external inputs, the sequential control goes to the next state to initiate other operations. The digital circuits

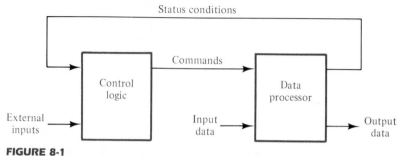

FIGURE 8-1

Control and data-processor interaction

that act as the control logic provide a time sequence of signals for initiating the operations in the data processor and also determine the next state of the control subsystem itself.

The control sequence and data-processing tasks of a digital system are specified by means of a hardware algorithm. An algorithm consists of a finite number of procedural steps that specify how to obtain a solution to a problem. A hardware algorithm is a procedure for implementing the problem with a given piece of equipment. The most challenging and creative part of digital design is the formulation of hardware algorithms for achieving required objectives.

A flow chart is a convenient way to specify the sequence of procedural steps and decision paths for an algorithm. A flow chart for a hardware algorithm translates the word statement to an information diagram that enumerates the sequence of operations together with the conditions necessary for their execution. A special flow chart that has been developed specifically to define digital hardware algorithms is called an *algorithmic state machine* (ASM) chart. A *state machine* is another term for a sequential circuit, which is the basic structure of a digital system.

The ASM chart resembles a conventional flow chart, but is interpreted somewhat differently. A conventional flow chart describes the sequence of procedural steps and decision paths for an algorithm without concern for their time relationship. The ASM chart describes the sequence of events as well as the timing relationship between the states of a sequential controller and the events that occur while going from one state to the next. It is specifically adapted to specify accurately the control sequence and data-processing operations in a digital system, taking into consideration the constraints of digital hardware.

This chapter presents a method of digital logic design using the ASM chart. The various blocks that make up the chart are first defined. The timing relationship between the blocks is then explained by example. Various ways of implementing the control logic are discussed together with examples of ASM charts and the corresponding digital systems that they represent.

8-2   ASM CHART

The ASM chart is a special type of flow chart suitable for describing the sequential operations in a digital system. The chart is composed of three basic elements: the state box, the decision box, and the conditional box. A state in the control sequence is indicated by a state box, as shown in Fig. 8-2. The shape of the state box is a rectangle within which are written register operations or output signal names that the control generates while being in this state. The state is given a symbolic name, which is placed at the upper left corner of the box. The binary code assigned to the state is placed at the upper right corner. Figure 8-2(b) shows a specific example of a state box. The state has the symbolic name T_3, and the binary code assigned to it is 011. Inside the box is written the register operation $R \leftarrow 0$, which indicates that register R is to be cleared to 0 when the system is in state T_3. The START name inside the box may indicate, for example, an output signal that starts a certain operation.

The decision box describes the effect of an input on the control subsystem. It has a diamond-shaped box with two or more exit paths, as shown in Fig. 8-3. The input condition to be tested is written inside the box. One exit path is taken if the condition is true and another when the condition is false. When an input condition is assigned a binary value, the two paths are indicated by 1 and 0.

The state and decision boxes are familiar from use in conventional flow charts. The third element, the conditional box, is unique to the ASM chart. The oval shape of the conditional box is shown in Fig. 8-4. The rounded corners differentiate it from the state box. The input path to the conditional box must come from one of the exit paths of a decision box. The register operations or outputs listed inside the conditional box are generated during a given state provided that the input condition is satisfied. Figure 8-5 shows an example with a conditional box. The control generates a START output signal when in state T_1. While in state T_1, the control checks the status of input E. If $E = 1$, then R is cleared to 0; otherwise, R remains unchanged. In either case, the next state is T_2.

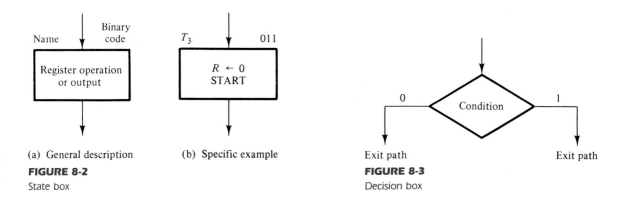

(a) General description

FIGURE 8-2

State box

(b) Specific example

Exit path

FIGURE 8-3

Decision box

Exit path

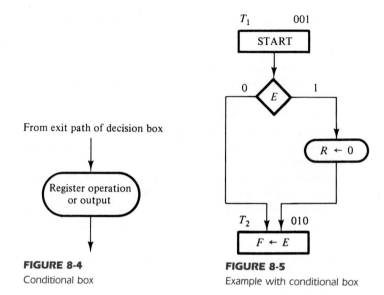

From exit path of decision box

FIGURE 8-4

Conditional box

FIGURE 8-5

Example with conditional box

ASM Block

An ASM block is a structure consisting of one state box and all the decision and conditional boxes connected to its exit path. An ASM block has one entrance and any number of exit paths represented by the structure of the decision boxes. An ASM chart consists of one or more interconnected blocks. An example of an ASM block is shown in Fig. 8-6. Associated with state T_1 are two decision boxes and one conditional box. The diagram distinguishes the block with dashed lines around the entire structure, but

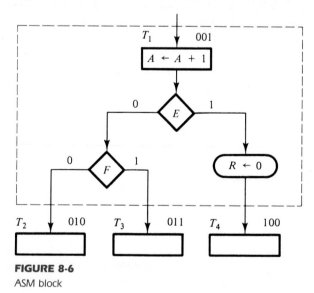

FIGURE 8-6

ASM block

this is not usually done, since the ASM chart uniquely defines each block from its structure. A state box without any decision or conditional boxes constitutes a simple block.

Each block in the ASM chart describes the state of the system during one clock-pulse interval. The operations within the state and conditional boxes in Fig. 8-6 are executed with a common clock pulse while the system is in state T_1. The same clock pulse also transfers the system controller to one of the next states, T_2, T_3, or T_4, as dictated by the binary values of E and F.

The ASM chart is very similar to a state diagram. Each state block is equivalent to a state in a sequential circuit. The decision box is equivalent to the binary information written along the directed lines that connect two states in a state diagram. As a consequence, it is sometimes convenient to convert the chart into a state diagram and then use sequential-circuit procedures to design the control logic. As an illustration, the ASM chart of Fig. 8-6 is drawn as a state diagram in Fig. 8-7. The three states are symbolized by circles, with their binary values written inside the circles. The directed lines indicate the conditions that determine the next state. The unconditional and conditional operations that must be performed are not indicated in the state diagram.

Register Operations

A digital system is quite often defined by the registers it contains and the operations that are performed on the data stored in them. A *register* in its broader sense includes storage registers, shift registers, counters, and single flip-flops. Examples of register operations are shift, increment, add, clear, and data transfer. It is sometimes convenient to adopt a suitable notation to describe the operations performed among the registers.

Table 8-1 gives examples of symbolic notation for some register operations. A register is designated by one or more capital letters such as A, B, or RA. The individual cells or flip-flops within an n-bit register are numbered in sequence from 1 to n or from 0 to $n - 1$. A single flip-flop is considered a 1-bit register. The transfer of data from one register to another is symbolized by a directed arrow that denotes a transfer of contents

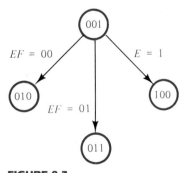

FIGURE 8-7
State-diagram equivalent to the ASM chart of Fig. 8-6

TABLE 8-1
Symbolic Notation for Register Operations

Symbolic Notation	Description
$A \leftarrow B$	Transfer contents of register B into register A
$R \leftarrow 0$	Clear register R
$F \leftarrow 1$	Set flip-flop F to 1
$A \leftarrow A + 1$	Increment register A by 1 (count-up)
$A \leftarrow A - 1$	Decrement register A by 1 (count-down)
$A \leftarrow A + B$	Add contents of register B to register A

from the source register to the destination register. The register-clear operation is symbolized by a transfer of 0 into the register. A single flip-flop can be set to 1 or cleared to 0. To increment a register by 1, it is necessary that the register be able to count up as in a binary counter. The decrement operation requires a count-down counter. The contents of two registers can be added by means of an adder circuit. Some operations, such as the shift operation, do not have a known symbol. In such a case, we will use the words "shift right R" to denote a shift right of register R.

8-3   TIMING CONSIDERATIONS

The timing for all registers and flip-flops in a digital system is controlled by a master-clock generator. The clock pulses are applied not only to the registers of the data-processor subsection, but also to all the flip-flops in the control logic. Inputs are also synchronized with the clock pulses because they are normally generated as outputs of another circuit that uses the same clock signals. If the input signal changes at an arbitrary time independent of the clock, we call it an asynchronous input. Asynchronous inputs may cause a variety of problems, as discussed in Chapter 9. To simplify the design, we will assume that all inputs are synchronized with the clock and change state in response to an edge transition of the clock pulse. Similarly, any output that is a function of the present state and a synchronous input will also be synchronous.

The major difference between a conventional flow chart and an ASM chart is in interpreting the time relationship among the various operations. For example, if Fig. 8-6 were a conventional flow chart, then the listed operations would be considered to follow one after another in time sequence: Register A is first incremented and only then is E evaluated. If $E = 1$, then register R is cleared and control goes to state T_4. Otherwise, if $E = 0$, the next step is to evaluate F and go to state T_2 or T_3. In contrast, an ASM chart considers the entire block as one unit. All the operations that are specified within the block must occur in synchronism during the edge transition of the same clock pulse while the system changes from T_1 to the next state. This is presented pictorially in Fig. 8-8. We assume positive-edge triggering of all flip-flops. The first positive transition of the clock transfers the control circuit into state T_1. While in state T_1, the control circuits check inputs E and F and generate appropriate signals accordingly. The

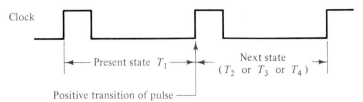

FIGURE 8-8

Transition between states

following operations occur simultaneously during the next positive transition of the clock pulse:

1. Register A is incremented.
2. If $E = 1$, register R is cleared.
3. Depending on the values of E and F, control is transferred to next state, T_2 or T_3 or T_4.

Note that the operations in the data-processor subsection and the change of state in the control logic occur at the same time.

We will now demonstrate the time relationship between the components of an ASM chart by going over a specific design example. The example does not have any known application and is merely formulated to show the usefulness of the ASM Chart. We start from the initial specifications and proceed with the development of an appropriate ASM chart from which the digital hardware can be derived.

Design Example

We wish to design a digital system with two flip-flops, E and F, and one 4-bit binary counter, A. The individual flip-flops in A are denoted by A_4, A_3, A_2, and A_1, with A_4 holding the most significant bit of the count. A start signal S initiates the system operation by clearing the counter A and flip-flop F. The counter is then incremented by 1 starting from the next clock pulse and continues to increment until the operations stop. Counter bits A_3 and A_4 determine the sequence of operations:

If $A_3 = 0$, E is cleared to 0 and the count continues.

If $A_3 = 1$, E is set to 1; then if $A_4 = 0$, the count continues, but if $A_4 = 1$, F is set to 1 on the next clock pulse and the system stops counting.

ASM Chart

The ASM chart is shown in Fig. 8-9. When no operations are performed, the system is in the initial state T_0, waiting for the start signal S. When input S is equal to 1, counter A and flip-flop F are cleared to 0 and the controller goes to state T_1. Note the conditional box that follows the decision box for S. This means that the counter and flip-flop will be cleared during T_0 if $S = 1$, and at the same time, control transfers to state T_1.

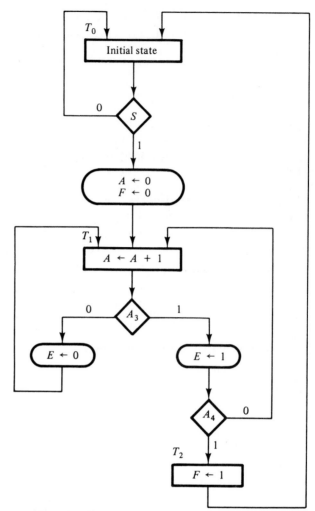

FIGURE 8-9
ASM chart for design example

The block associated with state T_1 has two decision boxes and two conditional boxes. The counter is incremented with every clock pulse. At the same time, one of three possible operations occur during the same clock pulse transition:

Either E is cleared and control stays in state T_1 $(A_3 = 0)$; or
E is set and control stays in state T_1 $(A_3 A_4 = 10)$; or
E is set and control goes to state T_2 $(A_3 A_4 = 11)$.

When the control is in state T_2, flip-flop F is set to 1 and the circuit goes back to its initial state, T_0.

The ASM chart consists of three states and three blocks. The block associated with T_0 consists of the state box, one decision box, and one conditional box. The block associated with T_2 consists of only the state box. The control logic has one external input, S, and two status inputs, A_3 and A_4.

Timing Sequence

Every block in an ASM chart specifies the operations that are to be performed during one common clock pulse. The operations specified within the state and conditional boxes in the block are performed in the data-processor subsection. The change from one state to the next is performed in the control logic. In order to appreciate the timing relationship involved, we will list the step-by-step sequence of operations after each clock pulse from the time the start signal occurs until the system goes back to its initial state.

Table 8-2 shows the binary values of the counter and the two flip-flops after every clock pulse. The table also shows separately the status of A_3 and A_4 as well as the present state of the controller. We start with state T_1 right after the input signal S has caused the counter and flip-flop F to be cleared. The value of E is assumed to be 1, because E is equal to 1 at T_0 (as shown at the end of the table) and because E does not change during the transition from T_0 to T_1. The system stays in state T_1 during the next thirteen clock pulses. Each pulse increments the counter and either clears or sets E. Note the relationship between the time at which A_3 becomes a 1 and the time at which E

TABLE 8-2
Sequence of Operations for Design Example

Counter				Flip-Flops			
A_4	A_3	A_2	A_1	E	F	Conditions	State
0	0	0	0	1	0	$A_3 = 0, A_4 = 0$	T_1
0	0	0	1	0	0		
0	0	1	0	0	0		
0	0	1	1	0	0		
0	1	0	0	0	0	$A_3 = 1, A_4 = 0$	
0	1	0	1	1	0		
0	1	1	0	1	0		
0	1	1	1	1	0		
1	0	0	0	1	0	$A_3 = 0, A_4 = 1$	
1	0	0	1	0	0		
1	0	1	0	0	0		
1	0	1	1	0	0		
1	1	0	0	0	0	$A_3 = 1, A_4 = 1$	
1	1	0	1	1	0		T_2
1	1	0	1	1	1		T_0

is set to 1. When $A = 0011$, the next clock pulse increments the counter to 0100, but that same clock pulse sees the value of A_3 as 0, so E is cleared. The next pulse changes the counter from 0100 to 0101, and now A_3 is initially equal to 1, so E is set to 1. Similarly, E is cleared to 0 not when the count goes from 0111 to 1000, but when it goes from 1000 to 1001, which is when A_3 is 0 in the present value of the counter.

When the count reaches 1100, both A_3 and A_4 are equal to 1. The next clock pulse increments A by 1, sets E to 1, and transfers control to state T_2. Control stays in T_2 for only one clock period. The pulse transition associated with T_2 sets flip-flop F to 1 and transfers control to state T_0. The system stays in the initial state T_0 as long as S is equal to 0.

From observation of Table 8-2, it may seem that the operations performed on E are delayed by one clock pulse. This is the difference between an ASM chart and a conventional flow chart. If Fig. 8-9 were a conventional flow chart, we would assume that A is first incremented and the incremented value would have been used to check the status of A_3. The operations that are performed in the digital hardware as specified by a block in the ASM chart occur during the same clock period and not in a sequence of operations following each other in time, as is usually interpreted in a conventional flow chart. Thus, the value of A_3 to be considered in the decision box is taken from the value of the counter in the present state and before it is incremented. This is because the decision box for E belongs with the same block as state T_1. The digital circuits in the control generate the signals for all the operations specified in the present block prior to the arrival of the next clock pulse. The next clock transition executes all the operations in the registers and flip-flops, including the flip-flops in the controller that determine the next state.

Data Processor

The ASM chart gives all the information necessary to design the digital system. The requirements for the design of the data-processor subsystem are specified inside the state and conditional boxes. The control logic is determined from the decision boxes and the required state transitions. A diagram showing the hardware for the design example is shown in Fig. 8-10. The control subsystem is shown with only its inputs and outputs. The detailed design of the control is considered in the next section. The data processor consists of a 4-bit binary counter, two flip-flops, and a number of gates. The counter is similar to the one shown in Fig. 7-17 except that additional gates are required for the synchronous clear operation. The counter is incremented with every clock pulse when control is in state T_1. It is cleared only when control is at state T_0 and S is equal to 1. This conditional operation requires an AND gate to guarantee that both conditions are present. The other two conditional operations use two other AND gates for setting or clearing flip-flop E. Flip-flop F is set unconditionally during state T_2. Note that all flip-flops and registers including the flip-flops in the control use a common clock-pulse source.

This example demonstrates a method of digital design using the ASM chart. The design of the data-processor subsystem requires an interpretation of the register operations and their implementation by means of the components introduced in Chapters 5 and 7

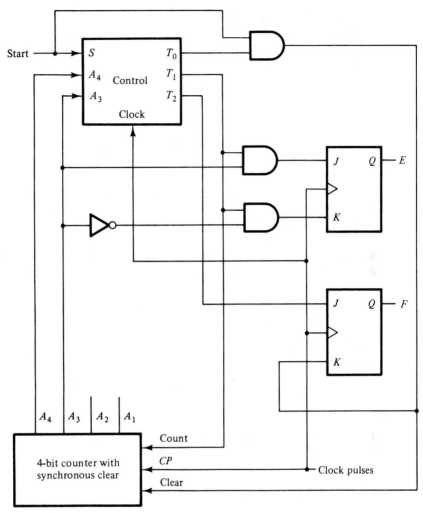

FIGURE 8-10
Data processor for design example

such as registers, counters, multiplexers, and adders. The design of the control subsystem requires the application of design procedures based on the theory of sequential logic. The next three sections present some of the alternatives that are available for designing the control logic.

8-4 CONTROL IMPLEMENTATION

The control section of a digital system is essentially a sequential circuit that can be designed by the procedure outlined in Chapter 6. However, in most cases, this method is impractical because of the large number of states and inputs that a typical control circuit may have. Except for very simple controllers, the design method that uses state and excitation tables is cumbersome and difficult to manage. Experienced digital designers use specialized methods for control logic design that may be considered an extension of the classical sequential method combined with other simplified assumptions. Two of these specialized methods are presented in this section, and a third method is explained in Section 8-5. Another alternative is to use a ROM or PLA to design the control logic. This is covered in Section 8-6.

State Table

As mentioned previously, the ASM chart resembles a state diagram, with each state box representing a state. The state diagram can be converted into a state table from which the sequential circuit of the controller can be designed. First, we must assign binary values to each state in the ASM chart. For n flip-flops in the control sequential circuit, the ASM chart can accommodate up to 2^n states. A chart with three or four states requires a sequential circuit with two flip-flops. With five to eight states, there is a need for three flip-flops. Each combination of flip-flop values represents a binary number for one of the states.

A state table for a controller is a list of present states and inputs and their corresponding next states and outputs. In most cases, there are many don't-care input conditions that must be included, so it is advisable to arrange the state table to take this into consideration. In order to clarify the procedure, we will illustrate by obtaining the state table of the controller defined in the example of the previous section.

The ASM chart of the design example is shown in Fig. 8-9. We assign the following binary values to the three states: $T_0 = 00$, $T_1 = 01$, $T_2 = 11$. Binary state 10 is not used and will be treated as a don't-care condition. The state table corresponding to the ASM chart is shown in Table 8-3. Two flip-flops are needed, and they are labeled G_1 and G_2. There are three inputs and three outputs. The inputs are taken from the conditions in the decision boxes. The outputs are equivalent to the present state of the control. Note that there is a row in the table for each possible transition between states. Initial state 00 goes to state 01 or stays in 00, depending on the value of input S. The

TABLE 8-3
State Table for Control of Fig. 8-10

Present-State Symbol	Present State		Inputs			Next State		Outputs		
	G_1	G_2	S	A_3	A_4	G_1	G_2	T_0	T_1	T_2
T_0	0	0	0	X	X	0	0	1	0	0
T_0	0	0	1	X	X	0	1	1	0	0
T_1	0	1	X	0	X	0	1	0	1	0
T_1	0	1	X	1	0	0	1	0	1	0
T_1	0	1	X	1	1	1	1	0	1	0
T_2	1	1	X	X	X	0	0	0	0	1

other two inputs are marked with don't-care X's, as they do not determine the next state in this case. While the system is in binary state 00, the control provides an output labeled T_0 to initiate the required register operations. The transition from binary state 01 depends on inputs A_3 and A_4. The system goes to binary state 11 only if $A_3 A_4 = 11$; otherwise, it remains in binary state 01. Finally, binary state 11 goes to 00 independently of the input variables.

This example demonstrates a state table for a sequential controller. Note again the large number of don't-care conditions under the inputs. The number of rows in the state table is equal to the number of distinct paths between the states in the ASM chart.

Logic Diagram with *JK* Flip-Flops

The procedure for designing a sequential circuit starting from a state table is presented in Section 6-7. This procedure requires that we obtain the excitation table of the flip-flop inputs and then simplify the combinational-circuit part of the sequential circuit. If we apply this procedure to Table 8-3, we will need to use five-variable maps (see Fig. 3-12) to simplify the input functions. This is because there are five variables listed under the "present-state" and "input" columns. Since this procedure was explained in Chapter 6, we will not show the detail work here. The flip-flop input functions obtained by this method, if we assume *JK* flip-flops, are

$$JG_1 = G_2 A_3 A_4 \qquad JG_2 = S$$

$$KG_1 = 1 \qquad KG_2 = G_1$$

To derive the three output functions, we can utilize the fact that binary state 10 is not used and obtain the following simplified functions:

$$T_0 = G_2'$$

$$T_1 = G_1' G_2$$

$$T_2 = G_1$$

The logic diagram of the control is shown in Fig. 8-11. This circuit replaces the control block in Fig. 8-10.

D Flip-Flops and Decoder

When the number of flip-flops plus inputs in a state table is greater than 5, it is necessary to use large maps to simplify the input functions. This is cumbersome and difficult to achieve, as explained in Chapter 3. Therefore, it is necessary to find alternative ways to design controllers except when they are very simple. One possibility is to use D-type flip-flops and obtain the input functions directly from the state table without the need of an excitation table. This is because the next state is the same as the input requirement for the D flip-flops (see Section 6-7). To design the sequential circuit with D flip-flops,

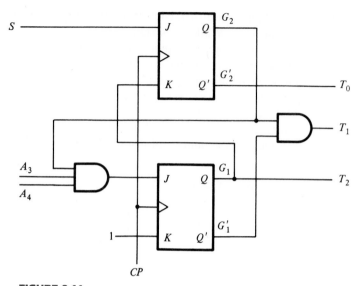

FIGURE 8-11
Logic diagram of control using JK flip-flops

it is necessary to go over the next state column in the state table and derive all the conditions that must set each flip-flop to 1. From Table 8-3, we note that the next state column of G_1 has a single 1 in the fifth row. The D input of flip-flop G_1 must be equal to 1 during present state $T_1 = G_1' G_2$ when both inputs A_3 and A_4 are equal to 1. This is expressed with the D flip-flop input function

$$DG_1 = G_1' G_2 A_3 A_4$$

Similarly, the next state column of G_2 has four 1's, and the condition for setting this flip-flop is

$$DG_2 = G_1' G_2' S + G_1' G_2$$

We can go one step further and insert a decoder at the output of the flip-flops to obtain the necessary three outputs, T_0, T_1, and T_2. Then, instead of using the flip-flop outputs as the present-state condition, we might as well use the outputs of the decoder to supply this information. The input functions to the D flip-flops can now be expressed as follows:

$$DG_1 = A_3 A_4 T_1$$

$$DG_2 = ST_0 + T_1$$

The alternative logic diagram is shown in Fig. 8-12. The decoder provides the three control outputs, and these outputs are also used to determine the next state of each flip-flop. The second control circuit requires more components than the first, but it has the advantage that it can be derived by inspection from the state table.

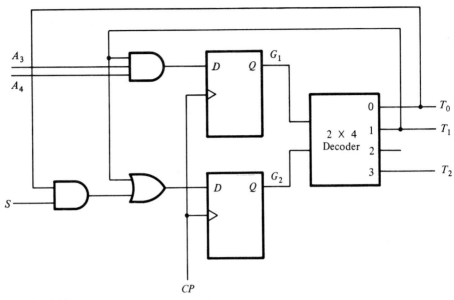

FIGURE 8-12
Alternate logic diagram of control using D flip-flops and a decoder

One Flip-Flop per State

Another possible method of control logic design is to use one flip-flop per state in the sequential circuit. Only one flip-flop is set at any particular time; all others are cleared to 0. The single bit is made to propagate from one flip-flop to the other under the control of decision logic. In such an array, each flip-flop represents a state that is activated only when the control bit is transferred to it.

It is obvious that this method does not use a minimum number of flip-flops for the sequential circuit. In fact, it uses a maximum number of flip-flops. For example, a sequential circuit with 12 states requires a minimum of four flip-flops. Yet by this method, the control circuit needs 12 flip-flops, one for each state.

A control organization that uses one flip-flop per state has the convenient characteristic that the circuit can be derived directly from the state diagram without the need of state or excitation tables. Consider, for example, the state diagram of Fig. 8-13. This diagram is equivalent to the ASM chart of the design example from Fig. 8-9 as far as the control-state transitions are concerned. Since the diagram has three states, we assign three flip-flops to the circuit and label them T_0, T_1, and T_2. The controller can be designed by inspection from the state diagram if D-type flip-flops are used. The Boolean function for setting the flip-flop is determined from the present-state and the input conditions along the directed lines. For example, flip-flop T_0 is set with the next clock pulse if present state $T_2 = 1$ or if present state $T_0 = 1$ and input $S = 0$. This condition is defined by the flip-flop input function:

$$DT_0 = T_2 + S'T_0$$

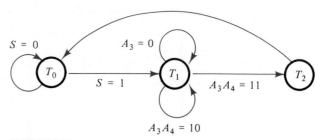

FIGURE 8-13

State diagram of a controller

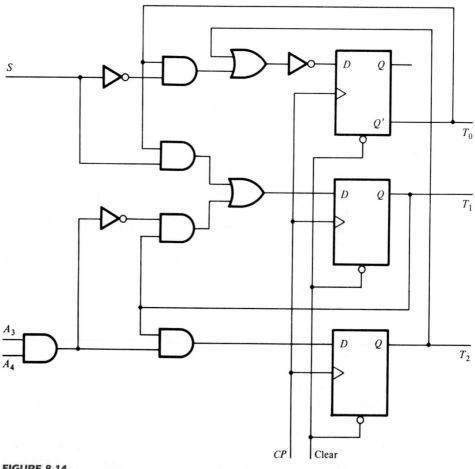

FIGURE 8-14

Third alternate logic diagram of control using one flip-flop per state

where DT_0 designates the D input of flip-flop T_0. In fact, the condition for setting a flip-flop to 1 is obtained directly from the state diagram from the condition specified in the directed lines going into the corresponding flip-flop state ANDed with the previous flip-flop state. If there is more than one directed line going into a state, all conditions must be ORed. Using this procedure for the other two flip-flops, we obtain the input functions:

$$DT_1 = ST_0 + A_3'T_1 + A_3 A_4' T_1 = ST_0 + (A_3 A_4)' T_1$$

$$DT_2 = A_3 A_4 T_1$$

The logic diagram is shown in Fig. 8-14. It consists of three D flip-flops, T_0, T_1, and T_2, and the associated gates specified by the input functions listed before.

Initially, flip-flop T_0 must be set to 1 and all other flip-flops cleared to 0 so that the flip-flop representing the initial state is equal to 1 and all other states equal to 0. Once started, the one-flip-flop-per-state controller will propagate itself from state to state in the proper manner. For a register with a common asynchronous clear input, as shown in Fig. 8-14, all flip-flops, including the Q output of T_0, are cleared to 0. Taking the output of T_0 from the complement output Q' provides the required initial 1 signal for T_0. In order to keep Q' as the output of T_0, it is necessary that the input function to the D input be complemented. This is done by the extra inverter that is placed at the D input of flip-flop T_0.

8-5 DESIGN WITH MULTIPLEXERS

One major goal of control-logic design is the development of a circuit that implements the desired control sequence in a logical and straightforward manner. The attempt to minimize the number of gates tends to produce an irregular network, making it difficult for anyone but the designer to identify the sequence of events the control undergoes. As a consequence, it is difficult to alter, service, or maintain the equipment after the initial design. The sequence of states in the control should be clearly evident from the circuit configuration even if this requires additional components and results in a non-minimal circuit. The multiplexer method is such an implementation.

The control circuit shown in Fig. 8-12 consists of three components: the flip-flops that hold the binary state value, the decoder that generates the control outputs, and the gates that determine the next state. We now replace the gates with multiplexers and use a register for the individual flip-flops. This design method results in a regular pattern of three levels of components. The first level consists of multiplexers that determine the next state of the register. The second level contains a register that holds the present binary state. The third level has the decoder that provides a separate output for each control state.

Consider, for example, the ASM chart of Fig. 8-15. It consists of four states and four control inputs. The state boxes are left empty in this case because we are interested only in the control sequence, which is independent of the register operations. The

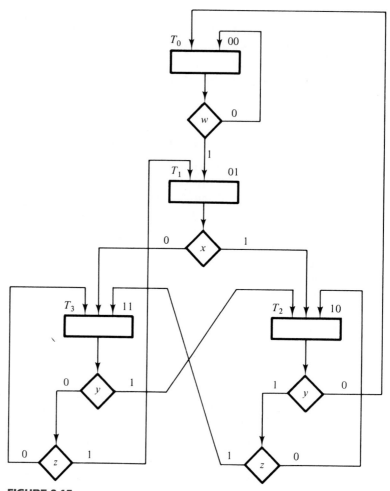

FIGURE 8-15

Example of ASM chart with four control inputs

binary assignment for each state is indicated at the upper right corner of the state boxes. The decision boxes specify the state transitions as a function of the four control inputs, w, x, y, and z. The three-level control implementation is shown in Fig. 8-16. It consists of two multiplexers, MUX1 and MUX2; a register with two flip-flops, G_1 and G_2; and a decoder with four outputs. The outputs of the register are applied to the decoder inputs and also to the select inputs of the multiplexers. In this way, the present state of the register is used to select one of the inputs from each multiplexer. The outputs of the multiplexers are then applied to the D inputs of G_1 and G_2. The purpose of each multiplexer is to produce an input to its corresponding flip-flop equal to the binary value of the next state.

The inputs of the multiplexers are determined from the decision boxes and state

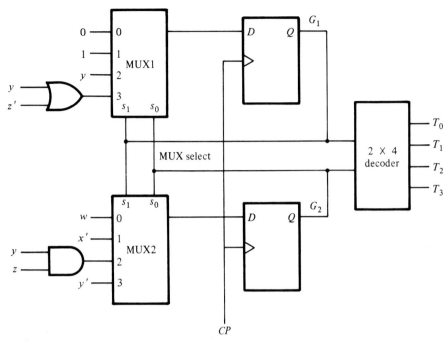

FIGURE 8-16
Control implementation with multiplexers

transitions given in the ASM chart. For example, state 00 stays at 00 or goes to 01, de-pending on the value of input w. Since the next state of G_1 is 0 in either case, we place a signal equivalent to logic-0 in MUX1 input 0. The next state of G_2 is 0 if $w = 0$ and 1 if $w = 1$. Since the next state of G_2 is equal to w, we apply control intput w to MUX2 input 0. What this means is that when the select inputs of the multiplexers are equal to present state 00, the outputs of the multiplexers provide the binary value that is trans-ferred to the register during the next clock pulse.

To facilitate the evaluation of the multiplexer inputs, we prepare a table showing the input conditions for each possible transition in the ASM chart. Table 8-4 gives this in-formation for the ASM chart of Fig. 8-15. There are two transitions from present state 00 or 01 and three transitions from present state 10 or 11. These are separated by hori-zontal lines across the table. The input conditions listed in the table are obtained from the decision boxes in the ASM chart. For example, from Fig. 8-15, we note that present state 01 will go to next state 10 if $x = 1$ or to next state 11 if $x = 0$. In the table, we mark these input conditions as x and x', respectively. The two columns under "multiplexer inputs" in the table specify the input values that must be applied to MUX1 and MUX2. The multiplexer input for each present state is determined from the input conditions when the next state of the flip-flop is equal to 1. Thus, after present state 01, the next state of G_1 is always equal to 1 and the next state of G_2 is equal to the comple-ment value of x. Therefore, the input of MUX1 is made equal to 1 and that of MUX2

TABLE 8-4
Multiplexer Input Conditions

Present State G_1	Present State G_2	Next State G_1	Next State G_2	Input Conditions	Multiplexer Inputs MUX1	Multiplexer Inputs MUX2
0	0	0	0	w'		
0	0	0	1	w	0	w
0	1	1	0	x		
0	1	1	1	x'	1	x'
1	0	0	0	y'		
1	0	1	0	yz'		
1	0	1	1	yz	$yz' + yz = y$	yz
1	1	0	1	$y'z$		
1	1	1	0	y		
1	1	1	1	$y'z'$	$y + y'z' = y + z'$	$y'z + y'z' = y'$

to x' when the present state of the register is 01. As another example, after present state 10, the next state of G_1 must be equal to 1 if the input conditions are yz' or yz. When these two Boolean terms are ORed together and then simplified, we obtain the single binary variable y, as indicated in the table. The next state of G_2 is equal to 1 if the input conditions are $yz = 11$. If the next state of G_1 remains at 0 after a given present state, we place a 0 in the multiplexer input as shown in present state 00 for MUX1. If the next state of G_1 is always 1, we place a 1 in the multiplexer input as shown in present state 01 for MUX1. The other entries for MUX1 and MUX2 are derived in a similar manner. The multiplexer inputs from the table are then used in the control implementation of Fig. 8-16. Note that if the next state of a flip-flop is a function of two or more control variables, the multiplexer may require one or more gates in its input. Otherwise, the multiplexer input is equal to the control variable, or the complement of the control variable, or 0, or 1.

Design Example

We will demonstrate the multiplexer control implementation by means of a second design example. The example will also demonstrate the formulation of the ASM chart and the implementation of the data-processor subsystem.

The digital system to be designed consists of two registers, $R1$ and $R2$, and a flip-flop, E. The system counts the number of 1's in the number loaded into register $R1$ and sets register $R2$ to that number. For example, if the binary number loaded into $R1$ is 10111001, the circuit counts the five 1's in $R1$ and sets register $R2$ to the binary count 101. This is done by shifting each bit from register $R1$ one at a time into flip-flop E. The value in E is checked by the control, and each time it is equal to 1, register $R2$ is incremented by 1.

The control subsystem uses one external input S to start the operation and two status inputs E and Z from the data processor. E is the output of the flip-flop. Z is the output of a circuit that checks the contents of register $R1$ for all 0's. The circuit produces an output $Z = 1$ when $R1$ is equal to 0.

The ASM chart for the design example is shown in Fig. 8-17. The binary number is loaded into $R1$, and register $R2$ is set to an all-1's value. Note that a number with all 1's in a register when incremented produces a number with all 0's. In state T_1, register $R2$ is incremented and the content of $R1$ is examined. If the content is zero, then $Z = 1$, and it signifies that there are no 1's stored in the register; so the operation terminates with $R2$ equal to 0. If the content of $R1$ is not zero, then $Z = 0$, and it indicates that

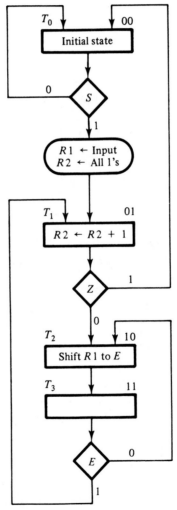

FIGURE 8-17
ASM chart for design example

there are some 1's stored in the register. The number in $R1$ is shifted and its leftmost bit transferred into E. This is done as many times as necessary until a 1 is transferred into E. For every 1 detected in E, register $R2$ is incremented and register $R1$ is checked again for more 1's. The major loop is repeated until all the 1's in $R1$ are counted. Note that the state box of T_3 has no register operations, but the block associated with it contains the decision box for E. Also note that the serial input to shift register $R1$ must be equal to 0 because we don't want to shift external 1's into $R1$.

The data-processor subsystem is shown in Fig. 8-18. The control has three inputs

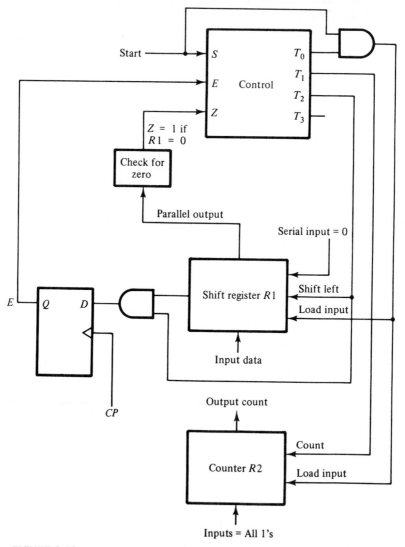

FIGURE 8-18

Data-processor subsystem for design example

TABLE 8-5
Multiplexer Input Conditions for Design Example

Present State		Next State		Input Conditions	Multiplexer Inputs	
G_1	G_2	G_1	G_2		MUX1	MUX2
0	0	0	0	S'		
0	0	0	1	S	0	S
0	1	0	0	Z		
0	1	1	0	Z'	Z'	0
1	0	1	1	None	1	1
1	1	1	0	E'		
1	1	0	1	E	E'	E

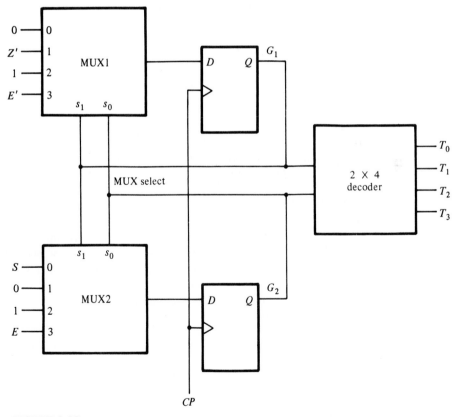

FIGURE 8-19
Control implementation of design example

and four outputs. Only three outputs are used by the data processor. Register $R1$ is a shift register similar to the one shown in Fig. 7-9. Register $R2$ is a counter with parallel load similar to the one shown in Fig. 7-19. In order not to complicate the diagram, the clock pulses are not shown, but they must be applied to the two registers, the E flip-flop, and the flip-flops in the control. The circuit that checks for zero is a NOR gate. For example, if $R1$ is a four-bit register with outputs R_1, R_2, R_3, R_4, then Z is generated with the Boolean function

$$Z = R_1' R_2' R_3' R_4' = (R_1 + R_2 + R_3 + R_4)'$$

which is the NOR function of all bits in the register.

The multiplexer input conditions for the control are determined from Table 8-5. The input conditions are obtained from the ASM chart for each possible binary state transition. The binary assignment to each state is written at the upper right corner of the state boxes. The transition from present state 00 depends on S, from present state 01 depends on Z, and from present state 11 on E. Present state 10 goes to next state 11 unconditionally. The values under MUX1 and MUX2 in the table are determined from the input Boolean conditions for the next state of G_1 and G_2, respectively.

The control implementation of the design example is shown in Fig. 8-19. This is a three-level implementation with the multiplexers in the first level. The inputs to the multiplexers are obtained from Table 8-5.

8-6 PLA CONTROL

We have seen from the examples presented in this chapter that the design of a control circuit is essentially a sequential-logic problem. In Section 7-2, we showed that a sequential circuit can be constructed by means of a register connected to a combinational circuit. In Section 5-8, we investigated the programmable logic array (PLA) and showed that it can be used to implement any combinational circuit. Since the control logic is a sequential circuit, it is then possible to design the control circuit with a register connected to a PLA. The design of a PLA control requires that we obtain the state table of the circuit. The PLA method should be used if the state table contains many don't-care entries; otherwise, it may be advantageous to use a ROM instead of a PLA.

The PLA control will be demonstrated by means of a third design example. This example is an arithmetic circuit that multiplies two unsigned binary numbers and produces their binary product.

Binary Multiplier

The multiplication of two binary numbers is done with paper and pencil by successive additions and shifting. This process is best illustrated with a numerical example. Let us multiply the two binary numbers 10111 and 10011.

23	10111	multiplicand
19	10011	multiplier
	10111	
	10111	
	00000	
	00000	
	10111	
437	110110101	product

The process consists of looking at successive bits of the multiplier, least significant bit first. If the multiplier bit is a 1, the multiplicand is copied down; otherwise, 0's are copied down. The numbers copied down in successive lines are shifted one position to the left from the previous number. Finally, the numbers are added and their sum forms the product. Note that the product obtained from the multiplication of two binary numbers of n bits each can be up to $2n$ bits long.

When the above process is implemented with digital hardware, it is convenient to change the process slightly. First, instead of providing digital circuits to store and add simultaneously as many binary numbers as there are 1's in the multiplier, it is convenient to provide circuits for the summation of only two binary numbers and successively accumulate the partial products in a register. Second, instead of shifting the multiplicand to the left, the partial product is shifted to the right, which results in leaving the partial product and the multiplicand in the required relative positions. Third, when the corresponding bit of the multiplier is a 0, there is no need to add all 0's to the partial product, since this will not alter its value.

The data-processor subsystem for the binary multiplier is shown in Fig. 8-20. The

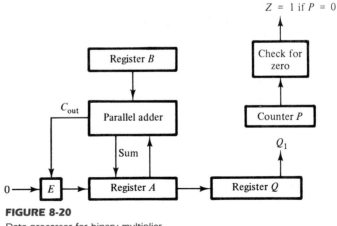

FIGURE 8-20

Data processor for binary multiplier

multiplicand is stored in register B, the multiplier is stored in register Q, and the partial product is formed in register A. A parallel adder similar to the circuit shown in Fig. 5-2 is used to add the contents of register B to register A. The E flip-flop stores the carry after the addition. The P counter is initially set to hold a binary number equal to the number of bits in the multiplier. This counter is decremented after the formation of each partial product. When the content of the counter reaches zero, the product is formed in the double register A and Q and the process stops.

The control logic stays in an initial state until the start signal S becomes a 1. The system then performs the multiplication. The sum of A and B forms a partial product, which is transferred to A. The output carry from the addition, whether 0 or 1, is transferred to E. Both the partial product in A and the multiplier in Q are shifted to the right. The least significant bit of A is shifted into the most significant position of Q; the

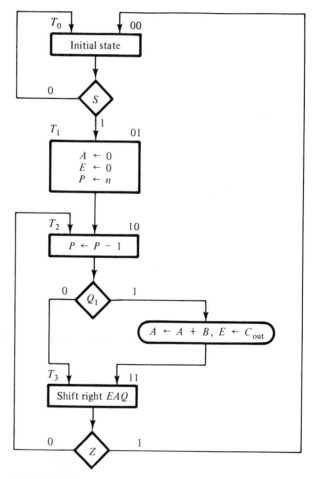

FIGURE 8-21
ASM chart for binary multiplier

carry from E is shifted into the most significant position of A; and 0 is shifted into E. After the shift-right operation, one bit of the partial product is transferred into Q while the multiplier bits in Q are shifted one position to the right. In this manner, the right-most bit of register Q, designated by Q_1, always holds the bit of the multiplier that must be inspected next.

ASM Chart

The ASM chart for the binary multiplier is shown in Fig. 8-21. Initially, the multiplicand is in B and the multiplier in Q. The multiplication process is initiated when $S = 1$. Register A and flip-flop E are cleared and the sequence counter P is set to a binary number n, which is equal to the number of bits in the multiplier.

Next we enter a loop that keeps forming the partial products. The multiplier bit in Q_1 is checked, and if it is equal to 1, the multiplicand in B is added to the partial product in A. The carry from the addition is transferred to E. The partial product in A is left unchanged if $Q_1 = 0$. The P counter is decremented by 1 regardless of the value of Q_1. Registers E, A, and Q are combined into one composite register EAQ, which is then shifted once to the right to obtain a new partial product.

The value in the P counter is checked after the formation of each partial product. If the content of P is not zero, control input Z is equal to 0 and the process is repeated to form a new partial product. The process stops when the P counter reaches 0 and the control input Z is equal to 1. Note that the partial product formed in A is shifted into Q one bit at a time and eventually replaces the multiplier. The final product is available in A and Q, with A holding the most significant bits and Q the least significant bits.

The previous numerical example is repeated in Fig. 8-22 to clarify the multiplication process. The procedure follows the steps outlined in the ASM chart.

Multiplicand B = 10111				
	E	A	Q	P
Multiplier in Q	0	00000	10011	101
$Q_1 = 1$; add B		10111		
First partial product	0	10111		100
Shift right EAQ	0	01011	11001	
$Q_1 = 1$: add B		10111		
Second partial product	1	00010		011
Shift right EAQ	0	10001	01100	
$Q_1 = 0$; shift right EAQ	0	01000	10110	010
$Q_1 = 0$; shift right EAQ	0	00100	01011	001
$Q_1 = 1$; add B		10111		
Fifth partial product	0	11011		000
Shift right EAQ	0	01101	10101	
Final product in AQ = 0110110101				

FIGURE 8-22

Example of binary multiplication

PLA Control

The control for the binary multiplier has four states and three inputs. The binary state assignment is shown in the ASM chart over each state box. The three control inputs are S, Q_1, and Z. The design of a control unit with a PLA is similar to the design using D flip-flops and a decoder. The only difference is in the way the combinational circuit part of the control is implemented. The PLA essentially replaces the decoder and all the gates in the inputs of the flip-flops.

The block diagram of the PLA control is shown in Fig. 8-23. The PLA is connected to a register with two flip-flops, G_1 and G_2. The inputs to the PLA are the values of the present state of the register and the three control inputs. The outputs of the PLA provide the values for the next state in the register and the control output variables. There is one output for each present state and an additional output for the conditional operation $D = Q_1 T_2$. Since the PLA implements the control combinational circuit, we might as well include within it the gates for all conditional operations. In the binary multiplier, there is a conditional operation to add B to A during state T_2 provided that $Q_1 = 1$. PLA output D will then activate this operation in the data processor.

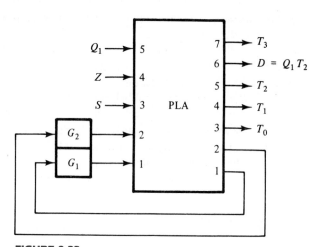

FIGURE 8-23

PLA control block diagram

At any given time, the present state of the register together with the input conditions determine the output values and the next state for the register. The next clock pulse initiates the register operations specified by the PLA outputs and transfers the next-state value into the register. This provides a new control state and possibly different input values. Thus, the PLA acts as the combinational-circuit part of the sequential circuit to generate the control outputs and the next-state values for the register.

PLA Program Table

The internal organization of the PLA was presented in Section 5-8. It was also shown there how to obtain the PLA program table. The reader is advised to review this section to make sure that the meaning of a PLA program table is understood. The internal paths inside the PLA are constructed according to the specifications given in the program table. The design of a PLA control requires that we obtain the state table for the circuit. The state table gives essentially all the information required for obtaining the PLA program table.

The state table for the control subsystem of the binary multiplier is shown in Table 8-6. The present state is determined from flip-flops G_1 and G_2. The input variables for the control are S, Z, and Q_1. The next state of G_1 and G_2 may be a function of one of the input control variables or it may be independent of any inputs. If an input variable does not influence the next state, we mark it with a don't-care condition X. If there are two different transitions from the same present state, the present state is repeated in the table, but the next states are assigned different binary values. The table also lists all control outputs as a function of the present state. Note that input Q_1 does not affect the next state, but only determines the value of output D during state T_2.

The PLA program table can be obtained directly from the state table without the need for simplification procedures. The PLA program table listed in Table 8-7 specifies seven product terms, one for each row in the state table. The input and output terminals are marked with numbers, and the variables applied to these numbered terminals are indicated in the block diagram of Fig. 8-23. The comments are not part of the table, but are included for clarification.

According to the rules established in Section 5-8, a no connection for a PLA path is indicated by a dash (–) in the table. The X's in the state table designate don't-care conditions and imply no connection for the PLA. The 0's in the output columns also indicate no connections to the OR gates within the PLA. The translation from the state table to a PLA program table is very simple. The X's in the "input" columns and the 0's in the "next-state" and "output" columns are changed to dashes, and all other en-

TABLE 8-6
State Table for Control Circuit

Present State		Inputs			Next State		Outputs				
G_1	G_2	S	Z	Q_1	G_1	G_2	T_0	T_1	T_2	D	T_3
0	0	0	X	X	0	0	1	0	0	0	0
0	0	1	X	X	0	1	1	0	0	0	0
0	1	X	X	X	1	0	0	1	0	0	0
1	0	X	X	0	1	1	0	0	1	0	0
1	0	X	X	1	1	1	0	0	1	1	0
1	1	X	0	X	1	0	0	0	0	0	1
1	1	X	1	X	0	0	0	0	0	0	1

TABLE 8-7
PLA Program Table

Product Term	Inputs					Outputs							Comments
	1	2	3	4	5	1	2	3	4	5	6	7	
1	0	0	0	–	–	–	–	1	–	–	–	–	$T_0 = 1,\quad S = 0$
2	0	0	1	–	–	–	1	1	–	–	–	–	$T_0 = 1,\quad S = 1$
3	0	1	–	–	–	1	–	–	1	–	–	–	$T_1 = 1$
4	1	0	–	–	0	1	1	–	–	1	–	–	$T_2 = 1,\quad Q_1 = 0$
5	1	0	–	–	1	1	1	–	–	1	1	–	$T_2 = 1,\quad D = 1$
6	1	1	–	0	–	1	–	–	–	–	–	1	$T_3 = 1,\quad Z = 0$
7	1	1	–	1	–	–	–	–	–	–	–	1	$T_3 = 1,\quad Z = 1$

tries remain the same. The inputs to the PLA are the same as the present state and inputs in the state table. The outputs of the PLA are the same as the next state and outputs in the state table.

The preceding example demonstrates the procedure for designing the control logic with a PLA. From the specifications of the system, we first obtain a state table for the controller. The number of states determines the number of flip-flops for the register. The PLA is then connected to the register and to the input and output variables. The PLA program table is obtained directly from the state table.

The examples introduced in this chapter demonstrate five methods of control-logic design. These should not be considered the only possible methods. A resourceful designer may be able to formulate a control configuration to suit a particular application. This configuration may consist of a combination of methods or may constitute a control organization other than the ones presented here.

REFERENCES

1. CLARE, C. R., *Designing Logic Systems Using State Machines.* New York: McGraw-Hill, 1971.

2. WINKEL, D., and F. PROSSER, *The Art of Digital Design,* 2nd Ed. Englewood Cliffs, NJ: Prentice-Hall, 1987.

3. PEATMAN, J. B., *Digital Hardware Design,* New York: McGraw-Hill, 1980.

4. WIATROWSKI, C. A., and C. H. HOUSE, *Logic Circuits and Microcomputer Systems.* New York: McGraw-Hill, 1980.

5. ROTH, C. H., *Fundamentals of Logic Design,* 3rd Ed. New York: West, 1985.

6. MANO, M. M., *Computer System Architecture,* 2nd Ed. Englewood Cliffs, NJ: Prentice-Hall, 1982.

7. SHIVA, S. G., *Introduction to Logic Design.* Glenview, IL: Scott, Foresman, 1988.

8. FLETCHER, W. I., *An Engineering Approach to Digital Design.* Englewood Cliffs, NJ: Prentice-Hall, 1979.

PROBLEMS

8-1 Draw the portion of an ASM chart that specifies a conditional operation to increment register R during state T_1 and transfer to state T_2 if control inputs z and y are equal to 1 and 0, respectively.

8-2 Show the eight exit paths in an ASM block emanating from the decision boxes that check the eight possible binary values of three control variables, x, y, and z.

8-3 Obtain the ASM charts for the following state transitions:
(a) If $x = 0$, control goes from state T_1 to state T_2; if $x = 1$, generate a conditional operation and go from T_1 to T_2.
(b) If $x = 1$, control goes from T_1 to T_2 and then to T_3; if $x = 0$, control goes from T_1 to T_3.
(c) Start from state T_1; then: if $xy = 00$, go to T_2; if $xy = 01$, go to T_3; if $xy = 10$, go to T_1; otherwise, go to T_3.

8-4 Construct an ASM chart for a digital system that counts the number of people in a room. People enter the room from one door with a photocell that changes a signal x from 1 to 0 when the light is interrupted. They leave the room from a second door with a similar photocell with a signal y. Both x and y are synchronized with the clock, but they may stay on or off for more than one clock-pulse period. The data-processor subsystem consists of an up–down counter with a display of its contents.

8-5 Explain how the ASM chart differs from a conventional flow chart. Using Fig. 8-5 as an illustration, show the difference in interpretation.

8-6 Design the 4-bit counter with synchronous clear specified in Fig. 8-10.

8-7 Using five-variable maps, derive the input Boolean functions to the JK flip-flops of Fig. 8-11.

8-8 Design the control whose state table is given in Table 8-3 using two multiplexers, a register, and a decoder.

8-9 Design the control of the design example of Section 8-5 by the method of one flip-flop per state.

8-10 The state diagram of a control unit is shown in Fig. P8-10. It has four states and two inputs, x and y.
(a) Draw the equivalent ASM chart, leaving the state boxes empty.
(b) Design the control with multiplexers.

8-11 Assume that $R1$ in Fig. 8-18 is the 4-bit shift register shown in Fig. 7-9. Show how the shift and load inputs in Fig. 8-18 are to be connected to the s_1 and s_0 inputs of the shift register.

8-12 Design a digital system with three 4-bit registers, A, B, and C, to perform the following operations:
1. Transfer two binary numbers to A and B when a start signal is enabled.
2. If $A < B$, shift left the contents of A and transfer the result to register C.
3. If $A > B$, shift right the contents of B and transfer the result to register C.
4. If $A = B$, transfer the number to register C unchanged.

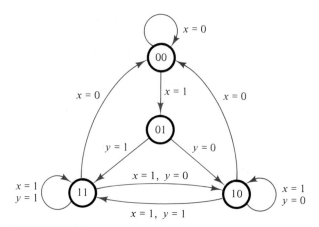

FIGURE P8-10

Control state diagram for Problem 8-10

8-13 Design a digital system that multiplies two binary numbers by the repeated addition method. For example, to multiply 5 × 4, the digital system evaluates the product by adding the multiplicand four times: 5 + 5 + 5 + 5 = 20. Let the multiplicand be in register *BR*, the multiplier in register *AR*, and the product in register *PR*. An adder circuit adds the contents of *BR* to *PR*. A zero-detection circuit *Z* checks when *AR* becomes 0 after each time that it is decremented.

8-14 Prove that the multiplication of two *n*-bit numbers gives a product of length less than or equal to 2*n* bits.

8-15 In Fig. 8-20, the *Q* register holds the multiplier and the *B* register holds the multiplicand. Assume that each number consists of 15 bits.
(a) How many bits can be expected in the product, and where is it available?
(b) How many bits are in the *P* counter, and what is the binary number loaded into it initially?
(c) Design the circuit that checks for zero in the *P* counter.

8-16 List the contents of registers *E*, *A*, *Q*, and *P* similar to Fig. 8-22 during the process of multiplying the two numbers 11111 (multiplicand) and 10101 (multiplier).

8-17 Determine the time it takes to process the multiplication operation in the binary multiplier described in Section 8-6. Assume that the *Q* register has *n* bits and the clock period is *t* nanoseconds.

8-18 Design the control circuit of the binary multiplier specified by the ASM chart of Fig. 8-21 using each of the following methods:
(a) *JK* flip-flops and gates.
(b) *D* flip-flops and a decoder.
(c) Input multiplexers and a register.
(d) One flip-flop per state.

8-19 Design the control whose state table is given in Table 8-3 using the PLA method.

8-20 Consider the ASM chart of Fig. P8-20. The register operations are not specified, because we are interested only in designing the control logic.
(a) Draw the equivalent state diagram.
(b) Design the control with one flip-flop per state.

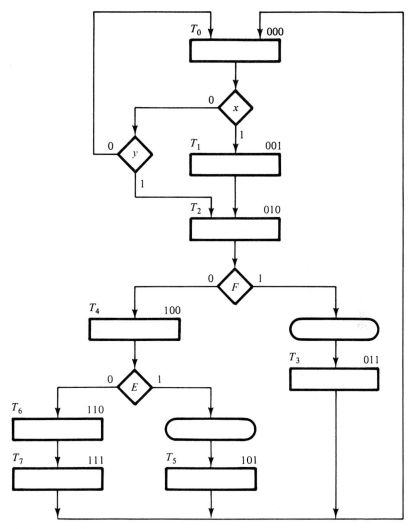

FIGURE P8-20
ASM chart for Problems 8-20 through 8-22

8-21 (a) Derive the state table for the ASM chart of Fig. P8-20.

(b) Design the control with three D flip-flops, a decoder, and gates.

(c) Design the control with a register and a PLA. List the PLA program table.

8-22 (a) Derive a table showing the multiplexer input conditions for the control specified in the ASM chart of Fig. P8-20.

(b) Design the control with three multiplexers, a register with three flip-flops, and a 3×8 decoder.

Asynchronous
Sequential Logic

9-1 INTRODUCTION

A sequential circuit is specified by a time sequence of inputs, outputs, and internal states. In synchronous sequential circuits, the change of internal state occurs in response to the synchronized clock pulses. Asynchronous sequential circuits do not use clock pulses. The change of internal state occurs when there is a change in the input variables. The memory elements in synchronous sequential circuits are clocked flip-flops. The memory elements in asynchronous sequential circuits are either unclocked flip-flops or time-delay elements. The memory capability of a time-delay device is due to the finite time it takes for the signal to propagate through digital gates. An asynchronous sequential circuit quite often resembles a combinational circuit with feedback.

The design of asynchronous sequential circuits is more difficult than that of synchronous circuits because of the timing problems involved in the feedback path. In a properly designed synchronous system, timing problems are eliminated by triggering all flip-flops with the pulse edge. The change from one state to the next occurs during the short time of the pulse transition. Since the asynchronous circuit does not use a clock, the state of the system is allowed to change immediately after the input changes. Care must be taken to ensure that each new state keeps the circuit in a stable condition even though a feedback path exists.

Asynchronous sequential circuits are useful in a variety of applications. They are used when speed of operation is important, especially in those cases where the digital system must respond quickly without having to wait for a clock pulse. They are more economical to use in small independent systems that require only a few components, as it may not be practical to go to the expense of providing a circuit for generating clock pulses. Asynchronous circuits are useful in applications where the input signals to the system may change at any time, independently of an internal clock. The communication between two units, with each unit having its own independent clock, must be done with asynchronous circuits. Digital designers often produce a mixed system where some part of the synchronous system has the characteristics of an asynchronous circuit. Knowledge of asynchronous sequential logic behavior is helpful in verifying that the total digital system is operating in the proper manner.

Figure 9-1 shows the block diagram of an asynchronous sequential circuit. It consists of a combinational circuit and delay elements connected to form feedback loops. There are n input variables, m output variables, and k internal states. The delay elements can be visualized as providing short-term memory for the sequential circuit. In a gate-type circuit, the propagation delay that exists in the combinational circuit path from input to output provides sufficient delay along the feedback loop so that no specific delay elements are actually inserted in the feedback path. The present-state and next-state variables in asynchronous sequential circuits are customarily called secondary variables and excitation variables, respectively. The excitation variables should not be confused with the excitable table used in the design of clocked sequential circuits.

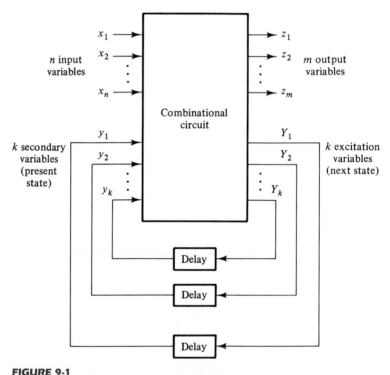

FIGURE 9-1
Block diagram of an asynchronous sequential circuit

When an input variable changes in value, the y secondary variables do not change instantaneously. It takes a certain amount of time for the signal to propagate from the input terminals through the combinational circuit to the Y excitation variables where new values are generated for the next state. These values propagate through the delay elements and become the new present state for the secondary variables. Note the distinction betwen the y's and the Y's. In the steady-state condition, they are the same, but during transition they are not. For a given value of input variables, the system is stable if the circuit reaches a steady-state condition with $y_i = Y_i$ for $i = 1, 2, \ldots, k$. Otherwise, the circuit is in a continuous transition and is said to be unstable. It is important to realize that a transition from one stable state to another occurs only in response to a change in an input variable. This is in contrast to synchronous systems, where the state transitions occur in response to the application of a clock pulse.

To ensure proper operation, asynchronous sequential circuits must be allowed to attain a stable state before the input is changed to a new value. Because of delays in the wires and the gate circuits, it is impossible to have two or more input variables change at exactly the same instant of time without an uncertainty as to which one changes first. Therefore, simultaneous changes of two or more variables are usually prohibited. This restriction means that only one input variable can change at any one time and the time between two input changes must be longer than the time it takes the circuit to reach a stable state. This type of operation is defined as *fundamental mode*. Fundamental-mode operation assumes that the input signals change one at a time and only when the circuit is in a stable condition.

9-2 ANALYSIS PROCEDURE

The analysis of asynchronous sequential circuits consists of obtaining a table or a diagram that describes the sequence of internal states and outputs as a function of changes in the input variables. A logic diagram manifests an asynchronous-sequential-circuit behavior if it has one or more feedback loops or if it includes unclocked flip-flops. In this section, we will investigate the behavior of asynchronous sequential circuits that have feedback paths without employing flip-flops. Unclocked flip-flops are called latches, and their use in asynchronous sequential circuits will be explained in the next section.

The analysis procedure will be presented by means of three specific examples. The first example introduces the transition table. The second example defines the flow table. The third example investigates the stability of asynchronous sequential circuits.

Transition Table

An example of an asynchronous sequential circuit with only gates is shown in Fig. 9-2. The diagram clearly shows two feedback loops from the OR-gate outputs back to the AND-gate inputs. The circuit consists of one input variable, x, and two internal states.

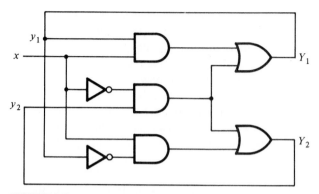

FIGURE 9-2
Example of an asynchronous sequential circuit

The internal states have two excitation variables, Y_1 and Y_2, and two secondary variables, y_1 and y_2. The delay associated with each feedback loop is obtained from the propagation delay between each y input and its corresponding Y output. Each logic gate in the path introduces a propagation delay of about 2 to 10 nanoseconds. The wires that conduct electrical signals introduce approximately one-nanosecond delay for each foot of wire. Thus, no additional external delay elements are necessary when the combinational circuit and the wires in the feedback path provide sufficient delay.

The analysis of the circuit starts by considering the excitation variables as outputs and the secondary variables as inputs. We then derive the Boolean expressions for the excitation variables as a function of the input and secondary variables. These can be readily obtained from the logic diagram.

$$Y_1 = xy_1 + x'y_2$$

$$Y_2 = xy_1' + x'y_2$$

The next step is to plot the Y_1 and Y_2 functions in a map, as shown in Fig. 9-3(a) and (b). The encoded binary values of the y variables are used for labeling the rows, and the input x variable is used to designate the columns. This configuration results in a slightly different three-variable map from the one used in previous chapters. However, it is still a valid map, and this type of configuration is more convenient when dealing with asynchronous sequential circuits. Note that the variables belonging to the appropriate squares are not marked along the sides of the map as done in previous chapters.

The transition table shown in Fig. 9-3(c) is obtained from the maps by combining the binary values in corresponding squares. The transition table shows the value of $Y = Y_1Y_2$ inside each square. The first bit of Y is obtained from the value of Y_1, and the second bit is obtained from the value of Y_2 in the same square position. For a state to be stable, the value of Y must be the same as that of $y = y_1y_2$. Those entries in the transition table where $Y = y$ are circled to indicate a stable condition. An uncircled entry represents an unstable state.

Now consider the effect of a change in the input variable. The square for $x = 0$ and

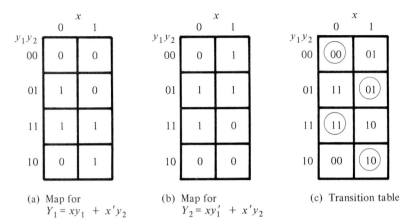

(a) Map for
$Y_1 = xy_1 + x'y_2$

(b) Map for
$Y_2 = xy_1' + x'y_2$

(c) Transition table

FIGURE 9-3
Maps and transition table for the circuit of Fig. 9-2

$y = 00$ in the transition table shows that $Y = 00$. Since Y represents the next value of y, this is a stable condition. If x changes from 0 to 1 while $y = 00$, the circuit changes the value of Y to 01. This represents a temporary unstable condition because Y is not equal to the present value of y. What happens next is that as soon as the signal propagates to make $Y = 01$, the feedback path in the circuit causes a change in y to 01. This is manifested in the transition table by a transition from the first row ($y = 00$) to the second row, where $y = 01$. Now that $y = Y$, the circuit reaches a stable condition with an input of $x = 1$. In general, if a change in the input takes the circuit to an unstable state, the value of y will change (while x remains the same) until it reaches a stable (circled) state. Using this type of analysis for the remaining squares of the transition table, we find that the circuit repeats the sequence of states 00, 01, 11, 10 when the input repeatedly alternates between 0 and 1.

Note the difference between a synchronous and an asynchronous sequential circuit. In a synchronous system, the present state is totally specified by the flip-flop values and does not change if the input changes while the clock pulse is inactive. In an asynchronous circuit, the internal state can change immediately after a change in the input. Because of this, it is sometimes convenient to combine the internal state with the input value together and call it the *total state* of the circuit. The circuit whose transition table is shown in Fig. 9-3(c) has four stable total states, $y_1 y_2 x = 000$, 011, 110, and 101, and four unstable total states, 001, 010, 111, and 100.

The transition table of asynchronous sequential circuits is similar to the state table used for synchronous circuits. If we regard the secondary variables as the present state and the excitation variables as the next state, we obtain the state table, as shown in Table 9-1. This table provides the same information as the transition table. There is one restriction that applies to the asynchronous case but does not apply to the synchronous case. In the asynchronous transition table, there usually is at least one next-

TABLE 9-1
State Table for the Circuit of Fig. 9-2

Present State		Next State			
		$x = 0$		$x = 1$	
0	0	0	0	0	1
0	1	1	1	0	1
1	0	0	0	1	0
1	1	1	1	1	0

state entry that is the same as the present-state value in each row. Otherwise, all the total states in that row will be unstable.

The procedure for obtaining a transition table from the circuit diagram of an asynchronous sequential circuit is as follows:

1. Determine all feedback loops in the circuit.
2. Designate the output of each feedback loop with variable Y_i and its corresponding input with y_i for $i = 1, 2, \ldots, k$, where k is the number of feedback loops in the circuit.
3. Derive the Boolean functions of all Y's as a function of the external inputs and the y's.
4. Plot each Y function in a map, using the y variables for the rows and the external inputs for the columns.
5. Combine all the maps into one table showing the value of $Y = Y_1 Y_2 \cdots Y_k$ inside each square.
6. Circle those values of Y in each square that are equal to the value of $y = y_1 y_2 \cdots y_k$ in the same row.

Once the transition table is available, the behavior of the circuit can be analyzed by observing the state transition as a function of changes in the input variables.

Flow Table

During the design of asynchronous sequential circuits, it is more convenient to name the states by letter symbols without making specific reference to their binary values. Such a table is called a *flow table*. A flow table is similar to a transition table except that the internal states are symbolized with letters rather than binary numbers. The flow table also includes the output values of the circuit for each stable state.

Examples of flow tables are shown in Fig. 9-4. The one in Fig. 9-4(a) has four states designated by the letters a, b, c, and d. It reduces to the transition table of Fig. 9-3(c) if we assign the following binary values to the states: $a = 00$, $b = 01$, $c = 11$, and $d = 10$. The table of Fig. 9-4(a) is called a *primitive* flow table because it has only one stable state in each row. Figure 9-4(b) shows a flow table with more than one stable state in the same row. It has two states, a and b; two inputs, x_1 and x_2; and one output,

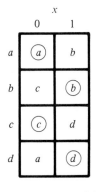

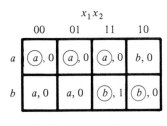

(b) Two states with two
inputs and one output

(a) Four states with
one input

FIGURE 9-4

Examples of flow tables

z. The binary value of the output variable is indicated inside the square next to the state symbol and is separated by a comma. From the flow table, we observe the following behavior of the circuit. If $x_1 = 0$, the circuit is in state a. If x_1 goes to 1 while x_2 is 0, the circuit goes to state b. With inputs $x_1 x_2 = 11$, the circuit may be either in state a or state b. If in state a, the output is 0, and if in state b, the output is 1. State b is maintained if the inputs change from 10 to 11. The circuit stays in state a if the inputs change from 01 to 11. Remember that in fundamental mode, two input variables cannot change simultaneously, and therefore we do not allow a change of inputs from 00 to 11.

In order to obtain the circuit described by a flow table, it is necessary to assign to each state a distinct binary value. This assignment converts the flow table into a transition table from which we can derive the logic diagram. This is illustrated in Fig. 9-5 for the flow table of Fig. 9-4(b). We assign binary 0 to state a and binary 1 to state b. The result is the transition table of Fig. 9-5(a). The output map shown in Fig. 9-5(b) is obtained directly from the output values in the flow table. The excitation function Y and the output function z are simplified by means of the two maps. The logic diagram of the circuit is shown in Fig. 9-5(c).

This example demonstrates the procedure for obtaining the logic diagram from a given flow table. This procedure is not always as simple as in this example. There are several difficulties associated with the binary state assignment and with the output assigned to the unstable states. These problems are discussed in detail in the following sections.

Race Conditions

A *race* condition is said to exist in an asynchronous sequential circuit when two or more binary state variables change value in response to a change in an input variable. When unequal delays are encountered, a race condition may cause the state variables to

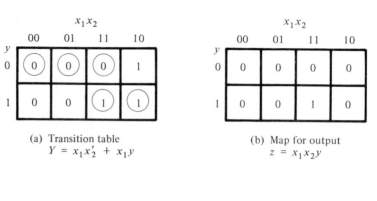

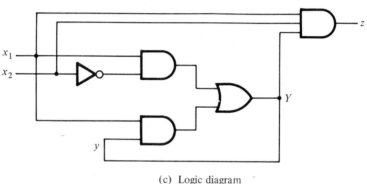

(c) Logic diagram

FIGURE 9-5
Derivation of a circuit specified by the flow table of Fig. 9-4(b)

change in an unpredictable manner. For example, if the state variables must change from 00 to 11, the difference in delays may cause the first variable to change faster than the second, with the result that the state variables change in sequence from 00 to 10 and then to 11. If the second variable changes faster than the first, the state variables will change from 00 to 01 and then to 11. Thus, the order by which the state variables change may not be known in advance. If the final stable state that the circuit reaches does not depend on the order in which the state variables change, the race is called a *noncritical* race. If it is possible to end up in two or more different stable states, depending on the order in which the state variables change, then it is a *critical* race. For proper operation, critical races must be avoided.

The two examples in Fig. 9-6 illustrate noncritical races. We start with the total stable state $y_1 y_2 x = 000$ and then change the input from 0 to 1. The state variables must change from 00 to 11, which defines a race condition. The listed transitions under each table show three possible ways that the state variables may change. They can either change simultaneously from 00 to 11, or they may change in sequence from 00 to 01

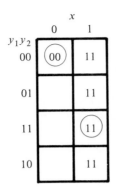

(a) Possible transitions:

    $00 \rightarrow 11$

    $00 \rightarrow 01 \rightarrow 11$

    $00 \rightarrow 10 \rightarrow 11$

(b) Possible transitions:

    $00 \rightarrow 11 \rightarrow 01$

    $00 \rightarrow 01$

    $00 \rightarrow 10 \rightarrow 11 \rightarrow 01$

FIGURE 9-6

Examples of noncritical races

and then to 11, or they may change in sequence from 00 to 10 and then to 11. In all cases, the final stable state is the same, which results in a noncritical race condition. In (a), the final total state is $y_1 y_2 x = 111$, and in (b), it is 011.

The transition tables of Fig. 9-7 illustrate critical races. Here again we start with the

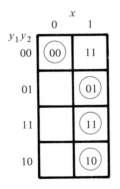

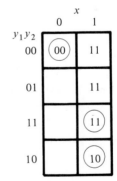

(a) Possible transitions:

    $00 \rightarrow 11$

    $00 \rightarrow 01$

    $00 \rightarrow 10$

(b) Possible transitions:

    $00 \rightarrow 11$

    $00 \rightarrow 01 \rightarrow 11$

    $00 \rightarrow 10$

FIGURE 9-7

Examples of critical races

total stable state $y_1 y_2 x = 000$ and then change the input from 0 to 1. The state variables must change from 00 to 11. If they change simultaneously, the final total stable state is 111. In the transition table of part (a), if Y_2 changes to 1 before Y_1 because of unequal propagation delay, then the circuit goes to the total stable state 011 and remains there. On the other hand, if Y_1 changes first, the internal state becomes 10 and the circuit will remain in the stable total state 101. Hence, the race is critical because the circuit goes to different stable states depending on the order in which the state variables change. The transition table of Fig. 9-7(b) illustrates another critical race, where two possible transitions result in one final total state, but the third possible transition goes to a different total state.

Races may be avoided by making a proper binary assignment to the state variables. The state variables must be assigned binary numbers in such a way that only one state variable can change at any one time when a state transition occurs in the flow table. The subject of race-free state assignment is discussed in Section 9-6.

Races can be avoided by directing the circuit through intermediate unstable states with a unique state-variable change. When a circuit goes through a unique sequence of unstable states, it is said to have a *cycle*. Figure 9-8 illustrates the occurrence of cycles. Again we start with $y_1 y_2 = 00$ and then change the input from 0 to 1. The transition table of part (a) gives a *unique* sequence that terminates in a total stable state 101. The table in (b) shows that even though the state variables change from 00 to 11, the cycle provides a unique transition from 00 to 01 and then to 11. Care must be taken when using a cycle that it terminates with a stable state. If a cycle does not terminate with a stable state, the circuit will keep going from one unstable state to another, making the entire circuit unstable. This is demonstrated in Fig. 9-8(c) and also in the following example.

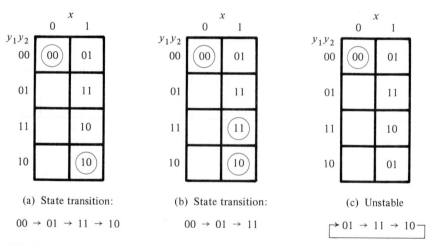

(a) State transition:

00 → 01 → 11 → 10

(b) State transition:

00 → 01 → 11

(c) Unstable

01 → 11 → 10

FIGURE 9-8
Examples of cycles

Stability Considerations

Because of the feedback connection that exists in asynchronous sequential circuits, care must be taken to ensure that the circuit does not become unstable. An unstable condition will cause the circuit to oscillate between unstable states. The transition-table method of analysis can be useful in detecting the occurrence of instability.

Consider, for example, the circuit of Fig. 9-9(a). The excitation function is

$$Y = (x_1 y)' x_2 = (x_1' + y') x_2 = x_1' x_2 + x_2 y'$$

The transition table for the circuit is shown in Fig. 9-9(b). Those values of Y that are equal to y are circled and represent stable states. The uncircled entries indicate unstable conditions. Note that column 11 has no stable states. This means that with input $x_1 x_2$ fixed at 11, the values of Y and y are never the same. If $y = 0$, then $Y = 1$, which causes a transition to the second row of the table with $y = 1$ and $Y = 0$. This causes a transition back to the first row, with the result that the state variable alternates between 0 and 1 indefinitely as long as the input is 11.

The instability condition can be detected directly from the logic diagram. Let $x_1 = 1$, $x_2 = 1$, and $y = 1$. The output of the NAND gate is equal to 0, and the output of the AND gate is equal to 0, making Y equal to 0, with the result that $Y \neq y$. Now if $y = 0$, the output of the NAND gate is 1, the output of the AND gate is 1, making Y equal to 1, with the result that $Y \neq y$. If it is assumed that each gate has a propagation delay of 5 ns (including the wires), we will find that Y will be 0 for 10 ns and 1 for the next 10 ns. This will result in a square-wave waveform with a period of 20 ns. The frequency of oscillation is the reciprocal of the period and is equal to 50 MHz. Unless one

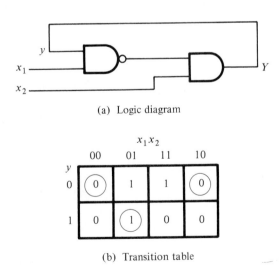

(a) Logic diagram

$$x_1 x_2$$

	00	01	11	10
y 0	⓪	1	1	⓪
1	0	①	0	0

(b) Transition table

FIGURE 9-9
Example of an unstable circuit

is designing a square-wave generator, the instability that may occur in asynchronous sequential circuits is undesirable and must be avoided.

9-3   CIRCUITS WITH LATCHES

Historically, asynchronous sequential circuits were known and used before synchronous circuits were developed. The first practical digital systems were constructed with relays, which are more adaptable to asynchronous-type operations. For this reason, the traditional method of asynchronous-circuit configuration has been with components that are connected to form one or more feedback loops. As electronic digital circuits were developed, it was realized that the flip-flop circuit could be used as a memory element in sequential circuits. Asynchronous sequential circuits can be implemented by employing a basic flip-flop commonly referred to as an *SR latch*. The use of *SR* latches in asynchronous circuits produces a more orderly pattern, which may result in a reduction of the circuit complexity. An added advantage is that the circuit resembles the synchronous circuit in having distinct memory elements that store and specify the internal states.

In this section, we will first explain the operation of the *SR* latch using the analysis technique introduced in the previous section. We will then proceed to give examples of analysis and implementation of asynchronous sequential circuits that employ *SR* latches.

SR Latch

The *SR* latch is a digital circuit with two inputs, S and R, and two cross-coupled NOR gates or two cross-coupled NAND gates. This circuit was introduced in Section 6-2 as a basic flip-flop from which other, more complicated flip-flop circuits were obtained. The cross-coupled NOR circuit is shown in Fig. 9-10(a). This circuit, and the truth table listed in Fig. 9-10(b), were taken directly from Fig. 6-2. In order to analyze the circuit by the transition-table method, we redraw the circuit, as shown in Fig. 9-10(c). Here we distinctly see a feedback path from the output of gate 1 to the input of gate 2. The output Q is identical to the excitation variable Y and the secondary variable y. The Boolean function for the output is

$$Y = [(S + y)' + R]' = (S + y)R' = SR' + R'y$$

Plotting Y as in Fig. 9-10(d), we obtain the transition table for the circuit.

We can now investigate the behavior of the *SR* latch from the transition table. With $SR = 10$, the output $Q = Y = 1$ and the latch is said to be set. Changing S to 0 leaves the circuit in the set state. With $SR = 01$, the output $Q = Y = 0$ and the latch is said to be reset. A change of R back to 0 leaves the circuit in the reset state. These conditions are also listed in the truth table. The circuit exhibits some difficulty when both S and R are equal to 1. From the truth table, we see that both Q and Q' are equal to 0, a condition that violates the requirement that these two outputs be the complement of

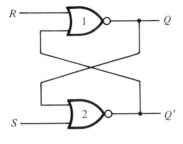

S	R	Q	Q'	
1	0	1	0	
0	0	1	0	(After SR = 10)
0	1	0	1	
0	0	0	1	(After SR = 01)
1	1	0	0	

(a) Crossed-coupled circuit   (b) Truth table

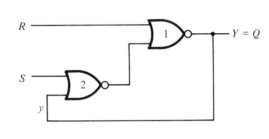

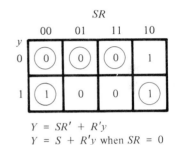

$Y = SR' + R'y$
$Y = S + R'y$ when $SR = 0$

(c) Circuit showing feedback   (d) Transition table

FIGURE 9-10
SR latch with NOR gates

each other. Moreover, from the transition table, we note that going from $SR = 11$ to $SR = 00$ produces an unpredictable result. If S goes to 0 first, the output remains at 0, but if R goes to 0 first, the output goes to 1. In normal operation, we must make sure that 1's are not applied to both the S and R inputs simultaneously. This condition can be expressed by the Boolean function $SR = 0$, which states that the ANDing of S and R must always result in a 0.

Coming back to the excitation function, we note that when we OR the Boolean expression SR' with SR, the result is the single variable S.

$$SR' + SR = S(R' + R) = S$$

From this we deduce that $SR' = S$ when $SR = 0$. Therefore, the excitation function derived previously,

$$Y = SR' + R'y$$

can be expressed as

$$Y = S + R'y \quad \text{when } SR = 0$$

To analyze a circuit with an SR latch, we must first check that the Boolean condition $SR = 0$ holds at all times. We then use the reduced excitation function to analyze the

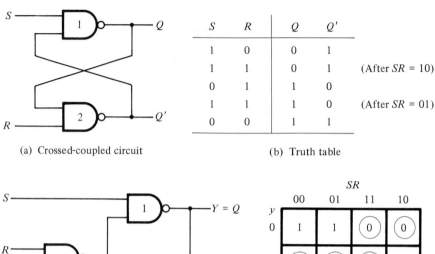

(a) Crossed-coupled circuit

(b) Truth table

S	R	Q	Q'	
1	0	0	1	
1	1	0	1	(After $SR = 10$)
0	1	1	0	
1	1	1	0	(After $SR = 01$)
0	0	1	1	

(c) Circuit showing feedback

(d) Transition table

$Y = S' + Ry$ with $S'R' = 0$

FIGURE 9-11
SR latch with NAND gates

circuit. However, if it is found that both S and R can be equal to 1 at the same time, then it is necessary to use the original excitation function.

The analysis of the SR latch with NAND gates is carried out in Fig. 9-11. The NAND latch operates with both inputs normally at 1 unless the state of the latch has to be changed. The application of 0 to R causes the output Q to go to 0, thus putting the latch in the reset state. After the R input returns to 1, a change of S to 0 causes a change to the set state. The condition to be avoided here is that both S and R not be 0 simultaneously. This condition is satisfied when $S'R' = 0$. The excitation function for the circuit is

$$Y = [S(Ry)']' = S' + Ry$$

Comparing it with the excitation function of the NOR latch, we note that S has been replaced with S' and R' with R. Hence, the input variables for the NAND latch require the complemented values of those used in the NOR latch. For this reason, the NAND latch is sometimes referred to as an $S'R'$ latch (or $\overline{S}$–$\overline{R}$ latch).

Analysis Example

Asynchronous sequential circuits can be constructed with the use of SR latches with or without external feedback paths. Of course, there is always a feedback loop within the latch itself. The analysis of a circuit with latches will be demonstrated by means of a specific example. From this example, it will be possible to generalize the procedural steps necessary to analyze other, similar circuits.

The circuit shown in Fig. 9-12 has two SR latches with outputs Y_1 and Y_2. There are two inputs, x_1 and x_2, and two external feedback loops giving rise to the secondary variables y_1 and y_2. Note that this circuit resembles a conventional sequential circuit with latches behaving like flip-flops without clock pulses. The analysis of the circuit requires that we first obtain the Boolean functions for the S and R inputs in each latch.

$$S_1 = x_1 y_2 \qquad S_2 = x_1 x_2$$
$$R_1 = x_1' x_2' \qquad R_2 = x_2' y_1$$

We then check whether the condition $SR = 0$ is satisfied to ensure proper operation.

$$S_1 R_1 = x_1 y_2 x_1' x_2' = 0$$
$$S_2 R_2 = x_1 x_2 x_2' y_1 = 0$$

The result is 0 because $x_1 x_1' = x_2 x_2' = 0$.

The next step is to derive the transition table of the circuit. Remember that the transition table specifies the value of Y as a function of y and x. The excitation functions are derived from the relation $Y = S + R'y$.

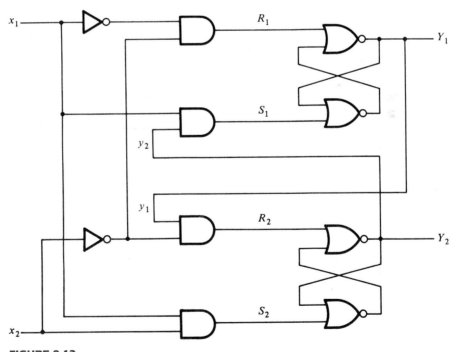

FIGURE 9-12
Example of a circuit with SR latches

y_1y_2 \ x_1x_2	00	01	11	10
00	(00)	(00)	01	(00)
01	(01)	(01)	11	11
11	00	(11)	(11)	10
10	00	(10)	11	(10)

FIGURE 9-13
Transition table for the circuit of Fig. 9-12

$$Y_1 = S_1 + R_1'y_1 = x_1y_2 + (x_1 + x_2)y_1 = x_1y_2 + x_1y_1 + x_2y_1$$

$$Y_2 = S_2 + R_2'y_2 = x_1x_2 + (x_2 + y_1')y_2 = x_1x_2 + x_2y_2 + y_1'y_2$$

We now develop a composite map for $Y = Y_1Y_2$. The y variables are assigned to the rows in the map, and the x variables are assigned to the columns, as shown in Fig. 9-13. The Boolean functions of Y_1 and Y_2 as expressed above are used to plot the composite map for Y. The entries of Y in each row that have the same value as that given to Y are circled and represent stable states. From investigation of the transition table, we deduce that the circuit is stable. There is a critical race condition when the circuit is initially in total state $y_1y_2x_1x_2 = 1101$ and x_2 changes from 1 to 0. If Y_1 changes to 0 before Y_2, the circuit goes to total state 0100 instead of 0000. However, with approximately equal delays in the gates and latches, this undesirable situation is not likely to occur.

The procedure for analyzing an asynchronous sequential circuit with SR latches can be summarized as follows:

1. Label each latch output with Y_i and its external feedback path (if any) with y_i for $i = 1, 2, \ldots, k$.
2. Derive the Boolean functions for the S_i and R_i inputs in each latch.
3. Check whether $SR = 0$ for each NOR latch or whether $S'R' = 0$ for each NAND latch. If this condition is not satisfied, there is a possibility that the circuit may not operate properly.
4. Evaluate $Y = S + R'y$ for each NOR latch or $Y = S' + Ry$ for each NAND latch.
5. Construct a map with the y's representing the rows and the x inputs representing the columns.
6. Plot the value of $Y = Y_1Y_2 \cdots Y_k$ in the map.
7. Circle all stable states where $Y = y$. The resulting map is then the transition table.

Implementation Example

The implementation of a sequential circuit with SR latches is a procedure for obtaining the logic diagram from a given transition table. The procedure requires that we determine the Boolean functions for the S and R inputs of each latch. The logic diagram is then obtained by drawing the SR latches and the logic gates that implement the S and R functions. To demonstrate the procedure, we will repeat the implementation example of Fig. 9-5. The output circuit remains the same and will not be repeated again.

The transition table from Fig. 9-5(a) is duplicated in Fig. 9-14(a). The latch excitation table is shown in Fig. 9-14(b). Remember that the transition table resembles a state table with y representing the present state and Y the next state. Moreover, the excitation table for the SR latch is exactly the same as that of an RS flip-flop as listed pre-

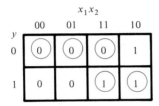

(a) Transition table
$Y = x_1 x_2' + x_1 y$

y	Y	S	R
0	0	0	X
0	1	1	0
1	0	0	1
1	1	X	0

(b) Latch excitation table

(c) Map for $S = x_1 x_2'$

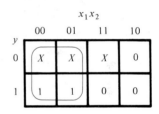

(d) Map for $R = x_1'$

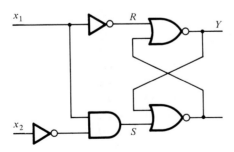

(e) Circuit with NOR latch

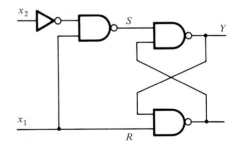

(f) Circuit with NAND latch

FIGURE 9-14
Derivation of a latch circuit from a transition table

viously in Table 6-10, except that y is replaced by $Q(t)$ and Y by $Q(t+1)$. Thus, the excitation table for the SR latch is used in the design of asynchronous sequential circuits just as the RS flip-flop excitation table is used in the design of synchronous sequential circuits as described in Section 6-7. From the information given in the transition table in Fig. 9-14(a) and from the latch excitation table conditions in Fig. 9-14(b), we can obtain the maps for the S and R inputs of the latch, as shown in Fig. 9-14(c) and (d). For example, the square in the second row and third column ($yx_1x_2 = 111$) in Fig. 9-14(a) requires a transition from $y = 1$ to $Y = 1$. The excitation table specifies $S = X$, $R = 0$ for this change. Therefore, the corresponding square in the S map is marked with an X and the one in the R map with a 0. All other squares are filled with values in a similar manner. The maps are then used to derive the simplified Boolean functions.

$$S = x_1 x_2' \qquad \text{and} \qquad R = x_1'$$

The logic diagram consists of an SR latch and the gates required to implement the S and R Boolean functions. The circuit is as shown in Fig. 9-14(e) when a NOR latch is used. With a NAND latch, we must use the complemented values for S and R.

$$S = (x_1 x_2')' \qquad \text{and} \qquad R = x_1$$

This circuit is shown in Fig. 9-14(f).

The general procedure for implementing a circuit with SR latches from a given transition table can now be summarized as follows:

1. Given a transition table that specifies the excitation function $Y = Y_1 Y_2 \cdots Y_k$, derive a pair of maps for S_i and R_i for each $i = 1, 2, \ldots, k$. This is done by using the conditions specified in the latch excitation table of Fig. 9-14(b).
2. Derive the simplified Boolean functions for each S_i and R_i. Care must be taken not to make S_i and R_i equal to 1 in the same minterm square.
3. Draw the logic diagram using k latches together with the gates required to generate the S and R Boolean functions. For NOR latches, use the S and R Boolean functions obtained in step 2. For NAND latches, use the complemented values of those obtained in step 2.

Another useful example of latch implementation can be found in Section 9-7 in conjunction with Fig. 9-38.

Debounce Circuit

Input binary information in a digital system can be generated manually by means of mechanical switches. One position of the switch provides a voltage equivalent to logic 1, and the other position provides a second voltage equivalent to logic 0. Mechanical switches are also used to start, stop, or reset the digital system. When testing digital circuits in the laboratory, the input signals will normally come from switches. A common characteristic of a mechanical switch is that when the arm is thrown from one position to the other, the switch contact vibrates or bounces several times before coming to a final rest. In a typical switch, the contact bounce may take several milliseconds to

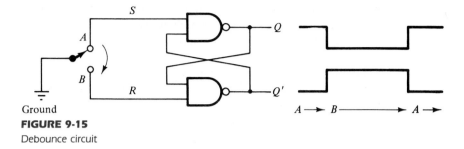

FIGURE 9-15
Debounce circuit

die out. This may cause the signal to oscillate between 1 and 0 because the switch contact is vibrating.

A debounce circuit is one that removes the series of pulses that result from a contact bounce and produces a single smooth transition of the binary signal from 0 to 1 or from 1 to 0. One such circuit consists of a single-pole double-throw switch connected to an *SR* latch, as shown in Fig. 9-15. The center contact is connected to ground that provides a signal equivalent to logic-0. When one of the two contacts, *A* or *B*, is not connected to ground through the switch, it behaves like a logic-1 signal. A resistor is sometimes connected from each contact to a fixed voltage to provide a firm logic-1 signal. When the switch is thrown from position *A* to position *B* and back, the outputs of the latch produce a single pulse as shown, negative for *Q* and positive for *Q'*. The switch is usually a pushbutton whose contact rests in position *A*. When the pushbutton is depressed, it goes to position *B* and when released, it returns to position *A*.

The operation of the debounce circuit is as follows. When the switch rests in position *A*, we have the condition $S = 0$, $R = 1$ and $Q = 1$, $Q' = 0$ (see Fig. 9-11(b)). When the switch is moved to position *B*, the ground connection causes *R* to go to 0 while *S* becomes a 1 because contact *A* is open. This condition causes output *Q* to go to 0 and *Q'* to go to 1. After the switch makes an initial contact with *B*, it bounces several times, but for proper operation, we must assume that it does not bounce back far enough to reach point *A*. The output of the latch will be unaffected by the contact bounce because *Q'* remains 1 (and *Q* remains 0) whether *R* is equal to 0 (contact with ground) or equal to 1 (no contact with ground). When the switch returns to position *A*, *S* becomes 0 and *Q* returns to 1. The output again will exhibit a smooth transition even if there is a contact bounce in position *A*.

9-4 DESIGN PROCEDURE

The design of an asynchronous sequential circuit starts from the statement of the problem and culminates in a logic diagram. There are a number of design steps that must be carried out in order to minimize the circuit complexity and to produce a stable circuit without critical races. Briefly, the design steps are as follows. A primitive flow table is obtained from the design specifications. The flow table is reduced to a minimum number of states. The states are then given a binary assignment from which we obtain the transition table. From the transition table, we derive the logic diagram as a combinational circuit with feedback or as a circuit with *SR* latches.

The design process will be demonstrated by going through a specific example. Once this example is mastered, it will be easier to understand the design steps that are enumerated at the end of this section. Some of the steps require the application of formal procedures, and these are discussed in greater detail in the following sections.

Design Example

It is necessary to design a gated latch circuit with two inputs, *G* (gate) and *D* (data), and one output, *Q*. Binary information present at the *D* input is transferred to the *Q* output when *G* is equal to 1. The *Q* output will follow the *D* input as long as *G* = 1. When *G* goes to 0, the information that was present at the *D* input at the time the transition occurred is retained at the *Q* output. The gated latch is a memory element that accepts the value of *D* when *G* = 1 and retains this value after *G* goes to 0. Once *G* = 0, a change in *D* does not change the value of the output *Q*.

Primitive Flow Table

As defined previously, a primitive flow table is a flow table with only one stable total state in each row. Remember that a total state consists of the internal state combined with the input. The derivation of the primitive flow table can be facilitated if we first form a table with all possible total states in the system. This is shown in Table 9-2 for the gated latch. Each row in the table specifies a total state, which consists of a letter designation for the internal state and a possible input combination for *D* and *G*. The output *Q* is also shown for each total state. We start with the two total states that have *G* = 1. From the design specifications, we know that *Q* = 0 if *DG* = 01 and *Q* = 1 if *DG* = 11 because *D* must be equal to *Q* when *G* = 1. We assign these conditions to states *a* and *b*. When *G* goes to 0, the output depends on the last value of *D*. Thus, if the transition of *DG* is from 01 to 00 to 10, then *Q* must remain 0 because *D* is 0 at the time of the transition from 1 to 0 in *G*. If the transition of *DG* is from 11 to 10 to 00, then *Q* must remain 1. This information results in six different total states, as shown in

TABLE 9-2
Gated-Latch Total States

State	Inputs		Output	Comments
	D	*G*	*Q*	
a	0	1	0	*D* = *Q* because *G* = 1
b	1	1	1	*D* = *Q* because *G* = 1
c	0	0	0	After state *a* or *d*
d	1	0	0	After state *c*
e	1	0	1	After state *b* or *f*
f	0	0	1	After state *e*

the table. Note that simultaneous transitions of two input variables, such as from 01 to 10 or from 11 to 00, are not allowed in fundamental-mode operation.

The primitive flow table for the gated latch is shown in Fig. 9-16. It has one row for each state and one column for each input combination. First, we fill in one square in each row belonging to the stable state in that row. These entries are determined from Table 9-2. For example, state a is stable and the output is 0 when the input is 01. This information is entered in the flow table in the first row and second column. Similarly, the other five stable states together with their output are entered in the corresponding input columns.

Next we note that since both inputs are not allowed to change simultaneously, we can enter dash marks in each row that differs in two or more variables from the input variables associated with the stable state. For example, the first row in the flow table shows a stable state with an input of 01. Since only one input can change at any given time, it can change to 00 or 11, but not to 10. Therefore, we enter two dashes in the 10 column of row a. This will eventually result in a don't-care condition for the next state and output in this square. Following this procedure, we fill in a second square in each row of the primitive flow table.

Next it is necessary to find values for two more squares in each row. The comments listed in Table 9-2 may help in deriving the necessary information. For example, state c is associated with input 00 and is reached after an input change from state a or d. Therefore, an unstable state c is shown in column 00 and rows a and d in the flow table. The output is marked with a dash to indicate a don't-care condition. The interpretation of this is that if the circuit is in stable state a and the input changes from 01 to 00, the circuit first goes to an unstable next state c, which changes the present state value from a to c, causing a transition to the third row and first column of the flow

DG

	00	01	11	10
a	$c,-$	ⓐ, 0	$b,-$	$-,-$
b	$-,-$	$a,-$	ⓑ, 1	$e,-$
c	ⓒ, 0	$a,-$	$-,-$	$d,-$
d	$c,-$	$-,-$	$b,-$	ⓓ, 0
e	$f,-$	$-,-$	$b,-$	ⓔ, 1
f	ⓕ, 1	$a,-$	$-,-$	$e,-$

FIGURE 9-16
Primitive flow table

table. The unstable state values for the other squares are determined in a similar manner. All outputs associated with unstable states are marked with a dash to indicate don't-care conditions. The assignment of actual values to the outputs is discussed further after the design example is completed.

Reduction of the Primitive Flow Table

The primitive flow table has only one stable state in each row. The table can be reduced to a smaller number of rows if two or more stable states are placed in the same row of the flow table. The grouping of stable states from separate rows into one common row is called *merging*. Merging a number of stable states in the same row means that the binary state variable that is ultimately assigned to the merged row will not change when the input variable changes. This is because in a primitive flow table, the state variable changes every time the input changes, but in a reduced flow table, a change of input will not cause a change in the state variable if the next stable state is in the same row.

A formal procedure for reducing a flow table is given in the next section. In order to complete the design example without going through the formal procedure, we will apply the merging process by using a simplified version of the merging rules. Two or more rows in the primitive flow table can be merged into one row if there are non-conflicting states and outputs in each of the columns. Whenever one state symbol and don't-care entries are encountered in the same column, the state is listed in the merged row. Moreover, if the state is circled in one of the rows, it is also circled in the merged row. The output value is included with each stable state in the merged row.

We now apply these rules to the primitive flow table of Fig. 9-16. To see how this is done, the primitive flow table is separated into two parts of three rows each, as shown in Fig. 9-17(a). Each part shows three stable states that can be merged because there are no conflicting entries in each of the four columns. The first column shows state c in all the rows and 0 or a dash for the output. Since a dash represents a don't-care condition, it can be associated with any state or output. The two dashes in the first column can be taken as 0 output to make all three rows identical to a stable state c with a 0 output. The second column shows that the dashes can be assigned to correspond to a stable state a with a 0 output. Note that if the state is circled in one of the rows, it is also circled in the merged row. Similarly, the third column can be merged into an unstable state b with a don't-care output and the fourth column can be merged into stable state d and a 0 output. Thus, the three rows, a, c, and d, can be merged into one row with three stable states and one unstable state, as shown in the first row of Fig. 9-17(b). The second row of the reduced table results from the merging of rows b, e, and f of the primitive flow table. There are two ways that the reduced table can be drawn. The letter symbols for the states can be retained to show the relationship between the reduced and primitive flow tables. The other alternative is to define a common letter symbol for all the stable states of the merged rows. Thus, states c and d are replaced by state a, and states e and f are replaced by state b. Both alternatives are shown in Fig. 9-17(b).

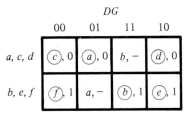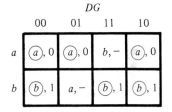

(a) States that are candidates for merging

DG

	00	01	11	10
a, c, d	(c), 0	(a), 0	b, −	(d), 0
b, e, f	(f), 1	a, −	(b), 1	(e), 1

DG

	00	01	11	10
a	(a), 0	(a), 0	b, −	(a), 0
b	(b), 1	a, −	(b), 1	(b), 1

(b) Reduced table (two alternatives)

FIGURE 9-17
Reduction of the primitive flow table

Transition Table and Logic Diagram

In order to obtain the circuit described by the reduced flow table, it is necessary to assign to each state a distinct binary value. This assignment converts the flow table into a transition table. In the general case, a binary state assignment must be made to ensure that the circuit will be free of critical races. The state-assignment problem in asynchronous sequential circuits and ways to solve it are discussed in Section 9-6. Fortunately, there can be no critical races in a two-row flow table, and, therefore, we can finish the design of the gated latch prior to studying Section 9-6. Assigning 0 to state a and 1 to state b in the reduced flow table of Fig. 9-17(b), we obtain the transition table of Fig. 9-18(a). The transition table is, in effect, a map for the excitation variable Y. The simplified Boolean function for Y is then obtained from the map.

$$Y = DG + G'y$$

There are two don't-care outputs in the final reduced flow table. If we assign values to the output, as shown in Fig. 9-18(b), it is possible to make output Q equal to the excitation function Y. If we assign the other possible values to the don't-care outputs, we can make output Q equal to y. In either case, the logic diagram of the gated latch is as shown in Fig. 9-19.

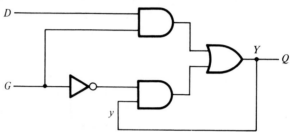

(a) $Y = DG + G'y$ (b) $Q = Y$

FIGURE 9-18

Transition table and output map for gated latch

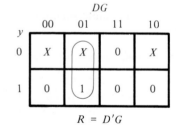

FIGURE 9-19

Gated-latch logic diagram

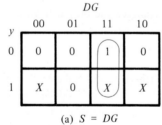

(a) $S = DG$ $R = D'G$

(a) Maps for S and R

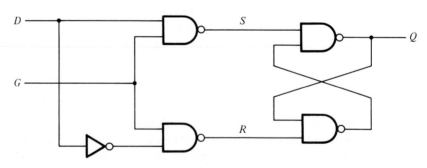

(b) Logic diagram

FIGURE 9-20

Circuit with *SR* latch

The diagram can be implemented also by means of an *SR* latch. Using the procedure outlines in Section 9-3, we first obtain the Boolean functions for *S* and *R*, as shown in Fig. 9-20(a). The logic diagram with NAND gates is shown in Fig. 9-20(b). Note that the gated latch is a level-sensitive *D*-type flip-flop with the clock pulses applied to input *G* (see Fig. 6-5).

Assigning Outputs to Unstable States

The stable states in a flow table have specific output values associated with them. The unstable states have unspecified output entries designated by a dash. The output values for the unstable states must be chosen so that no momentary false outputs occur when the circuit switches between stable states. This means that if an output variable is not supposed to change as the result of a transition, then an unstable state that is a transient state between two stable states must have the same output value as the stable states. Consider, for example, the flow table of Fig. 9-21(a). A transition from stable state *a* to stable state *b* goes through the unstable state *b*. If the output assigned to the unstable *b* is a 1, then a momentary short pulse will appear on the output as the circuit shifts from an output of 0 in state *a* to an output of 1 for the unstable *b* and back to 0 when the circuit reaches stable state *b*. Thus the output corresponding to unstable state *b* must be specified as 0 to avoid a momentary false output.

If an output variable is to change value as a result of a state change, then this variable is assigned a don't-care condition. For example, the transition from stable state *b* to stable state *c* in Fig. 9-21(a) changes the output from 0 to 1. If a 0 is entered as the output value for unstable *c*, then the change in the output variable will not take place until the end of the transition. If a 1 is entered, the change will take place at the start of the transition. Since it makes no difference when the output change occurs, we place a don't-care entry for the output associated with unstable state *c*. Figure 9-21(b) shows the output assignment for the flow table. It demonstrates the four possible combinations in output change that can occur. The procedure for making the assignment to outputs associated with unstable states can be summarized as follows:

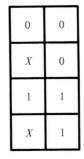

(a) Flow table

(b) Output assignment

FIGURE 9-21

Assigning output values to unstable states

1. Assign a 0 to an output variable associated with an unstable state that is a transient state between two stable states that have a 0 in the corresponding output variable.

2. Assign a 1 to an output variable associated with an unstable state that is a transient state between two stable states that have a 1 in the corresponding output variable.

3. Assign a don't-care condition to an output variable associated with an unstable state that is a transient state between two stable states that have different values (0 and 1 or 1 and 0) in the corresponding output variable.

Summary of Design Procedure

The design of asynchronous sequential circuits can be carried out by using the procedure illustrated in the previous example. Some of the design steps need further elaboration and are explained in the following sections. The procedural steps are as follows.

1. Obtain a primitive flow table from the given design specifications. This is the most difficult part of the design because it is necessary to use intuition and experience to arrive at the correct interpretation of the problem specifications.

2. Reduce the flow table by merging rows in the primitive flow table. A formal procedure for merging rows in the flow table is given in Section 9-5.

3. Assign binary state variables to each row of the reduced flow table to obtain the transition table. The procedure of state assignment that eliminates any possible critical races is given in Section 9-6.

4. Assign output values to the dashes associated with the unstable states to obtain the output maps. This procedure was explained previously.

5. Simplify the Boolean functions of the excitation and output variables and draw the logic diagram, as shown in Section 9-2. The logic diagram can be drawn using *SR* latches, as shown in Section 9-3 and also at the end of Section 9-7.

9-5   REDUCTION OF STATE AND FLOW TABLES

The procedure for reducing the number of internal states in an asynchronous sequential circuit resembles the procedure that is used for synchronous circuits. An algorithm for state reduction of a completely specified state table is given in Section 6-5. We will review this algorithm and apply it to a state-reduction method that uses an implication table. The algorithm and the implication table will then be modified to cover the state reduction of incompletely specified state tables. This modified algorithm will be used to explain the procedure for reducing the flow table of asynchronous sequential circuits.

Implication Table

The state-reduction procedure for completely specified state tables is based on the algorithm that two states in a state table can be combined into one if they can be shown

TABLE 9-3
State Table to Demonstrate Equivalent States

Present State	Next State		Output	
	$x = 0$	$x = 1$	$x = 0$	$x = 1$
a	c	b	0	1
b	d	a	0	1
c	a	d	1	0
d	b	d	1	0

to be equivalent. Two states are equivalent if for each possible input, they give exactly the same output and go to the same next states or to equivalent next states. Table 6-6 shows an example of equivalent states that have the same next states and outputs for each combination of inputs. There are occasions when a pair of states do not have the same next states, but, nonetheless, go to equivalent next states. Consider, for example, the state table shown in Table 9-3. The present states a and b have the same output for the same input. Their next states are c and d for $x = 0$ and b and a for $x = 1$. If we can show that the pair of states (c, d) are equivalent, then the pair of states (a, b) will also be equivalent because they will have the same or equivalent next states. When this relationship exists, we say that (a, b) *imply* (c, d). Similarly, from the last two rows of Table 9-3, we find that the pair of states (c, d) imply the pair of states (a, b). The characteristic of equivalent states is that if (a, b) imply (c, d) and (c, d) imply (a, b), then both pairs of states are equivalent; that is, a and b are equivalent as well as c and d. As a consequence, the four rows of Table 9-3 can be reduced to two rows by combining a and b into one state and c and d into a second state.

The checking of each pair of states for possible equivalence in a table with a large number of states can be done systematically by means of an implication table. The implication table is a chart that consists of squares, one for every possible pair of states, that provide spaces for listing any possible implied states. By judicious use of the table, it is possible to determine all pairs of equivalent states. The state table of Table 9-4 will

TABLE 9-4
State Table to Be Reduced

Present State	Next State		Output	
	$x = 0$	$x = 1$	$x = 0$	$x = 1$
a	d	b	0	0
b	e	a	0	0
c	g	f	0	1
d	a	d	1	0
e	a	d	1	0
f	c	b	0	0
g	a	e	1	0

be used to illustrate this procedure. The implication table is shown in Fig. 9-22. On the left side along the vertical are listed all the states defined in the state table except the first, and across the bottom horizontally are listed all the states except the last. The result is a display of all possible combinations of two states with a square placed in the intersection of a row and a column where the two states can be tested for equivalence.

Two states that are not equivalent are marked with a cross ($\times$) in the corresponding square, whereas their equivalence is recorded with a check mark ($\sqrt{}$). Some of the squares have entries of implied states that must be further investigated to determine whether they are equivalent or not. The step-by-step procedure of filling in the squares is as follows. First, we place a cross in any square corresponding to a pair of states whose outputs are not equal for every input. In this case, state c has a different output than any other state, so a cross is placed in the two squares of row c and the four squares of column c. There are nine other squares in this category in the implication table.

Next, we enter in the remaining squares the pairs of states that are implied by the pair of states representing the squares. We do that starting from the top square in the left column and going down and then proceeding with the next column to the right. From the state table, we see that pair (a, b) imply (d, e), so (d, e) is recorded in the square defined by column a and row b. We proceed in this manner until the entire table is completed. Note that states (d, e) are equivalent because they go to the same next state and have the same output. Therefore, a check mark is recorded in the square defined by column d and row e, indicating that the two states are equivalent and independent of any implied pair.

The next step is to make successive passes through the table to determine whether any additional squares should be marked with a cross. A square in the table is crossed out if it contains at least one implied pair that is not equivalent. For example, the square

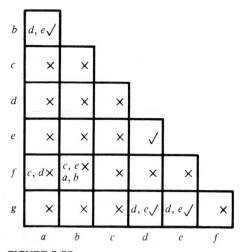

FIGURE 9-22
Implication table

TABLE 9-5
Reduced State Table

Present State	Next State		Output	
	$x = 0$	$x = 1$	$x = 0$	$x = 1$
a	d	a	0	0
c	d	f	0	1
d	a	d	1	0
f	c	a	0	0

defined by a and f is marked with a cross next to c, d because the pair (c, d) defines a square that contains a cross. This procedure is repeated until no additional squares can be crossed out. Finally, all the squares that have no crosses are recorded with check marks. These squares define pairs of equivalent states. In this example, the equivalent states are

$$(a, b) \quad (d, e) \quad (d, g) \quad (e, g)$$

We now combine pairs of states into larger groups of equivalent states. The last three pairs can be combined into a set of three equivalent states (d, e, g) because each one of the states in the group is equivalent to the other two. The final partition of the states consists of the equivalent states found from the implication table together with all the remaining states in the state table that are not equivalent to any other state.

$$(a, b) \quad (c) \quad (d, e, g) \quad (f)$$

This means that Table 9-4 can be reduced from seven states to four states, one for each member of the above partition. The reduced table is obtained by replacing state b by a and states e and g by d. The reduced state table is shown in Table 9-5.

Merging of the Flow Table

There are occasions when the state table for a sequential circuit is incompletely specified. This happens when certain combinations of inputs or input sequences may never occur because of external or internal constraints. In such a case, the next states and outputs that should have occurred if all inputs were possible are never attained and are regarded as don't-care conditions. Although synchronous sequential circuits may sometimes be represented by incompletely specified state tables, our interest here is with asynchronous sequential circuits where the primitive flow table is always incompletely specified.

Incompletely specified states can be combined to reduce the number of states in the flow table. Such states cannot be called equivalent, because the formal definition of equivalence requires that all outputs and next states be specified for all inputs. Instead, two incompletely specified states that can be combined are said to be *compatible*. Two states are compatible if for each possible input they have the same output whenever specified and their next states are compatible whenever they are specified. All don't-care conditions marked with dashes have no effect when searching for compatible states as they represent unspecified conditions.

The process that must be applied in order to find a suitable group of compatibles for the purpose of merging a flow table can be divided into three procedural steps.

1. Determine all compatible pairs by using the implication table.
2. Find the maximal compatibles using a merger diagram.
3. Find a minimal collection of compatibles that covers all the states and is closed.

The minimal collection of compatibles is then used to merge the rows of the flow table. We will now proceed to show and explain the three procedural steps using the primitive flow table from the design example in the previous section.

Compatible Pairs

The procedure for finding compatible pairs is illustrated in Fig. 9-23. The primitive flow table in (a) is the same as Fig. 9-16. The entries in each square represent the next state and output. The dashes represent the unspecified states or outputs. The implication table is used to find compatible states just as it is used to find equivalent states in the completely specified case. The only difference is that when comparing rows, we are at liberty to adjust the dashes to fit any desired condition.

Two states are compatible if in every column of the corresponding rows in the flow table, there are identical or compatible states and if there is no conflict in the output values. For example, rows *a* and *b* in the flow table are found to be compatible, but rows *a* and *f* will be compatible only if *c* and *f* are compatible. However, rows *c* and *f*

	00	01	11	10
a	*c*, –	(*a*), 0	*b*, –	–, –
b	–, –	*a*, –	(*b*), 1	*e*, –
c	(*c*), 0	*a*, –	–, –	*d*, –
d	*c*, –	–, –	*b*, –	(*d*), 0
e	*f*, –	–, –	*b*, –	(*e*), 1
f	(*f*), 1	*a*, –	–, –	*e*, –

(a) Primitive flow table

	a	*b*	*c*	*d*	*e*
b	✓				
c	✓	*d*, *e* ×			
d	✓	*d*, *e* ×	✓		
e	*c*, *f* ×	✓	*d*, *e* × *c*, *f* ×	×	
f	*c*, *f* ×	✓	×	*d*, *e* × *c*, *f* ×	✓

(b) Implication table

FIGURE 9-23
Flow and implication tables

are not compatible because they have different outputs in the first column. This information is recorded in the implication table. A check mark designates a square whose pair of states are compatible. Those states that are not compatible are marked with a cross. The remaining squares are recorded with the implied pairs that need further investigation.

Once the initial implication table has been filled, it is scanned again to cross out the squares whose implied states are not compatible. The remaining squares that contain check marks define the compatible pairs. In the example of Fig. 9-23, the compatible pairs are

$$(a, b) \quad (a, c) \quad (a, d) \quad (b, e) \quad (b, f) \quad (c, d) \quad (e, f)$$

Maximal Compatibles

Having found all the compatible pairs, the next step is to find larger sets of states that are compatible. The *maximal compatible* is a group of compatibles that contains all the possible combinations of compatible states. The maximal compatible can be obtained from a merger diagram, as shown in Fig. 9-24. The merger diagram is a graph in which each state is represented by a dot placed along the circumference of a circle. Lines are drawn between any two corresponding dots that form a compatible pair. All possible compatibles can be obtained from the merger diagram by observing the geometrical patterns in which states are connected to each other. An isolated dot represents a state that is not compatible to any other state. A line represents a compatible pair. A triangle constitutes a compatible with three states. An *n*-state compatible is represented in the merger diagram by an *n*-sided polygon with all its diagonals connected.

The merger diagram of Fig. 9-24(a) is obtained from the list of compatible pairs derived from the implication table of Fig. 9-23. There are seven straight lines connect-

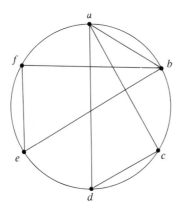

(a) Maximal compatible:
   $(a, b) (a, c, d) (b, e, f)$

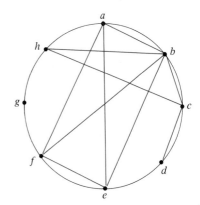

(b) Maximal compatible:
   $(a, b, e, f) (b, c, h) (c, d) (g)$

FIGURE 9-24
Merger diagrams

ing the dots, one for each compatible pair. The lines form a geometrical pattern consisting of two triangles connecting (a, c, d) and (b, e, f) and a line (a, b). The maximal compatibles are

$$(a, b) \quad (a, c, d) \quad (b, e, f)$$

Figure 9-24(b) shows the merger diagram of an 8-state flow table. The geometrical patterns are a rectangle with its two diagonals connected to form the 4-state compatible (a, b, e, f), a triangle (b, c, h), a line (c, d), and a single state g that is not compatible to any other state. The maximal compatibles are

$$(a, b, e, f) \quad (b, c, h) \quad (c, d) \quad (g)$$

The maximal compatible set can be used to merge the flow table by assigning one row in the reduced table to each member of the set. However, quite often the maximal compatibles do not necessarily constitute the set of compatibles that is minimal. In many cases, it is possible to find a smaller collection of compatibles that will satisfy the condition for row merging.

Closed Covering Condition

The condition that must be satisfied for row merging is that the set of chosen compatibles must *cover* all the states and must be *closed*. The set wll cover all the states if it includes all the states of the original state table. The closure condition is satisfied if there are no implied states or if the implied states are included within the set. A closed set of compatibles that covers all the states is called a *closed covering*. The closed-covering condition will be explained by means of two examples.

Consider the maximal compatibles from Fig. 9-24(a). If we remove (a, b), we are left with a set of two compatibles:

$$(a, c, d) \quad (b, e, f)$$

All six states from the flow table in Fig. 9-23 are included in this set. This satisfies the covering condition. There are no implied states for (a, c), (a, d), (c, d), (b, e), (b, f), and (e, f), as seen from the implication table of Fig. 9-23(b), so the closure condition is also satisfied. Therefore, the primitive flow table can be merged into two rows, one for each of the compatibles. The detailed construction of the reduced table for this particular example was done in the previous section and is shown in Fig. 9-17(b).

The second example is from a primitive flow table (not shown) whose implication table is given in Fig. 9-25(a). The compatible pairs derived from the implication table are

$$(a, b) \quad (a, d) \quad (b, c) \quad (c, d) \quad (c, e) \quad (d, e)$$

From the merger diagram of Fig. 9-25(b), we determine the maximal compatibles:

$$(a, b) \quad (a, d) \quad (b, c) \quad (c, d, e)$$

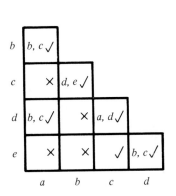

(a) Implication table

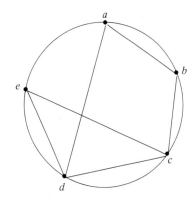

(b) Merger diagram

Compatibles	(a, b)	(a, d)	(b, c)	(c, d, e)
Implied states	(b, c)	(b, c)	(d, e)	(a, d)
				(b, c)

(c) Closure table

FIGURE 9-25

Choosing a set of compatibles

If we choose the two compatibles

$$(a, b) \quad (c, d, e)$$

the set will cover all five states of the original table. The closure condition can be checked by means of a closure table, as shown in Fig. 9-25(c). The implied pairs listed for each compatible are taken directly from the implication table. The implied states for (a, b) are (b, c). But (b, c) is not included in the chosen set of (a, b) (c, d, e), so this set of compatibles is not closed. A set of compatibles that will satisfy the closed covering condition is

$$(a, d) \quad (b, c) \quad (c, d, e)$$

The set is covered because it contains all five states. Note that the same state can be repeated more than once. The closure condition is satisfied because the implied states are (b, c) (d, e) and (a, d), which are included in the set. The original flow table (not shown here) can be reduced from five rows to three rows by merging rows a and d, b and c, and c, d, and e. Note that an alternative satisfactory choice of closed-covered compatibles would be (a, b) (b, c) (d, e). In general, there may be more than one possible way of merging rows when reducing a primitive flow table.

9-6 RACE-FREE STATE ASSIGNMENT

Once a reduced flow table has been derived for an asynchronous sequential circuit, the next step in the design is to assign binary variables to each stable state. This assignment results in the transformation of the flow table into its equivalent transition table. The primary objective in choosing a proper binary state assignment is the prevention of critical races. The problem of critical races was demonstrated in Section 9-2 in conjunction with Fig. 9-7.

Critical races can be avoided by making a binary state assignment in such a way that only one variable changes at any given time when a state transition occurs in the flow table. To accomplish this, it is necessary that states between which transitions occur be given adjacent assignments. Two binary values are said to be adjacent if they differ in only one variable. For example, 010 and 011 are adjacent because they only differ in the third bit.

In order to ensure that a transition table has no critical races, it is necessary to test each possible transition between two stable states and verify that the binary state variables change one at a time. This is a tedious process, especially when there are many rows and columns in the table. To simplify matters, we will explain the procedure of binary state assignment by going through examples with only three and four rows in the flow table. These examples will demonstrate the general procedure that must be followed to ensure a race-free state assignment. The procedure can then be applied to flow tables with any number of rows and columns.

Three-Row Flow-Table Example

The assignment of a single binary variable to a flow table with two rows does not impose critical race problems. A flow table with three rows requires an assignment of two binary variables. The assignment of binary values to the stable states may cause critical races if not done properly. Consider, for example, the reduced flow table of Fig. 9-26(a). The outputs have been omitted from the table for simplicity. Inspection of row *a*

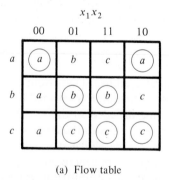

(a) Flow table

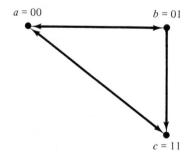

(b) Transition diagram

FIGURE 9-26
Three-row flow-table example

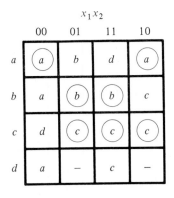

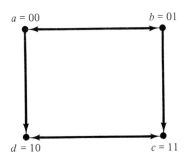

(a) Flow table

(b) Transition diagram

FIGURE 9-27

Flow table with an extra row

reveals that there is a transition from state a to state b in column 01 and from state a to state c in column 11. This information is transferred into a *transition diagram,* as shown in Fig. 9-26(b). The directed lines from a to b and from a to c represent the two transitions just mentioned. Similarly, the transitions from the other two rows are represented by directed lines in the transition diagram. The transition diagram is a pictorial representation of all required transitions between rows.

To avoid critical races, we must find a binary state assignment such that only one binary variable changes during each state transition. An attempt to find such assignment is shown in the transition diagram. State a is assigned binary 00, and state c is assigned binary 11. This assignment will cause a critical race during the transition from a to c because there are two changes in the binary state variables. Note that the transition from c to a also causes a race condition, but it is noncritical.

A race-free assignment can be obtained if we add an extra row to the flow table. The use of a fourth row does not increase the number of binary state variables, but it allows the formation of cycles between two stable states. Consider the modified flow table in Fig. 9-27. The first three rows represent the same conditions as the original three-row table. The fourth row, labeled d, is assigned the binary value 10, which is adjacent to both a and c. The transition from a to c must now go through d, with the result that the binary variables change from $a = 00$ to $d = 10$ to $c = 11$, thus avoiding a critical race. This is accomplished by changing row a, column 11 to d and row d, column 11 to c. Similarly, the transition from c to a is shown to go through unstable state d even though column 00 constitutes a noncritical race.

The transition table corresponding to the flow table with the indicated binary state assignment is shown in Fig. 9-28. The two dashes in row d represent unspecified states that can be considered dont't-care conditions. However, care must be taken not to assign 10 to these squares in order to avoid the possibility of an unwanted stable state being established in the fourth row.

This example demonstrates the use of an extra row in the flow table for the purpose

$x_1 x_2$

	00	01	11	10
$a = 00$	(00)	01	10	(00)
$b = 01$	00	(01)	(01)	11
$c = 11$	10	(11)	(11)	(11)
$d = 10$	00	–	11	–

FIGURE 9-28
Transition table

of achieving a race-free assignment. The extra row is not assigned to any specific stable state, but instead is used to convert a critical race into a cycle that goes through adjacent transitions between two stable states. Sometimes, just one extra row may not be sufficient to prevent critical races, and it may be necessary to add two or more extra rows in the flow table. This is demonstrated in the next example.

Four-Row Flow-Table Example

A flow table with four rows requires a minimum of two state variables. Although race-free assignment is sometimes possible with only two binary state variables, in many cases, the requirement of extra rows to avoid critical races will dictate the use of three binary state variables. Consider, for example, the flow table and its corresponding transition diagram, shown in Fig. 9-29. If there were no transitions in the diagonal direction (from b to d or from c to a), it would be possible to find an adjacent assignment for the remaining four transitions. With one or two diagonal transitions, there is no way of

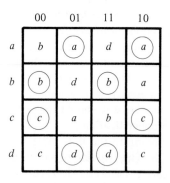

(a) Flow table

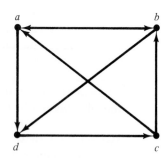

(b) Transition diagram

FIGURE 9-29
Four-row flow-table example

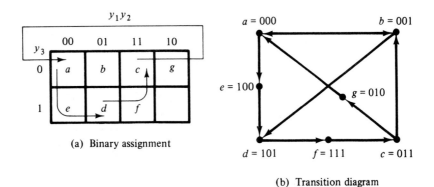

(a) Binary assignment

(b) Transition diagram

FIGURE 9-30
Choosing extra rows for the flow table

assigning two binary variables that satisfy the adjacency requirement. Therefore, at least three binary state variables are needed.

Figure 9-30 shows a state assignment map that is suitable for any four-row flow table. States a, b, c, and d are the original states, and e, f, and g are extra states. States placed in adjacent squares in the map will have adjacent assignments. State b is assigned binary 001 and is adjacent to the other three original states. The transition from a to d must be directed through the extra state e to produce a cycle so that only one binary variable changes at a time. Similarly, the transition from c to a is directed through

	00	01	11	10
000 = a	b	a	e	a
001 = b	b	d	b	a
011 = c	c	g	b	c
010 = g	$-$	a	$-$	$-$
110 $-$	$-$	$-$	$-$	$-$
111 = f	c	$-$	$-$	c
101 = d	f	d	d	f
100 = e	$-$	$-$	d	$-$

FIGURE 9-31
State assignment to modified flow table

g and the transition from d to c goes through f. By using the assignment given by the map, the four-row table can be expanded to a seven-row table that is free of critical races, as shown in Fig. 9-31. Note that although the flow table has seven rows, there are only four stable states. The uncircled states in the three extra rows are there merely to provide a race-free transition between the stable states.

This example demonstrates a possible way of selecting extra rows in a flow table in order to achieve a race-free assignment. A state-assignment map similar to the one used in Fig. 9-30(a) can be helpful in most cases. Sometimes it is possible to take advantage of unspecified entries in the flow table. Instead of adding rows to the table, it may be possible to eliminate critical races by directing some of the state transitions through the don't-care entries. The actual assignment is done by trial and error until a satisfactory assignment is found that resolves all critical races.

Multiple-Row Method

The method for making race-free state assignment by adding extra rows in the flow table, as demonstrated in the previous two examples, is sometimes referred to as the *shared-row* method. There is a second method that is not as efficient, but is easier to apply, called the *multiple-row* method. In the multiple-row assignment, each state in the original flow table is replaced by two or more combinations of state variables. The state-assignment map of Fig.9-32(a) shows a multiple-row assignment that can be used with any four-row flow table. There are two binary state variables for each stable state, each being the logical complement of each other. For example, the original state a is replaced with two equivalent states $a_1 = 000$ and $a_2 = 111$. The output values, not shown here, must be the same in a_1 and a_2. Note that a_1 is adjacent to b_1, c_2, and d_1 and a_2 is adjacent to c_1, b_2, and d_2, and, similarly, each state is adjacent to three states of different letter designation. The behavior of the circuit is the same whether the internal state is a_1 or a_2, and so on for the other states.

Figure 9-32(b) shows the multiple-row assignment for the original flow table of Fig. 9-29(a). The expanded table is formed by replacing each row of the original table with two rows. For example, row b is replaced by rows b_1 and b_2 and stable state b is entered in columns 00 and 11 in both rows b_1 and b_2. After all the stable states have been entered, the unstable states are filled in by reference to the assignment specified in the map of part (a). When choosing the next state for a given present state, a state that is adjacent to the present state is selected from the map. In the original table, the next states of b are a and d for inputs 10 and 01, respectively. In the expanded table, the next states for b_1 are a_1 and d_2 because these are the states adjacent to b_1. Similarly, the next states for b_2 are a_2 and d_1 because they are adjacent to b_2.

In the multiple-row assignment, the change from one stable state to another will always cause a change of only one binary state variable. Each stable state has two binary assignments with exactly the same output. At any given time, only one of the assignments is in use. For example, if we start with state a_1 and input 01 and then change the input to 11, 01, 00, and back to 01, the sequence of internal states will be a_1, d_1, c_1, and a_2. Although the circuit starts in state a_1 and terminates in state a_2, as far as the

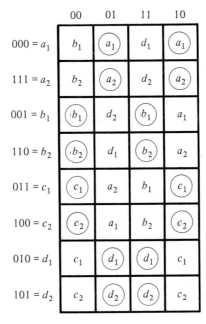

(a) Binary assignment

(b) Flow table

FIGURE 9-32
Multiple-row assignment

input–output relationship is concerned, the two states, a_1 and a_2, are equivalent to state a of the original flow table.

9-7 HAZARDS

When designing asynchronous sequential circuits, care must be taken to conform with certain restrictions and precautions to ensure proper operation. The circuit must be operated in fundamental mode with only one input changing at any time and must be free of critical races. In addition, there is one more phenomenon, called *hazard*, that may cause the circuit to malfunction. Hazards are unwanted switching transients that may appear at the output of a circuit because different paths exhibit different propagation delays. Hazards occur in combinational circuits, where they may cause a temporary false-output value. When this condition occurs in asynchronous sequential circuits, it may result in a transition to a wrong stable state. It is therefore necessary to check for possible hazards and determine whether they cause improper operations. Steps must then be taken to eliminate their effect.

Hazards in Combinational Circuits

A hazard is a condition where a single variable change produces a momentary output change when no ouput change should occur. The circuit of Fig. 9-33(a) demonstrates the occurrence of a hazard. Assume that all three inputs are initially equal to 1. This causes the output of gate 1 to be 1, that of gate 2 to be 0, and the output of the circuit to be equal to 1. Now consider a change of x_2 from 1 to 0. The output of gate 1 changes to 0 and that of gate 2 changes to 1, leaving the output at 1. However, the output may momentarily go to 0 if the propagation delay through the inverter is taken into consideration. The delay in the inverter may cause the output of gate 1 to change to 0 before the output of gate 2 changes to 1. In that case, both inputs of gate 3 are momentarily equal to 0, causing the output to go to 0 for the short interval of time that the input signal from x_2 is delayed while it is propagating through the inverter circuit.

The circuit of Fig. 9-33(b) is a NAND implementation of the same Boolean function. It has a hazard for the same reason. Because gates 1 and 2 are NAND gates, their outputs are the complement of the outputs of the corresponding AND gates. When x_2 changes from 1 to 0, both inputs of gate 3 may be equal to 1, causing the output to produce a momentary change to 0 when it should have stayed at 1.

The two circuits shown in Fig. 9-33 implement the Boolean function in sum of products.

$$Y = x_1 x_2 + x_2' x_3$$

This type of implementation may cause the output to go to 0 when it should remain a 1. If the circuit is implemented in product of sums (see Section 3-5),

$$Y = (x_1 + x_2')(x_2 + x_3)$$

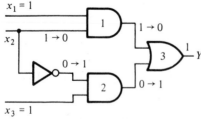

(a) AND-OR circuit   (b) NAND circuit

FIGURE 9-33
Circuits with hazards

(a) Static 1-hazard   (b) Static 0-hazard   (c) Dynamic hazard

FIGURE 9-34

Types of hazards

then the output may momentarily go to 1 when it should remain 0. The first case is referred to as *static 1-hazard* and the second case as *static 0-hazard*. A third type of hazard, known as *dynamic hazard,* causes the output to change three or more times when it should change from 1 to 0 or from 0 to 1. Figure 9-34 demonstrates the three types of hazards. When a circuit is implemented in sum of products with AND–OR gates or with NAND gates, the removal of static 1-hazard guarantees that no static 0-hazards or dynamic hazards will occur.

The occurrence of the hazard can be detected by inspecting the map of the particular circuit. To illustrate, consider the map in Fig. 9-35(a), which is a plot of the function implemented in Fig. 9-33. The change in x_2 from 1 to 0 moves the circuit from minterm 111 to minterm 101. The hazard exists because the change of input results in a different product term covering the two minterms. Minterm 111 is covered by the product term implemented in gate 1, and minterm 101 is covered by the product term implemented in gate 2 of Fig. 9-33. Whenever the circuit must move from one product term to another, there is a possibility of a momentary interval when neither term is equal to 1, giving rise to an undesirable 0 output.

The remedy for eliminating a hazard is to enclose the two minterms in question with another product term that overlaps both groupings. This is shown in the map of Fig. 9-35(b), where the two minterms that cause the hazard are combined into one product term. The hazard-free circuit obtained by this configuration is shown in Fig. 9-36. The extra gate in the circuit generates the product term $x_1 x_3$. In general, hazards in combinational circuits can be removed by covering any two minterms that may produce a hazard with a product term common to both. The removal of hazards requires the addition of redundant gates to the circuit.

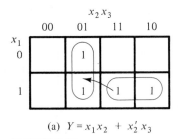

(a) $Y = x_1 x_2 + x_2' x_3$

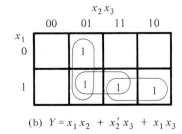

(b) $Y = x_1 x_2 + x_2' x_3 + x_1 x_3$

FIGURE 9-35

Maps demonstrating a hazard and its removal

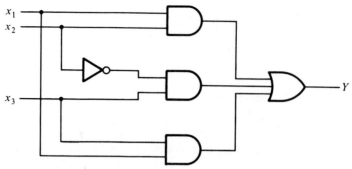

FIGURE 9-36
Hazard-free circuit

Hazards in Sequential Circuits

In normal combinational-circuit design associated with synchronous sequential circuits, hazards are not of concern, since momentary erroneous signals are not generally troublesome. However, if a momentary incorrect signal is fed back in an asynchronous sequential circuit, it may cause the circuit to go to the wrong stable state. This is illustrated in the example of Fig. 9-37. If the circuit is in total stable state $yx_1x_2 = 111$ and input x_2 changes from 1 to 0, the next total stable state should be 110. However, because of the hazard, output Y may go to 0 momentarily. If this false signal feeds back

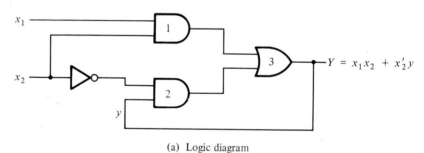

(a) Logic diagram

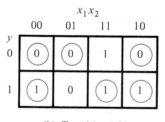

(b) Transition table

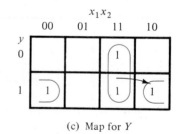

(c) Map for Y

FIGURE 9-37
Hazard in an asynchronous sequential circuit

into gate 2 before the output of the inverter goes to 1, the output of gate 2 will remain at 0 and the circuit will switch to the incorrect total stable state 010. This malfunction can be eliminated by adding an extra gate, as done in Fig. 9-36.

Implementation with *SR* Latches

Another way to avoid static hazards in asynchronous sequential circuits is to implement the circuit with *SR* latches. A momentary 0 signal applied to the *S* or *R* inputs of a NOR latch will have no effect on the state of the circuit. Similarly, a momentary 1 signal applied to the *S* and *R* inputs of a NAND latch will have no effect on the state of the latch. In Fig. 9-33(b), we observed that a two-level sum of product expression implemented with NAND gates may have a static 1-hazard if both inputs of gate 3 go to 1, changing the output from 1 to 0 momentarily. But if gate 3 is part of a latch, the momentary 1 signal will have no effect on the output because a third input to the gate will come from the complemented side of the latch that will be equal to 0 and thus maintain the output at 1. To clarify what was just said, consider a NAND *SR* latch with the following Boolean functions for *S* and *R*.

$$S = AB + CD$$

$$R = A'C$$

Since this is a NAND latch, we must apply the complemented values to the inputs.

$$S = (AB + CD)' = (AB)'(CD)'$$

$$R = (A'C)'$$

This implementation is shown in Fig. 9-38(a). *S* is generated with two NAND gates and one AND gate. The Boolean function for output *Q* is

$$Q = (Q'S)' = [Q'(AB)'(CD)']'$$

This function is generated in Fig. 9-38(b) with two levels of NAND gates. If output *Q* is equal to 1, then *Q'* is equal to 0. If two of the three inputs go momentarily to 1, the NAND gate associated with output *Q* will remain at 1 because *Q'* is maintained at 0.

Figure 9-38(b) shows a typical circuit that can be used to construct asynchronous sequential circuits. The two NAND gates forming the latch normally have two inputs. However, if the *S* or *R* functions contain two or more product terms when expressed in sum of products, then the corresponding NAND gate of the *SR* latch will have three or more inputs. Thus, the two terms in the original sum of products expression for *S* are *AB* and *CD* and each is implemented with a NAND gate whose output is applied to the input of the NAND latch. In this way, each state variable requires a two-level circuit of NAND gates. The first level consists of NAND gates that implement each product term in the original Boolean expression of *S* and *R*. The second level forms the cross-coupled connection of the *SR* latch with inputs that come from the outputs of each NAND gate in the first level.

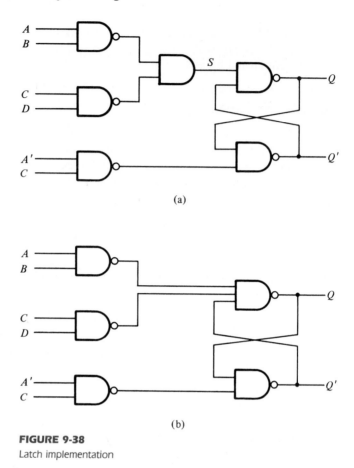

FIGURE 9-38
Latch implementation

Essential Hazards

Thus far we have considered what are known as static and dynamic hazards. There is another type of hazard that may occur in asynchronous sequential circuits, called *essential hazard*. An essential hazard is caused by unequal delays along two or more paths that originate from the same input. An excessive delay through an inverter circuit in comparison to the delay associated with the feedback path may cause such a hazard. Essential hazards cannot be corrected by adding redundant gates as in static hazards. The problem that they impose can be corrected by adjusting the amount of delay in the affected path. To avoid essential hazards, each feedback loop must be handled with individual care to ensure that the delay in the feedback path is long enough compared to delays of other signals that originate from the input terminals. This problem tends to be specialized, as it depends on the particular circuit used and the amount of delays that are encountered in its various paths.

9-8 DESIGN EXAMPLE

We are now in a position to examine a complete design example of an asynchronous sequential circuit. This example may serve as a reference for the design of other similar circuits. We will demonstrate the method of design by following the recommended procedural steps that were listed at the end of Section 9-4 and are repeated here:

1. State the design specifications.
2. Derive a primitive flow table.
3. Reduce the flow table by merging the rows.
4. Make a race-free binary state assignment.
5. Obtain the transition table and output map.
6. Obtain the logic diagram using *SR* latches.

Design Specifications

It is necessary to design a negative-edge-triggered *T* flip-flop. The circuit has two inputs, *T* (toggle) and *C* (clock), and one output, *Q*. The output state is complemented if $T = 1$ and the clock *C* changes from 1 to 0 (negative-edge triggering). Otherwise, under any other input condition, the output *Q* remains unchanged. Although this circuit can be used as a flip-flop in clocked sequential circuits, the internal design of the flip-flop (as is the case with all other flip-flops) is an asynchronous problem.

Primitive Flow Table

The derivation of the primitive flow table can be facilitated if we first derive a table that lists all the possible total states in the circuit. This is shown in Table 9-6. We start with the input condition $TC = 11$ and assign to it state *a*. The circuit goes to state *b* and the output *Q* complements from 0 to 1 when *C* changes from 1 to 0 while *T* remains a 1. Another change in the output occurs when the circuit goes from state *c* to state *d*. In

TABLE 9-6
Specification of Total States

State	Inputs		Output	Comments
	T	*C*	*Q*	
a	1	1	0	Initial output is 0
b	1	0	1	After state *a*
c	1	1	1	Initial output is 1
d	1	0	0	After state *c*
e	0	0	0	After state *d* or *f*
f	0	1	0	After state *e* or *a*
g	0	0	1	After states *b* or *h*
h	0	1	1	After states *g* or *c*

TC

	00	01	11	10
a	$-,-$	$f,-$	$\textcircled{a},0$	$b,-$
b	$g,-$	$-,-$	$c,-$	$\textcircled{b},1$
c	$-,-$	$h,-$	$\textcircled{c},1$	$d,-$
d	$e,-$	$-,-$	$a,-$	$\textcircled{d},0$
e	$\textcircled{e},0$	$f,-$	$-,-$	$d,-$
f	$e,-$	$\textcircled{f},0$	$a,-$	$-,-$
g	$\textcircled{g},1$	$h,-$	$-,-$	$b,-$
h	$g,-$	$\textcircled{h},1$	$c,-$	$-,-$

FIGURE 9-39
Primitive flow table

this case, $T = 1$, C changes from 1 to 0, and the output Q complements from 1 to 0. The other four states in the table do not change the output, because T is equal to 0. If Q is initially 0, it stays at 0, and if initially at 1, it stays at 1 even though the clock input changes. This information results in six total states. Note that simultaneous transitions of two input variables, such as from 01 to 10, are not included, as they violate the condition for fundamental-mode operation.

The primitive flow table is shown in Fig. 9-39. The information for the flow table can be obtained directly from the conditions listed in Table 9-6. We first fill in one square in each row belonging to the stable state in that row as listed in the table. Then we enter dashes in those squares whose input differs by two variables from the input corresponding to the stable state. The unstable conditions are then determined by utilizing the information listed under the comments in Table 9-6.

Merging of the Flow Table

The rows in the primitive flow table are merged by first obtaining all compatible pairs of states. This is done by means of the implication table shown in Fig. 9-40. The squares that contain check marks define the compatible pairs:

$$(a, f) \quad (b, g) \quad (b, h) \quad (c, h) \quad (d, e) \quad (d, f) \quad (e, f) \quad (g, h)$$

	a	b	c	d	e	f	g
b	a, c ✗						
c	✗	b, d ✗					
d	b, d ✗	✗	a, c ✗				
e	b, d ✗	e, g ✗ b, d ✗	f, h ✗	✓			
f	✓	e, g ✗ a, c ✗	f, h ✗ a, c ✗	✓	✓		
g	f, h ✗	✓	b, d ✗	e, g ✗ b, d ✗	✗	e, g ✗ f, h ✗	
h	f, h ✗ a, c ✗	✓	✓	e, g ✗ a, c ✗	e, g ✗ f, h ✗	✗	✓

FIGURE 9-40

Implication table

The maximal compatibles are obtained from the merger diagram shown in Fig. 9-41. The geometrical patterns that are recognized in the diagram consist of two triangles and two straight lines. The maximal compatible set is

$$(a, f) \quad (b, g, h) \quad (c, h) \quad (d, e, f)$$

In this particular example, the minimal collection of compatibles is also the maximal

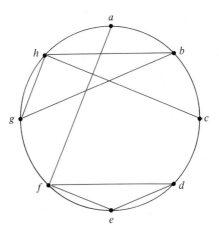

FIGURE 9-41

Merger diagram

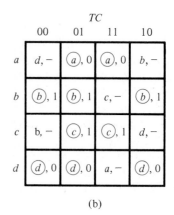

Table (a)

	TC 00	01	11	10
a, f	e, −	(f), 0	(a), 0	b, −
b, g, h	(g), 1	(h), 1	c, −	(b), 1
c, h	g, 1	(h), 1	(c), 1	d, −
d, e, f	(e), 0	(f), 0	a, −	(d), 0

(a)

Table (b)

	TC 00	01	11	10
a	d, −	(a), 0	(a), 0	b, −
b	(b), 1	(b), 1	c, −	(b), 1
c	b, −	(c), 1	(c), 1	d, −
d	(d), 0	(d), 0	a, −	(d), 0

(b)

FIGURE 9-42
Reduced flow table

compatible set. Note that the closed condition is satisfied because the set includes all the original eight states listed in the primitive flow table, although states h and f are repeated. The covering condition is also satisfied because all the compatible pairs have no implied states, as can be seen from the implication table.

The reduced flow table is shown in Fig. 9-42. The one shown in part (a) of the figure retains the original state symbols but merges the corresponding rows. For example, states a and f are compatible and are merged into one row that retains the original letter symbols of the states. Similarly, the other three compatible sets of states are used to merge the flow table into four rows, retaining the eight original letter symbols. The other alternative for drawing the merged flow table is shown in part (b) of the figure. Here we assign a common letter symbol to all the stable states in each merged row. Thus, the symbol f is replaced by a, and g and h are replaced by b, and similarly for the other two rows. The second alternative shows clearly a four-state flow table with only four letter symbols for the states.

State Assignment and Transition Table

The next step in the design is to find a race-free binary assignment for the four stable states in the reduced flow table. In order to find a suitable adjacent assignment, we draw the transition diagram, as shown in Fig. 9-43. For this example, it is possible to obtain a suitable adjacent assignment without the need of extra states. This is because there are no diagonal lines in the transition diagram.

Substituting the binary assignment indicated in the transition diagram into the reduced flow table, we obtain the transition table shown in Fig. 9-44. The output map is obtained from the reduced flow table. The dashes in the output section are assigned values according to the rules established in Section 9-4.

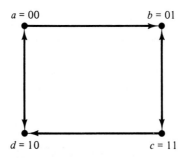

FIGURE 9-43

Transition diagram

	TC 00	01	11	10
y_1y_2				
$a = 00$	10	(00)	(00)	01
$b = 01$	(01)	(01)	11	(01)
$c = 11$	01	(11)	(11)	10
$d = 10$	(10)	(10)	00	(10)

(a) Transition table

	TC 00	01	11	10
y_1y_2				
00	0	0	0	X
01	1	1	1	1
11	1	1	1	X
10	0	0	0	0

(b) Output map $Q = y_2$

FIGURE 9-44

Transition table and output map

Logic Diagram

The circuit to be designed has two state variables, Y_1 and Y_2, and one output, Q. The output map in Fig. 9-44 shows that Q is equal to the state variable y_2. The implementation of the circuit requires two SR latches, one for each state variable. The maps for inputs S and R of the two latches are shown in Fig. 9-45. The maps are obtained from the information given in the transition table by using the conditions specified in the latch excitation table shown in Fig. 9-14(b). The simplified Boolean functions are listed under each map.

The logic diagram of the circuit is shown in Fig. 9-46. Here we use two NAND latches with two or three inputs in each gate. This implementation is according to the

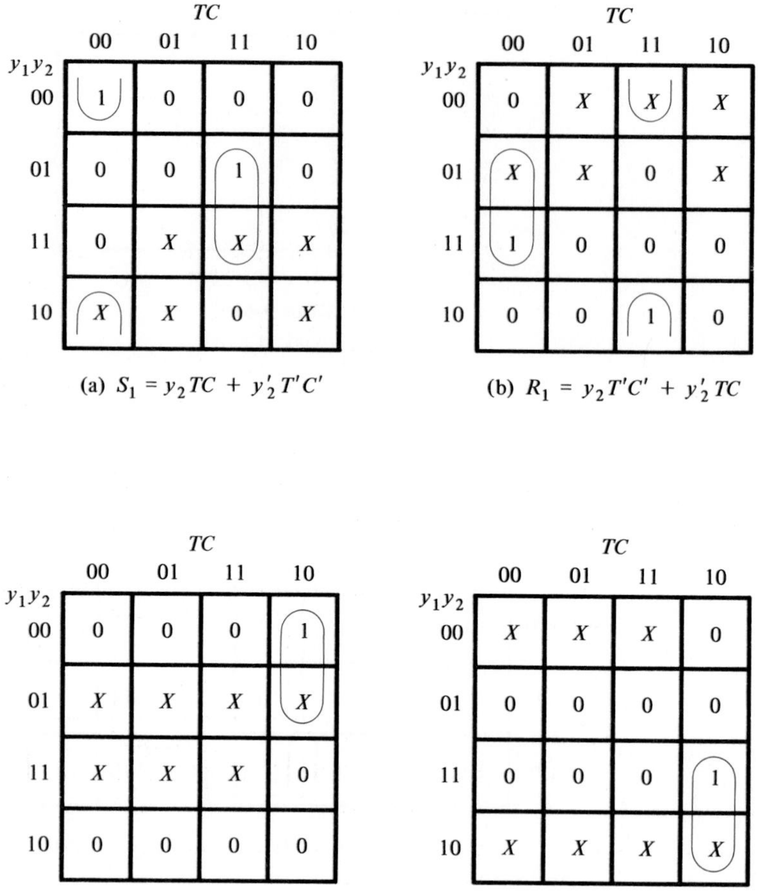

(a) $S_1 = y_2 TC + y_2' T'C'$

(b) $R_1 = y_2 T'C' + y_2' TC$

(c) $S_2 = y_1' TC'$

(d) $R_2 = y_1 TC'$

FIGURE 9-45

Maps for latch inputs

pattern established in Section 9-7 in conjunction with Fig. 9-38(b). The S and R input functions require six NAND gates for their implementation.

This example demonstrates the complexity involved in designing asynchronous sequential circuits. It was necessary to go through ten diagrams in order to obtain the final circuit diagram. Although most digital circuits are synchronous, there are occasions when one has to deal with asynchronous behavior. The basic properties presented in this chapter are essential to understand fully the internal behavior of digital circuits.

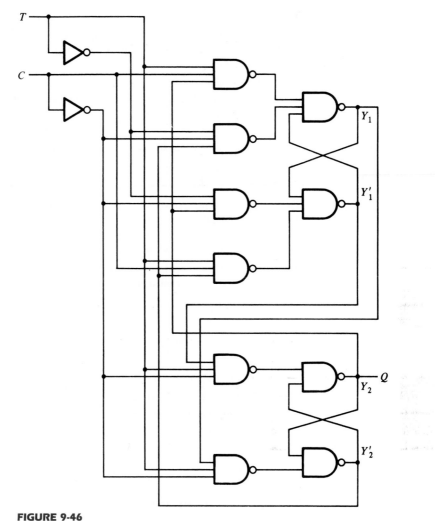

FIGURE 9-46
Logic diagram of negative-edge-triggered T flip-flop

REFERENCES

1. HILL, F. J., and G. R. PETERSON, *Introduction to Switching Theory and Logical Design,* 3rd Ed. New York: John Wiley, 1981.

2. KOHAVI, Z., *Switching and Automata Theory,* 2nd Ed. New York: McGraw-Hill, 1978.

3. MCCLUSKEY, E. J., *Logic Design Principles.* Englewood Cliffs, NJ: Prentice-Hall, 1986.

4. UNGER, S. H., *Asynchronous Sequential Switching Circuits*. New York: John Wiley, 1969.

5. BREEDING, K. J., *Digital Design Fundamentals*. Englewood Cliffs, NJ: Prentice-Hall, 1989.

6. FRIEDMAN, A. D., *Fundamentals of Logic Design and Switching Theory*. Rockville, MD: Computer Science Press, 1986.

7. GIVONE, D. D., *Introduction to Switching Circuit Theory*. New York: McGraw-Hill, 1970.

8. MANGE, D., *Analysis and Synthesis of Logic Systems*. Norwood, MA: Artech House, 1986.

9. ROTH, C. H., *Fundamentals of Logic Design*, 3rd Ed. New York: West, 1985.

PROBLEMS

9-1 (a) Explain the difference between asynchronous and synchronous sequential circuits.
(b) Define fundamental-mode operation.
(c) Explain the difference between stable and unstable states.
(d) What is the difference between an internal state and a total state?

9-2 Derive the transition table for the asynchronous sequential circuit shown in Fig. P9-2. Determine the sequence of internal states $Y_1 Y_2$ for the following sequence of inputs $x_1 x_2$: 00, 10, 11, 01, 11, 10, 00.

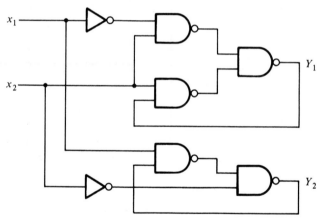

FIGURE P9-2

9-3 An asynchronous sequential circuit is described by the following excitation and output functions:

$$Y = x_1 x_2' + (x_1 + x_2')y$$

$$z = y$$

(a) Draw the logic diagram of the circuit.
(b) Derive the transition table and output map.

(c) Obtain a 2-state flow table.

(d) Describe in words the behavior of the circuit.

9-4 An asynchronous sequential circuit has two internal states and one output. The excitation and output functions describing the circuit are as follows:

$$Y_1 = x_1 x_2 + x_1 y_2' + x_2' y_1$$

$$Y_2 = x_2 + x_1 y_1' y_2 + x_1' y_1$$

$$z = x_2 + y_1$$

(a) Draw the logic diagram of the circuit.

(b) Derive the transition table and output map.

(c) Obtain a flow table for the circuit.

9-5 Convert the flow table of Fig. P9-5 into a transition table by assigning the following binary values to the states: $a = 00$, $b = 11$, and $c = 01$.

(a) Assign values to the extra fourth state to avoid critical races.

(b) Assign outputs to the don't-care states to avoid momentary false outputs.

(c) Derive the logic diagram of the circuit.

$x_1 x_2$

	00	01	11	10
a	$\textcircled{a}, 0$	$b, -$	$c, -$	$\textcircled{a}, 1$
b	$a, -$	$\textcircled{b}, 0$	$\textcircled{b}, 0$	$c, -$
c	$a, -$	$b, -$	$\textcircled{c}, 1$	$\textcircled{c}, 0$

FIGURE P9-5

9-6 Investigate the transition table of Fig. P9-6 and determine all race conditions and whether they are critical or noncritical. Also determine whether there are any cycles.

$x_1 x_2$

$y_1 y_2$	00	01	11	10
00	10	$\textcircled{00}$	11	10
01	$\textcircled{01}$	00	10	10
11	01	00	$\textcircled{11}$	$\textcircled{11}$
10	11	00	$\textcircled{10}$	$\textcircled{10}$

FIGURE P9-6

9-7   Analyze the T flip-flop shown in Fig. 6-7(a). Obtain the transition table and show that the circuit is unstable when both T and CP are equal to 1.

9-8   Convert the circuit of Fig. 6-30 into an asynchronous sequential circuit by removing the clock-pulse (CP) signal and changing the flip-flops into SR latches. Derive the transition table and output map of the modified circuit.

9-9   For the asynchronous sequential circuit shown in Fig. P9-9:
(a) Derive the Boolean functions for the outputs of the two SR latches Y_1 and Y_2. Note that the S input of the second latch is $x_1' y_1'$.
(b) Derive the transition table and output map of the circuit.

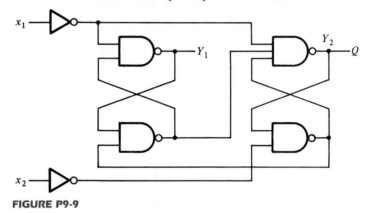

FIGURE P9-9

9-10 Implement the circuit defined in Problem 9-3 with a NOR SR latch. Repeat with a NAND SR latch.

9-11 Implement the circuit defined in Problem 9-4 with NAND SR latches.

9-12 Obtain a primitive flow table for a circuit with two inputs, x_1 and x_2, and two outputs, z_1 and z_2, that satisfy the following four conditions:
(a) When $x_1 x_2 = 00$, the output is $z_1 z_2 = 00$.
(b) When $x_1 = 1$ and x_2 changes from 0 to 1, the output is $z_1 z_2 = 01$.
(c) When $x_2 = 1$ and x_1 changes from 0 to 1, the output is $z_1 z_2 = 10$.
(d) Otherwise, the output does not change.

9-13 A traffic light is installed at a junction of a railroad and a road. The traffic light is controlled by two switches in the rails placed one mile apart on either side of the junction. A switch is turned on when the train is over it and is turned off otherwise. The traffic light changes from green (logic-0) to red (logic-1) when the beginning of the train is one mile from the junction. The light changes back to green when the end of the train is one mile away from the junction. Assume that the length of the train is less than two miles.
(a) Obtain a primitive flow table for the circuit.
(b) Show that the flow table can be reduced to four rows.

9-14 It is necessary to design an asynchronous sequential circuit with two inputs, x_1 and x_2, and one output, z. Initially, both inputs and output are equal to 0. When x_1 or x_2 becomes 1, z becomes 1. When the second input also becomes 1, the output changes to 0. The output stays at 0 until the circuit goes back to the initial state.

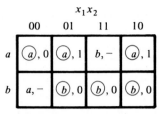

FIGURE P9-14

(a) Obtain a primitive flow table for the circuit and show that it can be reduced to the flow table shown in Fig. P9-14.

(b) Complete the design of the circuit.

9-15 Assign output values to the don't-care states in the flow tables of Fig. P9-15 in such a way as to avoid transient output pulses.

	00	01	11	10
a	(a), 0	b, −	−, −	d, −
b	a, −	(b), 1	(b), 1	c, −
c	b, −	−, −	b, −	(c), 0
d	c, −	(d), 1	c, −	(d), 1

(a)

	00	01	11	10
a	(a), 0	b, −	b, −	(a), 0
b	a, −	(b), 0	(b), 1	c, −
c	b, −	d, −	(c), 1	(c), 1
d	(d), 0	(d), 1	c, −	a, −

(b)

FIGURE P9-15

9-16 Using the implication-table method, show that the state table listed in Table 6-7 cannot be reduced any further.

9-17 Reduce the number of states in the state table listed in Problem 6-14. Use an implication table.

9-18 Merge each of the primitive flow tables shown in Fig. P9-18.
   Proceed as follows:
   (a) Find all compatible pairs by means of an implication table.
   (b) Find the maximal compatibles by means of a merger diagram.
   (c) Find a minimal set of compatibles that covers all the states and is closed.

	00	01	11	10
a	(a),0	b,-	-,-	e,-
b	a,-	(b),0	c,-	-,-
c	-,-	d,-	(c),0	h,-
d	a,-	(d),1	-,-	-,-
e	a,-	-,-	f,-	(e),0
f	-,-	g,-	(f),0	h,-
g	a,-	(g),0	-,-	-,-
h	a,-	-,-	-,-	(h),0

(a)

	00	01	11	10
a	(a),1	f,-	-,-	e,-
b	c,-	-,-	j,-	(b),0
c	(c),0	d,-	-,-	b,-
d	c,-	(d),0	g,-	-,-
e	a,-	-,-	g,-	(e),1
f	a,-	(f),1	g,-	-,-
g	-,-	d,-	(g),0	k,-
h	(h),0	d,-	-,-	k,-
j	-,-	f,-	(j),1	b,-
k	a,-	-,-	j,1	(k),0

(b)

FIGURE P9-18

9-19 (a) Obtain a binary state assignment for the reduced flow table shown in Fig. P9-19. Avoid critical race conditions.

(b) Obtain the logic diagram of the circuit using NAND latches and gates.

x_1x_2

	00	01	11	10
a	(a),0	(a),1	b,-	d,-
b	a,-	(b),0	(b),0	c,-
c	a,-	-,-	d,-	(c),0
d	a,-	a,-	(d),1	(d),1

FIGURE P9-19

9-20 Find a critical race-free state assignment for the reduced flow table shown in Fig. P9-20.

	00	01	11	10
a	(a)	d	(a)	c
b	a	(b)	(b)	d
c	d	(c)	b	(c)
d	(d)	(d)	e	(d)
e	f	c	(e)	c
f	(f)	b	a	(f)

FIGURE P9-20

9-21 Consider the reduced flow table shown in Fig. P9-21.
  (a) Obtain the transition diagram and show that three state variables are needed for a race-free binary state assignment.
  (b) Obtain the expanded flow table using the multiple-row-method assignment as specified in Fig. 9-32(a).

	00	01	11	10
a	(a)	c	(a)	d
b	a	(b)	c	(b)
c	(c)	(c)	(c)	d
d	(d)	b	a	(d)

FIGURE P9-21

9-22 Find a circuit that has no static hazards and implements the Boolean function:

$$F(A, B, C, D) = \sum (0, 2, 6, 7, 8, 10, 12)$$

9-23 Draw the logic diagram of the product of sums expression:

$$Y = (x_1 + x_2')(x_2 + x_3)$$

Show that there is a static 0-hazard when x_1 and x_3 are equal to 0 and x_2 goes from 0 to 1. Find a way to remove the hazard by adding one more OR gate.

9-24 The Boolean functions for the inputs of an SR latch are as follows. Obtain the circuit diagram using a minimum number of NAND gates.

$$S = x_1'x_2'x_3 + x_1x_2x_3$$

$$R = x_1x_2' + x_2x_3'$$

9-25 Complete the design of the circuit specified in Problem 9-13.

Digital Integrated Circuits

10-1 INTRODUCTION

The integrated circuit (IC) and the digital logic families were introduced in Section 2-8. This chapter presents the electronic circuits in each IC digital logic family and analyzes their electrical operation. A basic knowledge of electrical circuits is assumed.

The IC digital logic families to be considered here are

RTL	Resistor-transistor logic
DTL	Diode-transistor logic
TTL	Transistor-transistor logic
ECL	Emitter-coupled logic
MOS	Metal-oxide semiconductor
CMOS	Complementary metal-oxide semiconductor

The first two, RTL and DTL, have only historical significance since they are no longer used in the design of digital systems. RTL was the first commercial family to have been used extensively. It is included here because it represents a useful starting point for explaining the basic operation of digital gates. DTL circuits have been replaced by TTL. In fact, TTL is a modification of the DTL gate. The operation of the TTL gate will be easier to understand after the DTL gate is analyzed. TTL, ECL, and CMOS have a large number of SSI circuits, as well as MSI, LSI, and VLSI components. MOS is mostly used for LSI and VLSI components.

The basic circuit in each IC digital logic family is either a NAND or NOR gate. This basic circuit is the primary building block from which all other more complex digital components are obtained. Each IC logic family has available a data book that lists all the integrated circuits in that family. The differences in the logic functions available from each logic family are not so much in the functions that they achieve as in the specific electrical characteristics of the basic gate from which the circuit is constructed.

NAND and NOR gates are usually defined by the Boolean functions that they implement in terms of binary variables. When analyzing them as electronic circuits, it is necessary to investigate their input–output relationships in terms of two voltage levels: a *high* level designated by H and a *low* level designated by L. As mentioned in Section 2-8, the assignment of binary 1 to H results in a positive logic system and the assignment of binary 1 to L results in a negative logic system. The truth table in terms of H and L of a positive logic NAND gate is shown in Fig. 10-1. We notice that the output of the gate is high as long as one or more inputs are low. The output is low only when both inputs are high. The behavior of a positive logic NAND gate in terms of high and low signals can be stated as follows:

If *any* input of a NAND gate is low, the output is high.

If *all* inputs of a NAND gate are high, the output is low.

The corresponding truth table for a positive logic NOR gate is shown in Fig. 10-2. The output of the NOR gate is low when one or more inputs are high. The output is high when both inputs are low. The behavior of a positive logic NOR gate in terms of high and low signals can be stated as follows:

If *any* input of a NOR gate is high, the output is low.

If *all* inputs of a NOR gate are low, the output is high.

These statements for NAND and NOR gates must be remembered because they will be used during the analysis of the electronic gates in this chapter.

A bipolar junction transistor (BJT) can be either an *npn* or a *pnp* junction transistor. In contrast, the field-effect transistor (FET) is said to be unipolar. The operation of a bipolar transistor depends on the flow of two types of carriers: electrons and holes. A unipolar transistor depends on the flow of only one type of majority carrier, which may be electrons (n-channel) or holes (p-channel). The first four digital logic families listed, RTL, DTL, TTL, and ECL, use bipolar transistors. The last two families, MOS and

Inputs		Output
x	y	z
L	L	H
L	H	H
H	L	H
H	H	L

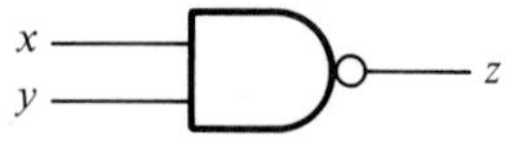

FIGURE 10-1
Positive logic NAND gate

Inputs		Output
x	y	z
L	L	H
L	H	L
H	L	L
H	H	L

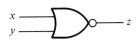

FIGURE 10-2
Positive logic NOR gate

CMOS, employ a type of unipolar transistor called a metal-oxide-semiconductor field-effect transistor, abbreviated MOSFET or MOS for short.

In this chapter, we first introduce the most common characteristics by which the digital logic families are compared. We then describe the properties of the bipolar transistor and analyze the basic gates in the bipolar logic families. We then explain the operation of the MOS transistor and introduce the basic gates of its two logic families.

10-2 SPECIAL CHARACTERISTICS

The characteristics of IC digital logic families are usually compared by analyzing the circuit of the basic gate in each family. The most important parameters that are evaluated and compared are fan-out, power dissipation, propagation delay, and noise margin. We first explain the properties of these parameters and then use them to compare the IC logic families.

Fan-Out

The fan-out of a gate specifies the number of standard loads that can be connected to the output of the gate without degrading its normal operation. A standard load is usually defined as the amount of current needed by an input of another gate in the same logic family. Sometimes the term *loading* is used instead of fan-out. This term is derived because the output of a gate can supply a limited amount of current, above which it ceases to operate properly and is said to be overloaded. The output of a gate is usually connected to the inputs of other gates. Each input consumes a certain amount of current from the gate output, so that each additional connection adds to the load of the gate. Loading rules are sometimes specified for a family of digital circuits. These rules give the maximum amount of loading allowed for each output of each circuit in the family. Exceeding the specified maximum load may cause a malfunction because the circuit cannot supply the power demanded from it. The fan-out is the maximum number of inputs that can be connected to the output of a gate, and is expressed by a number.

The fan-out is calculated from the amount of current available in the output of a gate and the amount of current needed in each input of a gate. Consider the connections shown in Fig. 10-3. The output of one gate is connected to one or more inputs of other gates. The output of the gate is in the high voltage level in Fig. 10-3(a). It provides a

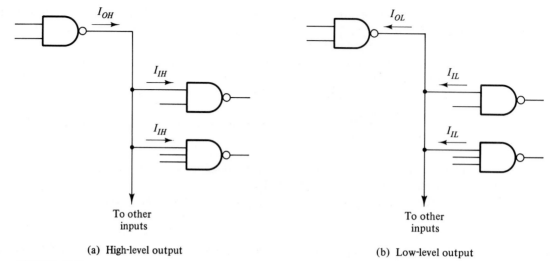

(a) High-level output

(b) Low-level output

FIGURE 10-3

Fan-out computation

current source I_{OH} to all the gate inputs connected to it. Each gate input requires a current I_{IH} for proper operation. Similarly, the output of the gate is in the low voltage level in Fig. 10-3(b). It provides a current sink I_{OL} for all the gate inputs connected to it. Each gate input supplies a current I_{IL}. The fan-out of the gate is calculated from the ratio I_{OH}/I_{IH} or I_{OL}/I_{IL}, whichever is smaller. For example, the standard TTL gates have the following values for the currents:

$$I_{OH} = 400 \ \mu A$$

$$I_{IH} = 40 \ \mu A$$

$$I_{OL} = 16 \ mA$$

$$I_{IL} = 1.6 \ mA$$

The two ratios give the same number in this case:

$$\frac{400 \ \mu A}{40 \ \mu A} = \frac{16 \ mA}{1.6 \ mA} = 10$$

Therefore, the fan-out of standard TTL is 10. This means that the output of a TTL gate can be connected to no more than ten inputs of other gates in the same logic family. Otherwise, the gate may not be able to drive or sink the amount of current needed from the inputs that are connected to it.

Power Dissipation

Every electronic circuit requires a certain amount of power to operate. The power dissipation is a parameter expressed in milliwatts (mW) and represents the amount of power needed by the gate. The number that represents this parameter does not include the power delivered from another gate; rather, it represents the power delivered to the gate from the power supply. An IC with four gates will require, from its power supply, four times the power dissipated in each gate.

The amount of power that is dissipated in a gate is calculated from the supply voltage V_{CC} and the current I_{CC} that is drawn by the circuit. The power is the product $V_{CC} \times I_{CC}$. The current drain from the power supply depends on the logic state of the gate. The current drawn from the power supply when the output of the gate is in the high-voltage level is termed I_{CCH}. When the output is in the low-voltage level, the current is I_{CCL}. The average current is

$$I_{CC}(\text{avg}) = \frac{I_{CCH} + I_{CCL}}{2}$$

and is used to calculate the average power dissipation:

$$P_D(\text{avg}) = I_{CC}(\text{avg}) \times V_{CC}$$

For example, a standard TTL NAND gate uses a supply voltage V_{CC} of 5 V and has current drains $I_{CCH} = 1$ mA and $I_{CCL} = 3$ mA. The average current is $(3 + 1)/2 = 2$ mA. The average power dissipation is $5 \times 2 = 10$ mW. An IC that has four NAND gates dissipates a total of $10 \times 4 = 40$ mW. In a typical digital system there will be many ICs, and the power required by each IC must be considered. The total power dissipation in the system is the sum total of the power dissipated in all ICs.

Propagation Delay

The propagation delay of a gate is the average transition-delay time for the signal to propagate from input to output when the binary signal changes in value. The signals through a gate take a certain amount of time to propagate from the inputs to the output. This interval of time is defined as the propagation delay of the gate. Propagation delay is measured in nanoseconds (ns). 1 ns is equal to 10^{-9} of a second.

The signals that travel from the inputs of a digital circuit to its outputs pass through a series of gates. The sum of the propagation delays through the gates is the total delay of the circuit. When speed of operation is important, each gate must have a short propagation delay and the digital circuit must have a minimum number of gates between inputs and outputs.

The average propagation delay time of a gate is calculated from the input and output waveforms, as shown in Fig. 10-4. The signal-delay time between the input and output when the output changes from the high to the low level is referred to as t_{PHL}. Similarly, when the output goes from the low to the high level, the delay is t_{PLH}. It is customary to measure the time between the 50 percent point on the input and output transitions. In general, the two delays are not the same, and both will vary with loading conditions. The average propagation-delay time is calculated as the average of the two delays.

As an example, the delays for a standard TTL gate are $t_{PHL} = 7$ ns and $t_{PLH} = 11$ ns. These quantities are given in the TTL data book and are measured with a load resistance of 400 ohms and a load capacitance of 15 pF. The average propagation delay of the TTL gate is $(11 + 7)/2 = 9$ ns.

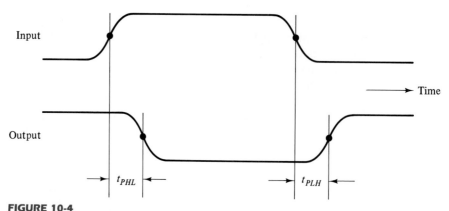

FIGURE 10-4
Measurement of propagation delay

Under certain conditions, it is more important to know the maximum delay time of a gate rather than the average value. The TTL data book lists the following maximum propagation delays for a standard NAND gate: $t_{PHL} = 15$ ns and $t_{PLH} = 22$ ns. When speed of operation is critical, it is necessary to take into account the maximum delay to ensure proper operation.

The input signals in most digital circuits are applied simultaneously to more than one gate. All the gates that are connected to external inputs constitute the first logic level of the circuit. Gates that receive at least one input from an output of a first-level gate are considered to be in the second logic level, and similarly for the third and higher logic levels. The total propagation delay of the circuit is equal to the propagation delay of a gate times the number of logic levels in the circuit. Thus, a reduction in the number of logic levels results in a reduction of signal delay and faster circuits. The reduction of the propagation delay in circuits may be more important than the reduction of the total number of gates if speed of operation is a major factor.

Noise Margin

Spurious electrical signals from industrial and other similar sources can induce undesirable voltages on the connecting wires between logic circuits. These unwanted signals are referred to as *noise*. There are two types of noise to be considered. DC noise is caused by a drift in the voltage levels of a signal. AC noise is a random pulse that may be created by other switching signals. Thus, noise is a term used to denote an undesirable signal that is superimposed upon the normal operating signal. *Noise margin* is the maximum noise voltage added to an input signal of a digital circuit that does not cause an undesirable change in the circuit output. The ability of circuits to operate reliably in a noise environment is important in many applications. Noise margin is expressed in volts and represents the maximum noise signal that can be tolerated by the gate.

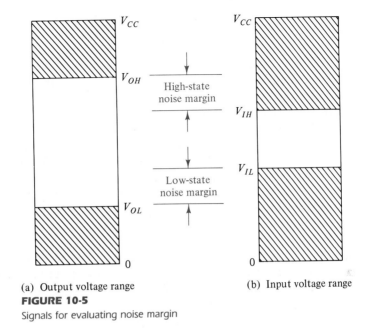

(a) Output voltage range   (b) Input voltage range

FIGURE 10-5

Signals for evaluating noise margin

The noise margin is calculated from knowledge of the voltage signal available in the output of the gate and the voltage signal required in the input of the gate. Figure 10-5 illustrates the signals for computing noise margin. Part (a) shows the range of output voltages that can occur in a typical gate. Any voltage in the gate output between V_{CC} and V_{OH} is considered as the high-level state and any voltage between 0 and V_{OL} in the gate output is considered as the low-level state. Voltages between V_{OL} and V_{OH} are indeterminate and do not appear under normal operating conditions except during transition between the two levels. The corresponding two voltage ranges that are recognized by the input of the gate are indicated in Fig. 10-5(b). In order to compensate for any noise signal, the circuit must be designed so that V_{IL} is greater than V_{OL} and V_{IH} is less than V_{OH}. The noise margin is the difference $V_{OH} - V_{IH}$ or $V_{IL} - V_{OL}$, whichever is smaller.

As illustrated in Fig. 10-5, V_{OL} is the maximum voltage that the output can be when in the low-level state. The circuit can tolerate any noise signal that is less than the noise margin ($V_{IL} - V_{OL}$) because the input will recognize the signal as being in the low-level state. Any signal greater than V_{OL} plus the noise-margin figure will send the input voltage into the indeterminate range, which may cause an error in the output of the gate. In a similar fashion, a negative-voltage noise greater than $V_{OH} - V_{IH}$ will send the input voltage into the indeterminate range.

The parameters for the noise margin in a standard TTL NAND gate are $V_{OH} = 2.4$ V, $V_{OL} = 0.4$ V, $V_{IH} = 2$ V, and $V_{IL} = 0.8$ V. The high-state noise margin is $2.4 - 2 = 0.4$ V, and the low-state noise margin is $0.8 - 0.4 = 0.4$ V. In this case, both values are the same.

10-3 BIPOLAR-TRANSISTOR CHARACTERISTICS

This section is devoted to a review of the bipolar transistor as applied to digital circuits. This information will be used for the analysis of the basic circuit in the four bipolar logic families. Bipolar transistors may be of the *npn* or *pnp* type. Moreover, they are constructed either with germanium or silicon semiconductor material. IC transistors, however, are made with silicon and are usually of the *npn* type.

The basic data needed for the analysis of digital circuits may be obtained from inspection of the typical characteristic curves of a common-emitter *npn* silicon transistor, shown in Fig. 10-6. The circuit in (a) is a simple inverter with two resistors and a transistor. The current marked I_C flows through resistor R_C and the collector of the transistor. Current I_B flows through resistor R_B and the base of the transistor. The emitter is connected to ground and its current $I_E = I_C + I_B$. The supply voltage is between V_{CC} and ground. The input is between V_i and ground, and the output is between V_o and ground.

We have assumed a positive direction for the currents as indicated. These are the directions in which the currents normally flow in an *npn* transistor. Collector and base currents, I_C and I_B, respectively, are positive when they flow into the transistor. Emit-

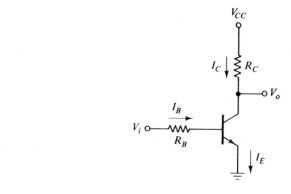

(a) Inverter circuit

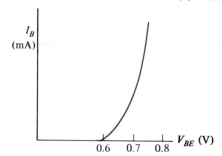

(b) Transistor-base characteristic

(c) Transistor-collector characteristic

FIGURE 10-6

Silicon *npn* transistor characteristics

ter current I_E is positive when it flows out of the transistor, as indicated by the arrow in the emitter terminal. The symbol V_{CE} stands for the voltage drop from collector to emitter and is always positive. Correspondingly, V_{BE} is the voltage drop across the base-to-emitter junction. This junction is forward biased when V_{BE} is positive. It is reverse biased when V_{BE} is negative.

The base–emitter graphical characteristic is shown in Fig. 10-6(b). This is a plot of V_{BE} versus I_B. If the base–emitter voltage is less than 0.6 V, the transistor is said to be *cut off* and no base current flows. When the base–emitter junction is forward biased with a voltage greater than 0.6 V, the transistor conducts and I_B starts rising very fast whereas V_{BE} changes very little. The voltage V_{BE} across a conducting transistor seldom exceeds 0.8 V.

The graphical collector–emitter characteristics, together with the load line, are shown in Fig. 10-6(c). When V_{BE} is less than 0.6 V, the transistor is cut off with $I_B = 0$ and a negligible current flows in the collector. The collector-to-emitter circuit then behaves like an open circuit. In the *active* region, collector voltage V_{CE} may be anywhere from about 0.8 V up to V_{CC}. Collector current I_C in this region can be calculated to be approximately equal to $I_B h_{FE}$, where h_{FE} is a transistor parameter called the *dc current gain*. The maximum collector current depends not on I_B, but rather on the external circuit connected to the collector. This is because V_{CE} is always positive and its lowest possible value is 0 V. For example, in the inverter shown, the maximum I_C is obtained by making $V_{CE} = 0$ to obtain $I_C = V_{CC}/R_C$.

It was stated that $I_C = h_{FE} I_B$ in the active region. The parameter h_{FE} varies widely over the operating range of the transistor, but still it is useful to employ an average value for the purpose of analysis. In a typical operating range, h_{FE} is about 50, but under certain conditions, it could go down to as low as 20. It must be realized that the base current I_B may be increased to any desirable value, but the collector current I_C is limited by external circuit parameters. As a consequence, a situation can be reached where $h_{FE} I_B$ is greater than I_C. If this condition exists, then the transistor is said to be in the *saturation* region. Thus, the condition for saturation is determined from the relationship

$$I_B \geqslant \frac{I_{CS}}{h_{FE}}$$

where I_{CS} is the maximum collector current flowing during saturation. V_{CE} is not exactly zero in the saturation region, but is normally about 0.2 V.

The basic data needed for analyzing bipolar transistor digital circuits are listed in Table 10-1. In the cutoff region, V_{BE} is less than 0.6 V, V_{CE} is considered as an open circuit, and both currents are negligible. In the active region, V_{BE} is about 0.7 V, V_{CE} may vary over a wide range, and I_C can be calculated as a function of I_B. In the saturation region, V_{BE} hardly changes, but V_{CE} drops to 0.2 V. The base current must be large enough to satisfy the inequality listed. To simplify the analysis, we will assume that $V_{BE} = 0.7$ V if the transistor is conducting, whether in the active or saturation region.

The analysis of digital circuits may be undertaken using a prescribed procedure: For each transistor in the circuit, determine if its V_{BE} is less than 0.6 V. If so, then the tran-

TABLE 10-1
Typical *npn* Silicon Transistor Parameters

Region	V_{BE} (V)*	V_{CE} (V)	Current Relationship
Cutoff	< 0.6	Open circuit	$I_B = I_C = 0$
Active	0.6–0.7	> 0.8	$I_C = h_{FE} I_B$
Saturation	0.7–0.8	0.2	$I_B \geqslant I_{CS}/h_{FE}$

*V_{BE} will be assumed to be 0.7 V if the transistor is conducting, whether in the active or saturation region.

sistor is cut off and the collector-to-emitter circuit is considered an open circuit. If V_{BE} is greater than 0.6 V, the transistor may be in the active or saturation region. Calculate the base current, assuming that $V_{BE} = 0.7$ V. Then calculate the maximum possible value of collector current I_{CS}, assuming $V_{CE} = 0.2$ V. These calculations will be in terms of voltages applied and resistor values. Then, if the base current is large enough that $I_B \geqslant I_{CS}/h_{FE}$, we deduce that the transistor is in the saturation region with $V_{CE} = 0.2$ V. However, if the base current is smaller and the above relationship is not satisfied, the transistor is in the active region and we recalculate collector current I_C using the equation $I_C = h_{FE} I_B$.

To demonstrate with an example, consider the inverter circuit of Fig. 10-6(a) with the following parameters:

$$R_C = 1 \text{ k}\Omega \qquad V_{CC} = 5 \text{ V (voltage supply)}$$

$$R_B = 22 \text{ k}\Omega \qquad H = 5 \text{ V (high-level voltage)}$$

$$h_{FE} = 50 \qquad L = 0.2 \text{ V (low-level voltage)}$$

With input voltage $V_i = L = 0.2$ V, we have that $V_{BE} < 0.6$ V and the transistor is cut off. The collector–emitter circuit behaves like an open circuit; so output voltage $V_o = 5$ V $= H$.

With input voltage $V_i = H = 5$ V, we deduce that $V_{BE} > 0.6$ V. Assuming that $V_{BE} = 0.7$, we calculate the base current:

$$I_B = \frac{V_i - V_{BE}}{R_B} = \frac{5 - 0.7}{22 \text{ k}\Omega} = 0.195 \text{ mA}$$

The maximum collector current, assuming $V_{CE} = 0.2$ V, is

$$I_{CS} = \frac{V_{CC} - V_{CE}}{R_C} = \frac{5 - 0.2}{1 \text{ k}\Omega} = 4.8 \text{ mA}$$

We then check for saturation:

$$0.195 = I_B \geqslant \frac{I_{CS}}{h_{FE}} = \frac{4.8}{50} = 0.096 \text{ mA}$$

and find that the inequality is satisfied since $0.195 > 0.096$. We conclude that the transistor is saturated and output voltage $V_o = V_{CE} = 0.2$ V $= L$. Thus, the circuit behaves as an inverter.

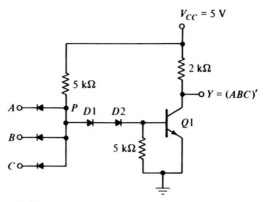

FIGURE 10-9
DTL basic NAND gate

The voltage at P now is equal to V_{BE} plus the two diode drops across $D1$ and $D2$, or $0.7 \times 3 = 2.1$ V. Since all inputs are high at 5 V and $V_P = 2.1$ V, the input diodes are reverse biased and off. The base current is equal to the difference of currents flowing in the two 5-kΩ resistors and is sufficient to drive the transistor into saturation (see Problem 10-3). With the transistor saturated, the output drops to V_{CE} of 0.2 V, which is the low level for the gate.

The power dissipation of a DTL gate is about 12 mW and the propagation delay averages 30 ns. The noise margin is about 1 V and a fan-out as high as 8 is possible. The fan-out of the DTL gate is limited by the maximum current that can flow in the collector of the saturated transistor (see Problem 10-4).

The fan-out of a DTL gate may be increased by replacing one of the diodes in the base circuit with a transistor, as shown in Fig. 10-10. Transistor $Q1$ is maintained in

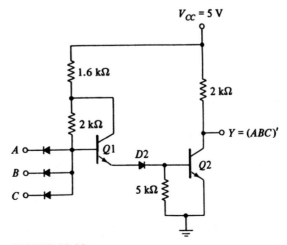

FIGURE 10-10
Modified DTL gate

the active region when output transistor $Q2$ is saturated. As a consequence, the modified circuit can supply a larger amount of base current to the output transistor. The output transistor can now draw a larger amount of collector current before it goes out of saturation. Part of the collector current comes from the conducting diodes in the loading gates when $Q2$ is saturated. Thus, an increase in allowable collector saturated current allows more loads to be connected to the output, which increases the fan-out capability of the gate.

10-5 TRANSISTOR-TRANSISTOR LOGIC (TTL)

The original basic TTL gate was a slight improvement over the DTL gate. As the TTL technology progressed, additional improvements were added to the point where this logic family became the most widely used family in the design of digital systems. There are several subfamilies or series of the TTL technology. The names and characteristics of seven TTL series appear in Table 10-2. Commercial TTL ICs have a number designation that starts with 74 and follows with a suffix that identifies the series type. Examples are 7404, 74S86, and 74ALS161. Fan-out, power dissipation and propagation delay were defined in Section 10-2. The speed–power product is an important parameter for comparing the various TTL series. This is the product of the propagation delay and power dissipation and is measured in picojoules (pJ). A low value for this parameter is desirable, because it indicates that a given propagation delay can be achieved without excessive power dissipation, and vice versa.

The standard TTL gate was the first version in the TTL family. This basic gate was then designed with different resistor values to produce gates with lower power dissipation or with higher speed. The propagation delay of a transistor circuit that goes into saturation depends mostly on two factors: storage time and RC time constants. Reducing the storage time decreases the propagation delay. Reducing resistor values in the circuit reduces the RC time constants and decreases the propagation delay. Of course, the trade-off is higher power dissipation because lower resistances draw more current

TABLE 10-2
TTL Series and Their Characteristics

TTL Series Name	Prefix	Fan-out	Power Dissipation (mW)	Propagation Delay (ns)	Speed–Power Product (pJ)
Standard	74	10	10	9	90
Low-power	74L	20	1	33	33
High-speed	74H	10	22	6	132
Schottky	74S	10	19	3	57
Low-power Schottky	74LS	20	2	9.5	19
Advanced Schottky	74AS	40	10	1.5	15
Advanced low-power Schottky	74ALS	20	1	4	4

from the power supply. The speed of the gate is inversely proportional to the propagation delay.

In the low-power TTL gate, the resistor values are higher than in the standard gate to reduce the power dissipation, but the propagation delay is increased. In the high-speed TTL gate, resistor values are lowered to reduce the propagation delay, but the power dissipation is increased. The Schottky TTL gate was the next improvement in the technology. The effect of the Schottky transistor is to remove the storage time delay by preventing the transistor from going into saturation. This series increases the speed of operation without an excessive increase in power dissipation. The low-power Schottky TTL sacrifices some speed for reduced power dissipation. It is equal to the standard TTL in propagation delay, but has only one-fifth the power dissipation. Recent innovations have led to the development of the advanced Schottky series. It provides an improvement in propagation delay over the Schottky series and also lowers the power dissipation. The advanced low-power Schottky has the lowest speed–power product and is the most efficient series. It is replacing all other low-power versions in new designs.

All TTL series are available in SSI and in more complex forms as MSI and LSI components. The differences in the TTL series are not in the digital logic that they perform, but rather in the internal construction of the basic NAND gate. In any case, TTL gates in all the available series come in three different types of output configuration:

1. Open-collector output
2. Totem-pole output
3. Three-state (or tristate) output

These three types of outputs will be considered in conjunction with the circuit description of the basic TTL gate.

Open-Collector Output Gate

The basic TTL gate shown in Fig. 10-11 is a modified circuit of the DTL gate. The multiple emitters in transistor $Q1$ are connected to the inputs. These emitters behave most of the time like the input diodes in the DTL gate since they form a *pn* junction with their common base. The base–collector junction of $Q1$ acts as another *pn* junction diode corresponding to $D1$ in the DTL gate (see Fig. 10-5). Transistor $Q2$ replaces the second diode, $D2$, in the DTL gate. The output of the TTL gate is taken from the open collector of $Q3$. A resistor connected to V_{CC} must be inserted external to the IC package for the output to "pull up" to the high voltage level when $Q3$ is off; otherwise, the output acts as an open circuit. The reason for not providing the resistor internally will be discussed later.

The two voltage levels of the TTL gate are 0.2 V for the low level and from 2.4 to 5 V for the high level. The basic circuit is a NAND gate. If any input is low, the corresponding base–emitter junction in $Q1$ is forward biased. The voltage at the base of $Q1$ is equal to the input voltage of 0.2 V plus a V_{BE} drop of 0.7 or 0.9 V. In order for $Q3$ to start conducting, the path from $Q1$ to $Q3$ must overcome a potential of one diode drop

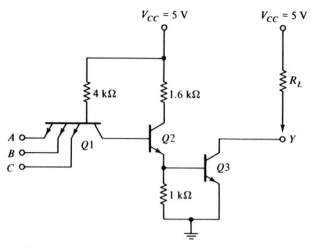

FIGURE 10-11
Open-collector TTL gate

in the base–collector *pn* junction of $Q1$ and two V_{BE} drops in $Q2$ and $Q3$, or $3 \times 0.6 = 1.8$ V. Since the base of $Q1$ is maintained at 0.9 V by the input signal, the output transistor cannot conduct and is cut off. The output level will be high if an external resistor is connected between the output and V_{CC} (or an open circuit if a resistor is not used).

If all inputs are high, both $Q2$ and $Q3$ conduct and saturate. The base voltage of $Q1$ is equal to the voltage across its base–collector *pn* junction plus two V_{BE} drops in $Q2$ and $Q3$, or about $0.7 \times 3 = 2.1$ V. Since all inputs are high and greater than 2.4 V, the base–emitter junctions of $Q1$ are all reverse biased. When output transistor $Q3$ saturates (provided it has a current path), the output voltage goes low to 0.2 V. This confirms the conditions of a NAND operation.

In this analysis, we said that the base–collector junction of $Q1$ acts like a *pn* diode junction. This is true in the steady-state condition. However, during the turn-off transition, $Q1$ does exhibit transistor action, resulting in a reduction in propagation delay. When all inputs are high and then one of the inputs is brought to a low level, both $Q2$ and $Q3$ start turning off. At this time, the collector junction of $Q1$ is reverse biased and the emitter is forward biased; so transistor $Q1$ goes momentarily into the active region. The collector current of $Q1$ comes from the base of $Q2$ and quickly removes the excess charge stored in $Q2$ during its previous saturation state. This causes a reduction in the storage time of the circuit as compared to the DTL type of input. The result is a reduction of the turn-off time of the gate.

The open-collector TTL gate will operate without the external resistor when connected to inputs of other TTL gates, although this is not recommended because of the low noise immunity encountered. Without an external resistor, the output of the gate will be an open circuit when $Q3$ is off. An open circuit to an input of a TTL gate behaves as if it has a high-level input (but a small amount of noise can change this to a low level). When $Q3$ conducts, its collector will have a current path supplied by the input of the loading gate through V_{CC}, the 4-kΩ resistor, and the forward-biased base-emitter junction.

Open-collector gates are used in three major applications: driving a lamp or relay, performing wired logic, and for the construction of a common-bus system. An open-collector output can drive a lamp placed in its output through a limiting resistor. When the output is low, the saturated transistor $Q3$ forms a path for the current that turns the lamp on. When the output transistor is off, the lamp turns off because there is no path for the current.

If the outputs of several open-collector TTL gates are tied together with a single external resistor, a wired-AND logic is performed. Remember that a positive-logic AND function gives a high level only if all variables are high; otherwise, the function is low. With outputs of open-collector gates connected together, the common output is high only when all output transistors are off (or high). If an output transistor conducts, it forces the output to the low state.

The wired logic performed with open-collector TTL gates is depicted in Fig. 10-12. The physical wiring in (a) shows how the outputs must be connected to a common resistor. The graphic symbol for such a connection is demonstrated in (b). The AND function formed by connecting together the two outputs is called a wired-AND function. The AND gate is drawn with the lines going through the center of the gate to distinguish it from a conventional gate. The wired-AND gate is not a physical gate, but only a symbol to designate the function obtained from the indicated connection. The Boolean function obtained from the circuit of Fig. 10-12 is the AND operation between the outputs of the two NAND gates:

$$Y = (AB)' \cdot (CD)' = (AB + CD)'$$

The second expression is preferred since it shows an operation commonly referred to as an AND-OR-INVERT function (see Section 3-7).

Open-collector gates can be tied together to form a common bus. At any time, all gate outputs tied to the bus, except one, must be maintained in their high state. The se-

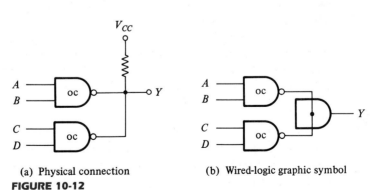

(a) Physical connection   (b) Wired-logic graphic symbol

FIGURE 10-12

Wired-AND of two open-collector (oc) gates, $Y = (AB + CD)'$

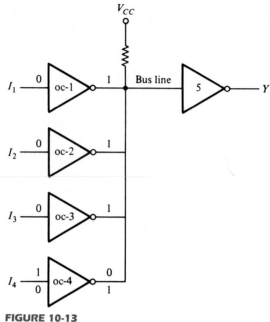

FIGURE 10-13
Open-collector gates forming a common bus line

lected gate may be either in the high or low state, depending on whether we want to transmit a 1 or 0 on the bus. Control circuits must be used to select the particular gate that drives the bus at any given time.

Figure 10-13 demonstrates the connection of four sources tied to a common bus line. Each of the four inputs drives an open-collector inverter, and the outputs of the inverters are tied together to form a single bus line. The figure shows that three of the inputs are 0, which produces a 1 or high level on the bus. The fourth input, I_4, can now transmit information through the common-bus line into inverter 5. Remember that an AND operation is performed in the wired logic. If $I_4 = 1$, the output of gate 4 is 0 and the wired-AND operation produces a 0. If $I_4 = 0$, the output of gate 4 is 1 and the wired-AND operation produces a 1. Thus, if all other outputs are maintained at 1, the selected gate can transmit its value through the bus. The value transmitted is the complement of I_4, but inverter 5 in the receiving end can easily invert this signal again to make $Y = I_4$.

Totem-Pole Output

The output impedance of a gate is normally a resistive plus a capacitive load. The capacitive load consists of the capacitance of the output transistor, the capacitance of the fan-out gates, and any stray wiring capacitance. When the output changes from the low to the high state, the output transistor of the gate goes from saturation to cutoff and the total load capacitance, C, charges exponentially from the low to the high voltage level with a time constant equal to RC. For the open-collector gate, R is the external resistor marked R_L. For a typical operating value of $C = 15$ pF and $R_L = 4$ kΩ, the propagation delay of a TTL open-collector gate during the turn-off time is 35 ns. With an *active pull-up* circuit replacing the passive pull-up resistor R_L, the propagation delay is reduced to 10 ns. This configuration, shown in Fig. 10-14, is called a *totem-pole* output because transistor $Q4$ "sits" upon $Q3$.

The TTL gate with the totem-pole output is the same as the open-collector gate, except for the output transistor $Q4$ and the diode $D1$. When the output Y is in the low state, $Q2$ and $Q3$ are driven into saturation as in the open-collector gate. The voltage in the collector of $Q2$ is $V_{BE}(Q3) + V_{CE}(Q2)$ or $0.7 + 0.2 = 0.9$ V. The output $Y = V_{CE}(Q3) = 0.2$ V. Transistor $Q4$ is cutoff because its base must be one V_{BE} drop plus one diode drop, or $2 \times 0.6 = 1.2$ V, to start conducting. Since the collector of $Q2$ is connected to the base of $Q4$, the latter's voltage is only 0.9 V instead of the required 1.2 V, and so $Q4$ is cut off. The reason for placing the diode in the circuit is to provide a diode drop in the output path and thus ensure that $Q4$ is cut off when $Q3$ is saturated.

When the output changes to the high state because one of the inputs drops to the low state, transistors $Q2$ and $Q3$ go into cutoff. However, the output remains momentarily low because the voltages across the load capacitance cannot change instantaneously. As soon as $Q2$ turns off, $Q4$ conducts because its base is connected to V_{CC} through the 1.6-kΩ resistor. The current needed to charge the load capacitance causes $Q4$ to momentarily saturate, and the output voltage rises with a time constant RC. But R in this case is equal to 130 Ω, plus the saturation resistance of $Q4$, plus the resistance of the diode, for a total of approximately 150 Ω. This value of R is much smaller than the

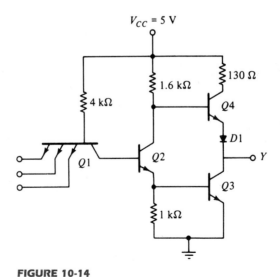

FIGURE 10-14

TTL gate with totem-pole output

passive pull-up resistance used in the open-collector circuit. As a consequence, the transition from the low to high level is much faster.

As the capacitive load charges, the output voltage rises and the current in $Q4$ decreases, bringing the transistor into the active region. Thus, in contrast to the other transistors, $Q4$ is in the *active* region when in a steady-state condition. The final value of the output voltage is then 5 V, minus a V_{BE} drop in $Q4$, minus a diode drop in $D1$ to about 3.6 V. Transistor $Q3$ goes into cutoff very fast, but during the initial transition time, both $Q3$ and $Q4$ are on and a peak current is drawn from the power supply. This current spike generates noise in the power-supply distribution system. When the change of state is frequent, the transient-current spikes increase the power-supply current requirement and the average power dissipation of the circuit increases.

The wired-logic connection is not allowed with totem-pole output circuits. When two totem-poles are wired together with the output of one gate high and the output of the second gate low, the excessive amount of current drawn can produce enough heat to damage the transistors in the circuit (see Problem 10-7). Some TTL gates are constructed to withstand the amount of current that flows under this condition. In any case, the collector current in the low gate may be high enough to move the transistor into the active region and produce an output voltage in the wired connection greater than 0.8 V, which is not a valid binary signal for TTL gates.

Schottky TTL Gate

As mentioned before, a reduction in storage time results in a reduction of propagation delay. This is because the time needed for a transistor to come out of saturation delays the switching of the transistor from the on condition to the off condition. Saturation can be eliminated by placing a Schottky diode between the base and collector of each saturated transistor in the circuit. The Schottky diode is formed by the junction of a metal and semiconductor, in contrast to a conventional diode, which is formed by the junction of p-type and n-type semiconductor material. The voltage across a conducting Schottky diode is only 0.4 V, as compared to 0.7 V in a conventional diode. The presence of a Schottky diode between the base and collector prevents the transistor from going into saturation. The resulting transistor is called a *Schottky transistor*. The use of Schottky transistors in a TTL decreases the propagation delay without a sacrifice of power dissipation.

The Schottky TTL gate is shown in Fig. 10-15. Note the special symbol used for the Schottky transistors and diodes. The diagram shows all transistors to be of the Schottky type except $Q4$. An exception is made of $Q4$ since it does not saturate, but stays in the active region. Note also that resistor values have been reduced to further decrease the propagation delay.

In addition to using Schottky transistors and lower resistor values, the circuit of Fig. 10-15 includes other modifications not available in the standard gate of Fig. 10-14. Two new transistors, $Q5$ and $Q6$ have been added, and Schottky diodes are inserted between each input terminal and ground. There is no diode in the totem-pole circuit.

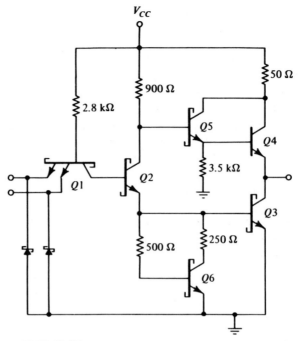

FIGURE 10-15
Schottky TTL gate

However, the new combination of $Q5$ and $Q4$ still gives the two V_{BE} drops necessary to prevent $Q4$ from conducting when the output is low. This combination comprises a double emitter-follower called a *Darlington pair*. The Darlington pair provides a very high current gain and extremely low resistance. This is exactly what is needed during the low-to-high swing of the output, resulting in a decrease of propagation delay.

The diodes in each input shown in the circuit help clamp any ringing that may occur in the input lines. Under transient switching conditions, signal lines appear inductive; this, along with stray capacitance, causes signals to oscillate or "ring." When the output of a gate switches from the high to the low state, the ringing waveform at the input may have excursions below ground as great as 2–3 V, depending on line length. The diodes connected to ground help clamp this ringing since they conduct as soon as the negative voltage exceeds 0.4 V. When the negative excursion is limited, the positive swing is also reduced. The success of the clamp diodes in limiting line effects has been so successful that all versions of TTL gates use them.

The emitter resistor of $Q2$ in Fig. 10-14 has been replaced in Fig. 10-15 by a circuit consisting of transistor $Q6$ and two resistors. The effect of this circuit is to reduce the turn-off current spikes discussed previously. The analysis of this circuit, which helps to reduce the propagation time of the gate, is too involved to present in this brief discussion.

Three-State Gate

As mentioned earlier, the outputs of two TTL gates with totem-pole structures cannot be connected together as in open-collector outputs. There is, however, a special type of totem-pole gate that allows the wired connection of outputs for the purpose of forming a common-bus system. When a totem-pole output TTL gate has this property, it is called a *three-state* (or tristate) gate.

A three-state gate exhibits three output states: (1) a low-level state when the lower transistor in the totem-pole is on and the upper transistor is off, (2) a high-level state when the upper transistor in the totem-pole is on and the lower transistor is off, and (3) a third state when both transistors in the totem-pole are off. The third state provides an open circuit or high-impedance state that allows a direct wire connection of many outputs to a common line. Three-state gates eliminate the need for open-collector gates in bus configurations.

Figure 10-16(a) shows the graphic symbol of a three-state buffer gate. When the control input C is high, the gate is enabled and behaves like a normal buffer with the output equal to the input binary value. When the control input is low, the output is an open circuit, which gives a high impedance (the third state) regardless of the value of input A. Some three-state gates produce a high-impedance state when the control input is high. This is shown symbolically in Fig. 10-16(b). Here we have two small circles, one for the inverter output and the other to indicate that the gate is enabled when C is low.

The circuit diagram of the three-state inverter is shown in Fig. 10-16(c). Transistors $Q6$, $Q7$, and $Q8$ associated with the control input form a circuit similar to the open-collector gate. Transistors $Q1$–$Q5$, associated with the data input, form a totem-pole TTL circuit. The two circuits are connected together through diode $D1$. As in an open-collector circuit, transistor $Q8$ turns off when the control input at C is in the low-level state. This prevents diode $D1$ from conducting, and also, the emitter in $Q1$ connected to $Q8$ has no conduction path. Under this condition, transistor $Q8$ has no effect on the operation of the gate and the output in Y depends only on the data input at A.

When the control input is high, transistor $Q8$ turns on, and the current flowing from V_{CC} through diode $D1$ causes transistor $Q8$ to saturate. The voltage at the base of $Q5$ is now equal to the voltage across the saturated transistor, $Q8$, plus one diode drop, or 0.9 V. This voltage turns off $Q5$ and $Q4$ since it is less than two V_{BE} drops. At the same time, the low input to one of the emitters of $Q1$ forces transistor $Q3$ (and $Q2$) to turn off. Thus, both $Q3$ and $Q4$ in the totem-pole are turned off and the output of the circuit behaves like an open circuit with a very high output impedance.

A three-state bus is created by wiring several three-state outputs together. At any given time, only one control input is enabled while all other outputs are in the high-impedance state. The single gate not in a high-impedance state can transmit binary information through the common bus. Extreme care must be taken that all except one of the outputs are in the third state; otherwise, we have the undesirable condition of having two active totem-pole outputs connected together.

An important feature of most three-state gates is that the output enable delay is longer than the output disable delay. If a control circuit enables one gate and disables

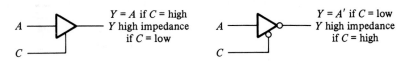

(a) Three-state buffer gate　　　　(b) Three-state inverter gate

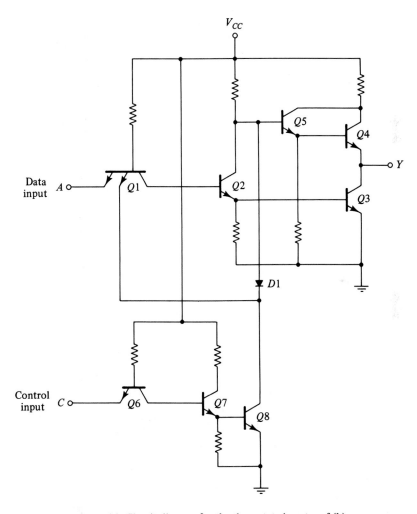

(c) Circuit diagram for the three-state inverter of (b)

FIGURE 10-16

Three-state TTL gate

another at the same time, the disabled gate enters the high-impedance state before the other gate is enabled. This eliminates the situation of both gates being active at the same time.

There is a very small leakage current associated with the high-impedance condition in a three-state gate. Nevertheless, this current is so small that as many as 100 three-state outputs can be connected together to form a common-bus line.

10-6   EMITTER-COUPLED LOGIC (ECL)

Emitter-coupled logic (ECL) is a nonsaturated digital logic family. Since transistors do not saturate, it is possible to achieve propagation delays of 2 ns and even below 1 ns. This logic family has the lowest propagation delay of any family and is used mostly in systems requiring very high-speed operation. Its noise immunity and power dissipation, however, are the worst of all the logic families available.

A typical basic circuit of the ECL family is shown in Fig. 10-17. The outputs

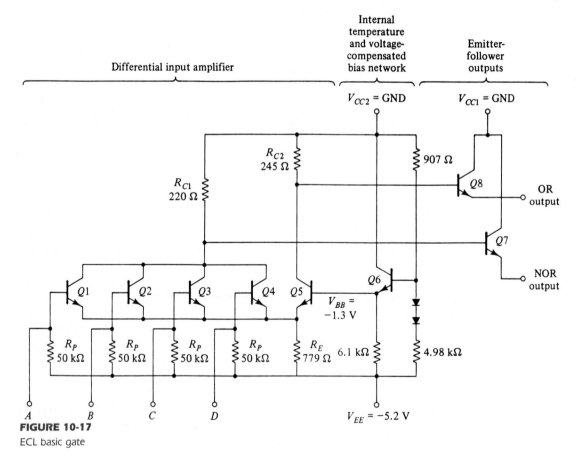

FIGURE 10-17
ECL basic gate

provide both the OR and NOR functions. Each input is connected to the base of a transistor. The two voltage levels are about -0.8 V for the high state and about -1.8 V for the low state. The circuit consists of a differential amplifier, a temperature- and voltage-compensated bias network, and an emitter-follower output. The emitter outputs require a pull-down resistor for current to flow. This is obtained from the input resistor, R_P, of another similar gate or from an external resistor connected to a negative voltage supply.

The internal temperature- and voltage-compensated bias circuit supplies a reference voltage to the differential amplifier. Bias voltage V_{BB} is set at -1.3 V, which is the midpoint of the signal logic swing. The diodes in the voltage divider, together with $Q6$, provide a circuit that maintains a constant V_{BB} value despite changes in temperature or supply voltage. Any one of the power supply inputs could be used as ground. However, the use of the V_{CC} node as ground and V_{EE} at -5.2 V results in best noise immunity.

If any input in the ECL gate is high, the corresponding transistor is turned on and $Q5$ is turned off. An input of -0.8 V causes the transistor to conduct and places -1.6 V on the emitters of all transistors (V_{BE} drop in ECL transistors is 0.8 V). Since $V_{BB} = -1.3$ V, the base voltage of $Q5$ is only 0.3 V more positive than its emitter. $Q5$ is cut off because its V_{BE} voltage needs at least 0.6 V to start conducting. The current in resistor R_{C2} flows into the base of $Q8$ (provided there is a load resistor). This current is so small that only a negligible voltage drop occurs across R_{C2}. The OR output of the gate is one V_{BE} drop below ground, or -0.8 V, which is the high state. The current flowing through R_{C1} and the conducting transistor causes a drop of about 1 V below ground (see Problem 10-9). The NOR output is one V_{BE} drop below this level, or at -1.8 V, which is the low state.

If all inputs are at the low level, all input transistors turn off and $Q5$ conducts. The voltage in the common-emitter node is one V_{BE} drop below V_{BB}, or -2.1 V. Since the base of each input is at a low level of -1.8 V, each base–emitter junction has only 0.3 V and all input transistors are cut off. R_{C2} draws current through $Q5$ that results in a voltage drop of about 1 V, making the OR output one V_{BE} drop below this, at -1.8 V or the low level. The current in R_{C1} is negligible and the NOR output is one V_{BE} drop below ground, at -0.8 V or the high level. This verifies the OR and NOR operations of the circuit.

The propagation delay of the ECL gate is 2 ns, and the power dissipation is 25 mW. This gives a speed–power product of 50, which is about the same as for the Schottky TTL. The noise margin is about 0.3 V and not as good as in the TTL gate. High fan-out is possible in the ECL gate because of the high input impedance of the differential amplifier and the low output impedance of the emitter-follower. Because of the extreme high speed of the signals, external wires act like transmission lines. Except for very short wires of a few centimeters, ECL outputs must use coaxial cables with a resistor termination to reduce line reflections.

The graphic symbol for the ECL gate is shown in Fig. 10-18(a). Two outputs are available: one for the NOR function and the other for the OR function. The outputs of two or more ECL gates can be connected together to form wired logic. As shown in

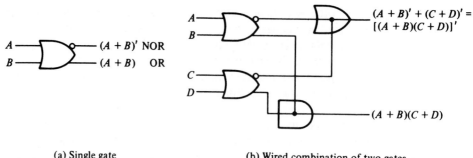

(a) Single gate

(b) Wired combination of two gates

FIGURE 10-18

Graphic symbols of ECL gates

Fig. 10-18(b), an *external* wired connection of two NOR outputs produces a wired-OR function. An *internal* wired connection of two OR outputs is employed in some ECL ICs to produce a wired-AND (sometimes called dot-AND) logic. This property may be utilized when ECL gates are used to form the OR-AND-INVERT and the OR-AND functions.

10-7   METAL-OXIDE SEMICONDUCTOR (MOS)

The field-effect transistor (FET) is a unipolar transistor, since its operation depends on the flow of only one type of carrier. There are two types of field-effect transistors: the junction field-effect transistor (JFET) and the metal-oxide semi-conductor (MOS). The former is used in linear circuits and the latter in digital circuits. MOS transistors can be fabricated in less area than bipolar transistors.

The basic structure of the MOS transistor is shown in Fig. 10-19. The p-channel MOS consists of a lightly doped substrate of n-type silicon material. Two regions are heavily doped by diffusion with p-type impurities to form the *source* and *drain*. The region between the two p-type sections serves as the *channel*. The *gate* is a metal plate separated from the channel by an insulated dielectric of silicon dioxide. A negative

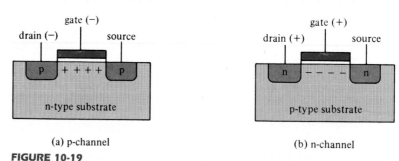

(a) p-channel

(b) n-channel

FIGURE 10-19

Basic structure of MOS transistor

voltage (with respect to the substrate) at the gate terminal causes an induced electric field in the channel that attracts p-type carriers from the substrate. As the magnitude of the negative voltage on the gate increases, the region below the gate accumulates more positive carriers, the conductivity increases, and current can flow from source to drain provided a voltage difference is maintained between these two terminals.

There are four basic types of MOS structures. The channel can be a p- or n-type, depending on whether the majority carriers are holes or electrons. The mode of operation can be enhancement or depletion, depending on the state of the channel region at zero gate voltage. If the channel is initially doped lightly with p-type impurity (diffused channel), a conducting channel exists at zero gate voltage and the device is said to operate in the *depletion* mode. In this mode, current flows unless the channel is depleted by an applied gate field. If the region beneath the gate is left initially uncharged, a channel must be induced by the gate field before current can flow. Thus, the channel current is enhanced by the gate voltage and such a device is said to operate in the *enhancement* mode.

The source is the terminal through which the majority carriers enter the bar. The drain is the terminal through which the majority carriers leave the bar. In a p-channel MOS, the source terminal is connected to the substrate and a negative voltage is applied to the drain terminal. When the gate voltage is above a threshold voltage V_T (about -2 V), no current flows in the channel and the drain-to-source path is like an open circuit. When the gate voltage is sufficiently negative below V_T, a channel is formed and p-type carriers flow from source to drain. P-type carriers are positive and correspond to a positive current flow from source to drain.

In the n-channel MOS, the source terminal is connected to the substrate and a positive voltage is applied to the drain terminal. When the gate voltage is below the threshold voltage V_T (about 2 V), no current flows in the channel. When the gate voltage is sufficiently positive above V_T to form the channel, n-type carriers flow from source to drain. n-type carriers are negative, which corresponds to a positive current flow from drain to source. The threshold voltage may vary from 1 to 4 V, depending on the particular process used.

The graphic symbols for the MOS transistors are shown in Fig. 10-20. The accepted symbol for the enhancement type is the one with the broken-line connection between source and drain. In this symbol, the substrate can be identified and is shown connected

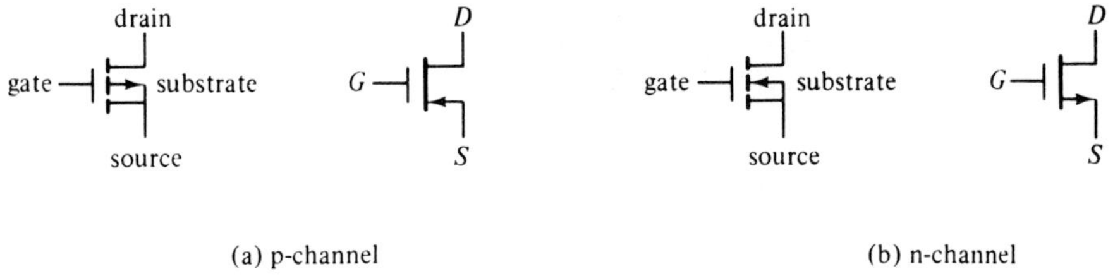

(a) p-channel

(b) n-channel

FIGURE 10-20

Symbols for MOS transistors

to the source. We will use an alternative symbol that omits the substrate; in this symbol, the arrow is placed in the source terminal to show the direction of *positive* current flow (from source to drain in the p-channel and from drain to source in the n-channel).

Because of the symmetrical construction of source and drain, the MOS transistor can be operated as a bilateral device. Although normally operated so that carriers flow from source to drain, there are circumstances when it is convenient to allow carrier flow from drain to source (see Problem 10-12).

One advantage of the MOS device is that it can be used not only as a transistor, but as a resistor as well. A resistor is obtained from the MOS by permanently biasing the

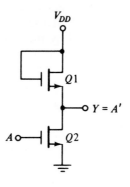

(a) Inverter

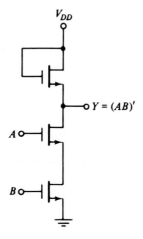

(b) NAND gate

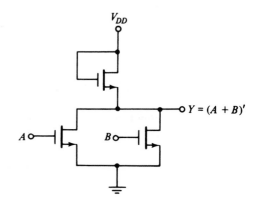

(c) NOR gate

FIGURE 10-21

n-channel MOS logic circuits

gate terminal for conduction. The ratio of the source–drain voltage to the channel current then determines the value of the resistance. Different resistor values may be constructed during manufacturing by fixing the channel length and width of the MOS device.

Three logic circuits using MOS devices are shown in Fig. 10-21. For an n-channel MOS, supply voltage V_{DD} is positive (about 5 V) to allow positive current flow from drain to source. The two voltage levels are a function of the threshold voltage V_T. The low level is anywhere from zero to V_T, and the high level ranges from V_T to V_{DD}. The n-channel gates usually employ positive logic. The p-channel MOS circuits use a negative voltage for V_{DD} to allow positive current flow from source to drain. The two voltage levels are both negative above and below the negative threshold voltage V_T. P-channel gates usually employ negative logic.

The inverter circuit shown in Fig. 10-21(a) uses two MOS devices. $Q1$ acts as the load resistor and $Q2$ as the active device. The load resistor MOS has its gate connected to V_{DD}, thus maintaining it always in the conduction state. When the input voltage is low (below V_T), $Q2$ turns off. Since $Q1$ is always on, the output voltage is at about V_{DD}. When the input voltage is high (above V_T), $Q2$ turns on. Current flows from V_{DD} through the load resistor $Q1$ and into $Q2$. The geometry of the two MOS devices must be such that the resistance of $Q2$, when conducting, is much less than the resistance of $Q1$ to maintain the output Y at a voltage below V_T.

The NAND gate shown in Fig. 10-21(b) uses transistors in series. Inputs A and B must both be high for all transistors to conduct and cause the output to go low. If either input is low, the corresponding transistor is turned off and the output is high. Again, the series resistance formed by the two active MOS devices must be much less than the resistance of the load-resistor MOS. The NOR gate shown in Fig. 10-21(c) uses transistors in parallel. If either input is high, the corresponding transistor conducts and the output is low. If all inputs are low, all active transistors are off and the output is high.

10-8   COMPLEMENTARY MOS (CMOS)

Complementary MOS circuits take advantage of the fact that both n-channel and p-channel devices can be fabricated on the same substrate. CMOS circuits consist of both types of MOS devices interconnected to form logic functions. The basic circuit is the inverter, which consists of one p-channel transistor and one n-channel transistor, as shown in Fig. 10-22(a). The source terminal of the p-channel device is at V_{DD}, and the source terminal of the n-channel device is at ground. The value of V_{DD} may be anywhere from +3 to +18 V. The two voltage levels are 0 V for the low level and V_{DD} for the high level.

To understand the operation of the inverter, we must review the behavior of the MOS transistor from the previous section:

1. The n-channel MOS conducts when its gate-to-source voltage is positive.

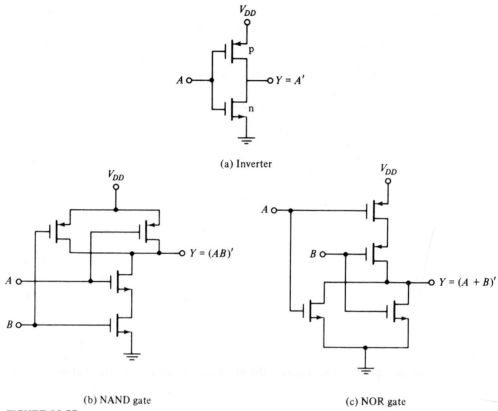

(a) Inverter

(b) NAND gate

(c) NOR gate

FIGURE 10-22
CMOS logic circuits

2. The p-channel MOS conducts when its gate-to-source voltage is negative.

3. Either type of device is turned off if its gate-to-source voltage is zero.

Now consider the operation of the inverter. When the input is low, both gates are at zero potential. The input is at $-V_{DD}$ relative to the source of the p-channel device and at 0 V relative to the source of the n-channel device. The result is that the p-channel device is turned on and the n-channel device is turned off. Under these conditions, there is a low-impedance path from V_{DD} to the output and a very high-impedance path from output to ground. Therefore, the output voltage approaches the high level V_{DD} under normal loading conditions. When the input is high, both gates are at V_{DD} and the situation is reversed: The p-channel device is off and the n-channel device is on. The result is that the output approaches the low level of 0 V.

Two other CMOS basic gates are shown in Fig. 10-22. A two-input NAND gate consists of two p-type units in parallel and two n-type units in series, as shown in Fig. 10-22(b). If all inputs are high, both p-channel transistors turn off and both n-channel transistors turn on. The output has a low impedance to ground and produces a low state. If any input is low, the associated n-channel transistor is turned off and the associated p-channel transistor is turned on. The output is coupled to V_{DD} and goes to the high state. Multiple-input NAND gates may be formed by placing equal numbers of p-type and n-type transistors in parallel and series, respectively, in an arrangement similar to that shown in Fig. 10-22(b).

A two-input NOR gate consists of two n-type units in parallel and two p-type units in series, as shown in Fig. 10-22(c). When all inputs are low, both p-channel units are on and both n-channel units are off. The output is coupled to V_{DD} and goes to the high state. If any input is high, the associated p-channel transistor is turned off and the associated n-channel transistor turns on. This connects the output to ground, causing a low-level output.

CMOS Characteristics

When a CMOS logic circuit is in a static state, its power dissipation is very low. This is because there is always an off transistor in the current path when the state of the circuit is not changing. As a result, a typical CMOS gate has a static power dissipation on the order of 0.01 mW. However, when the circuit is changing state at a rate of 1 MHz, the power dissipation increases to about 1 mW.

CMOS logic is usually specified for a single power-supply operation over the voltage range between 3 and 18 V with a typical V_{DD} value of 5 V. Operating CMOS at a larger value of supply voltage reduces the propagation delay time and improves the noise margin, but the power dissipation is increased. The propagation delay time with $V_{DD} = 5$ V ranges from 8 to 50 ns, depending on the type of CMOS used. The noise margin is usually about 40 percent of the V_{DD} supply voltage. The fan-out of CMOS gates is 50 when operated at a frequency of less than 1 MHz. The fan-out decreases with increase in frequency of operation.

There are several series of the CMOS digital logic family (see Table 2-10). The original design of CMOS ICs is recognized from the 4000 number designation. The 74C series are pin- and function-compatible with TTL devices having the same number. For example, CMOS IC type 74C04 has six inverters with the same pin configuration as TTL type 7404. The performance characteristics of the 74C series are about the same as the 4000 series. The high-speed CMOS 74HC series is an improvement of the 74C series with a tenfold increase in switching speed. The 74HCT series is electrically compatible with TTL ICs. This means that the circuits in this series can be connected to inputs and outputs of TTL ICs without the need of additional interfacing circuits.

The CMOS fabrication process is simpler than TTL and provides a greater packing density. This means that more circuits can be placed on a given area of silicon at a reduced cost per function. This property of CMOS, together with its low power dissipation, excellent noise immunity, and reasonable propagation delay, makes it a strong contender for a popular standard as a digital logic family.

10-9   CMOS TRANSMISSION GATE CIRCUITS

A special CMOS circuit that is not available in the other digital logic families is the *transmission gate*. The transmission gate is essentially an electronic switch that is controlled by an input logic level. It is used for simplifying the construction of various digital components when fabricated with CMOS technology.

Figure 10-23(a) shows the basic circuit of the transmission gate. It consists of one n-channel and one p-channel MOS transistor connected in parallel. The graphic symbol used here is the conventional symbol that shows the substrate, as indicated in Fig. 10-20. The n-channel substrate is connected to ground and the p-channel substrate is connected to V_{DD}. When the N gate is at V_{DD} and the P gate is at ground, both transistors conduct and there is a closed path between input X and output Y. When the N gate is at ground and the P gate at V_{DD}, both transistors are off and there is an open circuit between X and Y. Figure 10-23(b) shows the block diagram of the transmission gate. Note that the terminal of the p-channel gate is marked with the negation small-circle symbol. Figure 10-23(c) demonstrates the behavior of the switch in terms of positive-logic assignment with V_{DD} equivalent to logic-1 and ground equivalent to logic-0.

The transmission gate is usually connected to an inverter, as shown in Fig. 10-24.

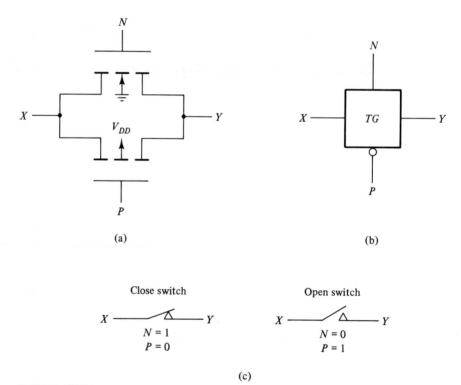

(a)

(b)

(c)

FIGURE 10-23

Transmission gate (*TG*)

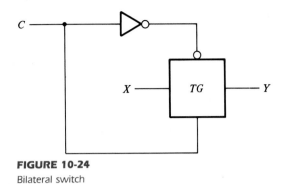

FIGURE 10-24
Bilateral switch

This type of arrangement is referred to as a *bilateral switch*. The control input C is connected directly to the n-channel gate and its inverse to the p-channel gate. When $C = 1$, the switch is closed, producing a path between X and Y. When $C = 0$, the switch is open, disconnecting the path between X and Y.

Various circuits can be constructed using the transmission gate. In order to demonstrate its usefulness as a component in the CMOS family, we will show three circuit examples.

The exclusive-OR gate can be constructed with two transmission gates and two inverters, as shown in Fig. 10-25. Input A controls the paths in the transmission gates and input B is connected to output Y through the gates. When input A is equal to 0, transmission gate $TG1$ is closed and output Y is equal to input B. When input A is equal to 1, $TG2$ is closed and output Y is equal to the complement of input B. This results in the exclusive-OR truth table, as indicated in the table of Fig. 10-25.

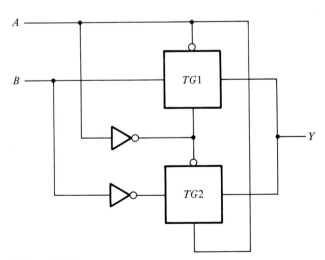

A	B	$TG1$	$TG2$	Y
0	0	close	open	0
0	1	close	open	1
1	0	open	close	1
1	1	open	close	0

FIGURE 10-25
Exclusive-OR constructed with transmission gates

Another circuit that can be constructed with transmission gates is the multiplexer. A 4-to-1-line multiplexer implemented with transmission gates is shown in Fig. 10-26. The *TG* circuit provides a transmission path between its horizontal input and output lines when the two vertical control inputs have the value of 1 in the uncircled terminal and 0 in the circled terminal. With an opposite polarity in the control inputs, the path disconnects and the circuit behaves like an open switch. The two selection inputs, S_1 and S_0, control the transmission path in the *TG* circuits. Inside each box is marked the

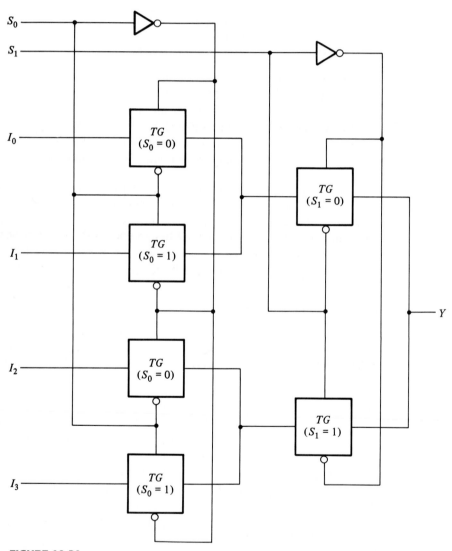

FIGURE 10-26
Multiplexer with transmission gates

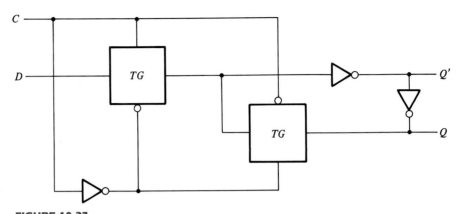

FIGURE 10-27
Gated *D* latch with transmission gates

condition for the transmission gate switch to be closed. Thus, if $S_0 = 0$ and $S_1 = 0$, there is a closed path from input I_0 to output Y through the two *TGs* marked with $S_0 = 0$ and $S_1 = 0$. The other three inputs are disconnected from the output by one of the other *TG* circuits.

The level-sensitive *D* flip-flop commonly referred to as gated *D* latch can be constructed with transmission gates, as shown in Fig. 10-27. The *C* input controls two transmission gates *TG*. When $C = 1$, the *TG* connected to input *D* has a closed path and the one connected to output *Q* has an open path. This produces an equivalent circuit from input *D* through two inverters to output *Q*. Thus, the output follows the data input as long as *C* remains active. When *C* switches to 0, the first *TG* disconnects input *D* from the circuit and the second *TG* produces a closed path between the two inverters at the output. Thus, the value that was present at input *D* at the time that *C* went from 1 to 0 is retained at the *Q* output.

A master–slave *D* flip-flop can be constructed with two circuits of the type shown in Fig. 10-27. The first circuit is the master and the second is the slave. Thus, a master–slave *D* flip-flop can be constructed with four transmission gates and six inverters.

REFERENCES

1. HODGES, D. A., and H. G. JACKSON, *Analysis and Design of Digital Integrated Circuits*, 2nd Ed. New York: McGraw-Hill, 1988.

2. TAUB, H., and D. SCHILLING, *Digital Integrated Electronics*. New York: McGraw-Hill, 1977.

3. GRINICH, V. H., and H. G. JACKSON, *Introduction to Integrated Circuits*. New York: McGraw-Hill, 1975.

4. TOCCI, R. J., *Digital Systems Principles and Applications*, 4th Ed., Englewood Cliffs, NJ: Prentice-Hall, 1988.

5. *The TTL Data Book*. Dallas: Texas Instruments, 1988.

6. *CMOS Logic Data Book*. Dallas: Texas Intruments, 1984.

7. MORRIS, R. L., and J. R. MILLER, *Designing with TTL Integrated Circuits*. New York: McGraw-Hill, 1971.

8. GLASER, A. B., and G. E. SUBAK-SHARPE, *Integrated Circuit Engineering*. Reading, MA: Addison-Wesley, 1977.

9. HAMILTON, D. J., and W. G. HOWARD, *Basic Integrated Circuit Engineering*. New York: McGraw-Hill, 1975.

PROBLEMS

10-1 The following are the specifications for the Schottky TTL 74S00 quadruple 2-input NAND gates. Calculate the fan-out, power dissipation, propagation delay, and noise margin of the Schottky NAND gate.

Parameter	Name	Value
V_{CC}	Supply voltage	5 V
I_{CCH}	High-level supply current (four gates)	10 mA
I_{CCL}	Low-level supply current (four gates)	20 mA
V_{OH}	High-level output voltage (min)	2.7 V
V_{OL}	Low-level output voltage (max)	0.5 V
V_{IH}	High-level input voltage (min)	2 V
V_{IL}	Low-level input voltage (max)	0.8 V
I_{OH}	High-level output current (max)	1 mA
I_{OL}	Low-level output current (max)	20 mA
I_{IH}	High-level input current (max)	0.05 mA
I_{IL}	Low-level input current (max)	2 mA
t_{PLH}	Low-to-high delay	3 ns
t_{PHL}	High-to-low delay	3 ns

10-2 (a) Determine the high-level output voltage of the RTL gate for a fan-out of 5. (b) Determine the minimum input voltage required to drive an RTL transistor to saturation when $h_{FE} = 20$. (c) From the results in (a) and (b), determine the noise margin of the RTL gate when the input is high and the fan-out is 5.

10-3 Show that the output transistor of the DTL gate of Fig. 10-9 goes into saturation when all inputs are high. Assume that $h_{FE} = 20$.

10-4 Connect the output Y of the DTL gate shown in Fig. 10-9 to N inputs of other similar gates. Assume that the output transistor is saturated and its base current is 0.44 mA. Let $h_{FE} = 20$.

(a) Calculate the current in the 2 kΩ resistor.
(b) Calculate the current coming from each input connected to the gate.
(c) Calculate the total collector current in the output transistor as a function of N.
(d) Find the value of N that will keep the transistor in saturation.
(e) What is the fan-out of the gate?

10-5 Let all inputs in the open-collector TTL gate of Fig. 10-11 be in the high state of 3 V.
  (a) Determine the voltages in the base, collector, and emitter of all transistors.
  (b) Determine the minimum h_{FE} of $Q2$ that ensures that this transistor saturates.
  (c) Calculate the base current of $Q3$.
  (d) Assume that the minimum h_{FE} of $Q3$ is 6.18. What is the maximum current that can be tolerated in the collector to ensure saturation of $Q3$?
  (e) What is the minimum value of R_L that can be tolerated to ensure saturation of $Q3$?

10-6 (a) Using the actual output transistors of two open-collector TTL gates, show (by means of a truth table) that when connected together to an external resistor and V_{CC}, the wired connection produces an AND function.
  (b) Prove that two open-collector TTL inverters when connected together produce the NOR function.

10-7 It was stated in Section 10-5 that totem-pole outputs should not be tied together to form wired logic. To see why this is prohibitive, connect two such circuits together and let the output of one gate be in the high state and the output of the other gate be in the low state. Show that the load current (which is the sum of the base and collector currents of the saturated transistor $Q4$ in Fig. 10-14) is about 32 mA. Compare this value with the recommended load current in the high state of 0.4 mA.

10-8 For the following conditions, list the transistors that are off and those that are conducting in the three-state TTL gate of Fig. 10-16(c). (For $Q1$ and $Q6$, it would be necessary to list the states in the base–emitter and base–collector junctions separately.)
  (a) When C is low and A is low.
  (b) When C is low and A is high.
  (c) When C is high.

  What is the state of the output in each case?

10-9 Calculate the emitter current I_E across R_E in the ECL gate of Fig. 10-17 when:
  (a) At least one input is high at -0.8 V.
  (b) All inputs are low at -1.8 V.
  Now assume that $I_C = I_E$. Calculate the voltage drop across the collector resistor in each case and show that it is about 1 V as required.

10-10 Calculate the noise margin of the ECL gate.

10-11 Using the NOR outputs of two ECL gates, show that when connected together to an external resistor and negative supply voltage, the wired connection produces an OR function.

10-12 The MOS transistor is bilateral, i.e., current may flow from source to drain or from drain to source. Using this property, derive a circuit that implements the Boolean function

$$Y = (AB + CD + AED + CEB)'$$

  using six MOS transistors.

10-13 (a) Show the circuit of a four-input NAND gate using CMOS transistors. (b) Repeat for a four-input NOR gate.

10-14 Construct an exclusive-NOR circuit with two inverters and two transmission gates.

10-15 Construct an 8-to-1-line multiplexer using transmission gates and inverters.

10-16 Draw the logic diagram of a master–slave D flip-flop using transmission gates and inverters.

Laboratory Experiments

11-0 INTRODUCTION TO EXPERIMENTS

This chapter presents 18 laboratory experiments in digital circuits and logic design. They provide hands-on experience for the student using this book. The digital circuits can be constructed by using standard integrated circuits (ICs) mounted on breadboards that are easily assembled in the laboratory. The experiments are ordered according to the material presented in the book.

A logic breadboard suitable for performing the experiments must have the following equipment:

1. LED (light-emitting diode) indicator lamps.
2. Toggle switches to provide logic-1 and -0 signals.
3. Pulsers with pushbuttons and debounce circuits to generate single pulses.
4. A clock-pulse generator with at least two frequencies, a low frequency of about one pulse per second to observe slow changes in digital signals and a higher frequency of about 10 kHz or higher for observing waveforms in an oscilloscope.
5. A power supply of 5 V for TTL ICs.
6. Socket strips for mounting the ICs.
7. Solid hookup wire and a pair of wire strippers for cutting the wires.

Digital logic trainers that include the required equipment are available from several manufacturers. A digital logic trainer contains LED lamps, toggle switches, pulsers, a variable clock, power supply, and IC socket strips. Some experiments may require additional switches, lamps, or IC socket strips. Extended breadboards with more solderless sockets and plug-in switches and lamps may be needed.

Additional equipment required are a dual-trace oscilloscope (for Experiments 1, 2, 8, and 15), a logic probe to be used for debugging, and a number of ICs. The ICs required for the experiments are of the transistor-transistor logic (TTL) type, series 7400.

The integrated circuits to be used in the experiments can be classified as small-scale integration (SSI) or medium-scale integration (MSI) circuits. SSI circuits contain individual gates or flip-flops, and MSI circuits perform specific digital functions. The eight SSI gate ICs needed for the experiments are shown in Fig. 11-1. They include 2-input NAND, NOR, AND, OR, and XOR gates, inverters, and 3-input and 4-input NAND gates. The pin assignment for the gates is indicated in the diagram. The pins are numbered from 1 to 14. Pin number 14 is marked V_{CC}, and pin number 7 is marked GND (ground). These are the supply terminals, which must be connected to a power supply of 5 V for proper operation. Each IC is recognized by its identification number; for example, the 2-input NAND gates are found inside the IC whose number is 7400.

Detailed descriptions of the MSI circuits can be found in data books published by the manufacturers. The best way to acquire experience with commercial MSI circuits is to study their description in a data book that provides complete information on the internal, external, and electrical characteristics of the integrated circuits. Various semiconductor companies publish data books for the TTL 7400 series. Examples are *The TTL Data Book*, published by Texas Instruments, and the *Logic Databook*, published by National Semiconductor Corp.

The MSI circuits that are needed for the experiments are introduced and explained when they are used for the first time. The operation of the circuit is explained by referring to similar circuits in previous chapters. The information given in this chapter about the MSI circuits should be sufficient for performing the experiments adequately. Nevertheless, a reference to a data book will always be preferable, as it gives more detailed description of the circuits.

We will now demonstrate the method of presentation of MSI circuits adopted here. This will be done by means of a specific example that introduces the ripple counter IC, type 7493. This IC is used in Experiment 1 and in subsequent experiments to generate a sequence of binary numbers for verifying the operation of combinational circuits.

The information about the 7493 IC that is found in a data book is shown in Figs. 11-2(a) and (b). Part (a) shows a diagram of the internal logic circuit and its connection to external pins. All inputs and outputs are given symbolic letters and assigned to pin numbers. Part (b) shows the physical layout of the IC with its 14-pin assignment to signal names. Some of the pins are not used by the circuit and are marked as NC (no connection). The IC is inserted into a socket, and wires are connected to the various pins through the socket terminals. When drawing schematic diagrams in this chapter, we will show the IC in a block diagram form as in Fig. 11-2(c). The IC number 7493 is written inside the block. All input terminals are placed on the left of the block and all output terminals on the right. The letter symbols of the signals, such as A, $R1$, and QA, are written inside the block, and the corresponding pin numbers, such as 14, 2, and 12, are written along the external lines. V_{CC} and GND are the power terminals connected to

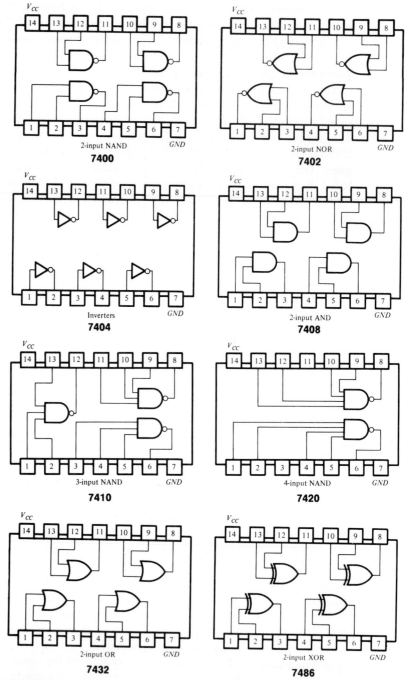

FIGURE 11-1
Digital gates in IC packages with identification numbers and pin assignments

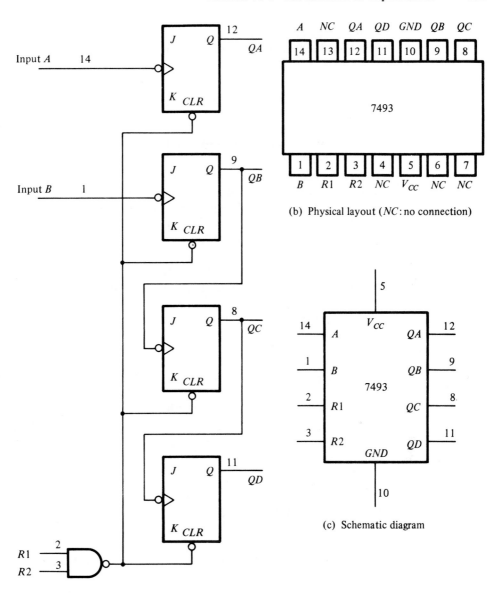

(b) Physical layout (*NC*: no connection)

(c) Schematic diagram

(a) Internal circuit diagram

FIGURE 11-2

IC type 7493 ripple counter

pins 5 and 10. The size of the block may vary to accommodate all input and output terminals. Inputs or outputs may sometimes be placed on the top or the bottom of the block for convenience.

The operation of the circuit is similar to the ripple counter shown in Fig. 7-12 with an asynchronous clear to each flip-flop, as shown in Fig. 6-15. When inputs $R1$ or $R2$

or both are equal to logic 0 (ground for TTL circuits), all asynchronous clears are equal to 1 and are disabled. To clear all four flip-flops to 0, the output of the NAND gate must be equal to 0. This is accomplished by having both inputs $R1$ and $R2$ at logic-1 (about 3 to 5 V in TTL circuits). Note that the J and K inputs show no connections. It is characteristic of TTL circuits that an input terminal with no external connections has the effect of producing a signal equivalent to logic-1. Also note that output QA is not connected to input B internally.

The 7493 IC can operate as a 3-bit counter using input B and flip-flops QB, QC, and QD. It can operate as a 4-bit counter using input A if output QA is connected to input B. Therefore, to operate the circuit as a 4-bit counter, it is necessary to have an external connection between pin 12 and pin 1. The reset inputs, $R1$ and $R2$, at pins 2 and 3, respectively, must be grounded. Pins 5 and 10 must be connected to a 5-V power supply. The input pulses must be applied to input A at pin 14, and the four flip-flop outputs of the counter are taken from QA, QB, QC, and QD, at pins 12, 9, 8, and 11, respectively, with QA being the least significant bit.

Figure 11-2(c) demonstrates the way that all MSI circuits will be symbolized graphically in this chapter. Only a block diagram similar to the one shown in this figure will be shown for each IC. The letter symbols for the inputs and outputs in the IC block diagram will be according to the symbols used in the data book. The operation of the circuit will be explained with reference to logic diagrams from previous chapters. The operation of the circuit will be specified by means of a truth table or a function table.

Other possible graphic symbols for the ICs are presented in Chapter 12. These are

TABLE 11-1
Integrated Circuits Required for the Experiments

		Graphic Symbol	
IC Number	Description	In Chap. 11	In Chap. 12
	Various gates	Fig. 11-1	Fig. 12-1
7447	BCD-to-seven-segment decoder	Fig. 11-8	
7474	Dual D-type flip-flops	Fig. 11-13	Fig. 12-9(b)
7476	Dual JK-type flip-flops	Fig. 11-12	Fig. 12-9(a)
7483	4-bit binary adder	Fig. 11-10	Fig. 12-2
7489	16 × 4 random-access memory	Fig. 11-18	Fig. 12-15
7493	4-bit ripple counter	Fig. 11-2	Fig. 12-13
74151	8 × 1 multiplexer	Fig. 11-9	Fig. 12-7(a)
74155	3 × 8 decoder	Fig. 11-7	Fig. 12-6
74157	Quadruple 2 × 1 multiplexers	Fig. 11-17	Fig. 12-7(b)
74161	4-bit synchronous counter	Fig. 11-15	Fig. 12-14
74194	Bidirectional shift register	Fig. 11-19	Fig. 12-12
74195	4-bit shift register	Fig. 11-16	Fig. 12-11
7730	Seven-segment LED display	Fig. 11-8	
72555	Timer (same as 555)	Fig. 11-21	

standard graphic symbols approved by the Institute of Electrical and Electronics Engineers and are given in IEEE standard 91-1984. The standard graphic symbols for SSI gates have rectangular shapes, as shown in Fig. 12-1. The standard graphic symbol for the 7493 IC is shown in Fig. 12-13. This symbol can be substituted in place of the one shown in Fig. 11-2(c). The standard graphic symbols of the other ICs that are needed to run the experiments are presented in Chapter 12. They can be used for drawing schematic diagrams of the logic circuits if the standard symbols are preferred.

Table 11-1 lists the ICs that are needed for the experiments together with the figure numbers where they are presented in this chapter. In addition, the table lists the figure numbers in Chapter 12 where the equivalent standard graphic symbols are drawn.

The rest of this chapter is divided into 18 sections, with each section covering one experiment. The section number designates the experiment number. Each experiment should take about two to three hours of laboratory time, except for Experiments 14, 16, and 17, which may take longer.

11-1 BINARY AND DECIMAL NUMBERS

This experiment demonstrates the count sequence of binary numbers and the binary-coded decimal (BCD) representation. It serves as an introduction to the breadboard used in the laboratory and acquaints the student with the cathode-ray oscilloscope. Reference material from the text that may be useful to know while performing the experiment can be found in Section 1-2, on binary numbers, and Section 1-7, on BCD numbers.

Binary Count.   IC type 7493 consists of four cells called flip-flops, as shown in Fig. 11-2. The cells can be connected to count in binary or in BCD. Connect the IC to operate as a 4-bit binary counter by wiring the external terminals, as shown in Fig. 11-3. This is done by connecting a wire from pin 12 (output QA) to pin 1 (input B). Input A at pin 14 is connected to a pulser that provides single pulses. The two reset inputs, $R1$ and $R2$, are connected to ground. The four outputs go to four indicator lamps with the low-order bit of the counter from QA connected to the rightmost indicator lamp. Do not forget to supply 5 V and ground to the IC. All connections should be made with the power supply in the off position.

Turn the power on and observe the four indicator lamps. The 4-bit number in the output is incremented by one for every pulse generated in the push-button pulser. The count goes to binary 15 and then back to 0. Disconnect the input of the counter at pin 14 from the pulser and connect it to a clock generator that produces a train of pulses at a low frequency of about one pulse per second. This will provide an automatic binary count. Note that the binary counter will be used in subsequent experiments to provide the input binary signals for testing combinational circuits.

Oscilloscope Display.   Increase the frequency of the clock to 10 kHz or higher and connect its output to an oscilloscope. Observe the clock output on the oscilloscope and

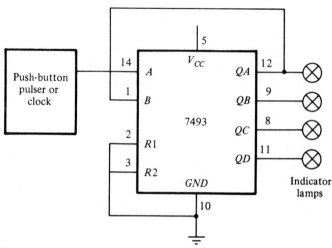

FIGURE 11-3
Binary counter

sketch its waveform. If a dual-trace oscilloscope is available, connect the output of QA to one channel and the output of the clock to the second channel. Note that the output of QA is complemented every time the clock pulse goes through a negative transition from 1 to 0. (The two waveforms should look similar to the timing diagram shown in Fig. 7-15 for the count pulses and Q_1.)

When the count pulses into the counter occur at constant frequency, the frequency at the output of the first cell, QA, is one-half that of the input clock frequency. Each cell in turn divides its incoming frequency by 2. The 4-bit counter divides the incoming frequency by 16 at output QD. Obtain a timing diagram showing the time relationship of the clock and the four outputs of the counter. Make sure that you include at least 16 clock pulses. The way to proceed with a dual-trace oscilloscope is as follows. First, observe the clock pulses and QA and record their timing waveforms. Then repeat by observing and recording the waveforms of QA together with QB, followed by the waveforms of QB with QC and then QC with QD. Your final result should be a diagram showing the time relationship of the clock and the four outputs in one composite diagram having at least 16 clock pulses.

BCD Count.   The BCD representation uses the binary numbers from 0000 to 1001 to represent the coded decimal digits from 0 to 9. IC type 7493 can be operated as a BCD counter by making the external connections shown in Fig. 11-4. Outputs QB and QD are connected to the two reset inputs, $R1$ and $R2$. When both $R1$ and $R2$ are equal to 1, all four cells in the counter clear to 0 irrespective of the input pulse. The counter starts from 0, and every input pulse increments it by 1 until it reaches the count of 1001. The next pulse changes the ouput to 1010, making QB and QD equal to 1. This momentary output cannot be sustained, because the four cells immediately clear to 0,

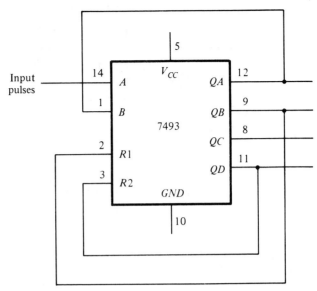

FIGURE 11-4
BCD counter

with the result that the output goes to 0000. Thus, the pulse after the count of 1001 changes the output to 0000, producing a BCD count.

Connect the IC to operate as a BCD counter. Connect the input to a pulser and the four outputs to indicator lamps. Verify that the count goes from 0000 to 1001.

Disconnect the input from the pulser and connect it to a clock generator with a frequency of 10 kHz or higher. Observe the clock waveform and the four outputs on the oscilloscope. Obtain an accurate timing diagram showing the time relationship between the clock and the four outputs. Make sure to include at least ten clock pulses in the oscilloscope display and in the composite timing diagram.

Output Pattern.   When the count pulses into the BCD counter are continuous, the counter keeps repeating the sequence from 0000 to 1001 and back to 0000 over and over. This means that each bit in the four outputs produces a fixed pattern of 1's and 0's, which is repeated every ten pulses. These patterns can be predicted from the list of the binary numbers from 0000 to 1001. The list will show that output QA, being the least significant bit, produces a pattern of alternate 1's and 0's. Output QD, being the most significant bit, produces a pattern of eight 0's followed by two 1's. Obtain the pattern for the other two outputs and then check all four patterns on the oscilloscope. This is done with a dual-trace oscilloscope by displaying the clock pulses in one channel and one of the output waveforms in the other channel. The pattern of 1's and 0's for the corresponding output is obtained by observing the output levels at the vertical positions where the pulses change from 1 to 0.

Other Counts. IC type 7493 can be connected to count from 0 to a variety of final counts. This is done by connecting one or two outputs to the reset inputs, $R1$ and $R2$. Thus, if $R1$ is connected to QA instead of QB in Fig. 11-4, the resulting count will be from 0000 to 1000, which is 1 less than 1001 ($QD = 1$ and $QA = 1$).

Utilizing your knowledge of how $R1$ and $R2$ affect the final count, connect the 7493 IC to count from 0000 to the following final counts.

(a) 0101

(b) 0111

(c) 1011

Connect each circuit and verify its count sequence by applying pulses from the pulser and observing the output count in the indicator lamps. If the initial count starts with a value greater than the final count, keep applying input pulses until the output clears to 0.

11-2 DIGITAL LOGIC GATES

In this experiment, you will investigate the logic behavior of various IC gates:

7400 Quadruple 2-input NAND gates

7402 Quadruple 2-input NOR gates

7404 Hex inverters

7408 Quadruple 2-input AND gates

7432 Quadruple 2-input OR gates

7486 Quadruple 2-input XOR gates

The pin assignments to the various gates are shown in Fig. 11-1. "Quadruple" means that there are four gates within the package. The TTL digital logic gates and their characteristics are discussed in Section 2-8. NAND implementation is discussed in Sections 3-6 and 4-7.

Truth Tables. Use one gate from each IC listed above and obtain the truth table of the gate. The truth table is obtained by connecting the inputs of the gate to switches and the output to an indicator lamp. Compare your results with the truth tables listed in Fig. 2-5.

Waveforms. For each gate listed above, obtain the input–output waveform relationship of the gate. The waveforms are to be observed in the oscilloscope. Use the two low-order outputs of a binary counter (Fig. 11-3) to provide the inputs to the gate. As an example, the circuit and waveforms for the NAND gate are illustrated in Fig. 11-5. The oscilloscope display will repeat this waveform, but you should record only the non-repetitive portion.

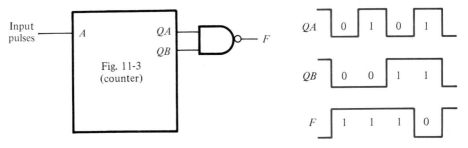

FIGURE 11-5
Waveforms for NAND gate

Propagation Delay.   Connect all six inverters inside the 7404 IC in cascade. The output will be the same as the input except that it will be delayed by the time it takes the signal to propagate through all six inverters. Apply clock pulses to the input of the first inverter. Using the oscilloscope, determine the delay from the input to the output of the sixth inverter during the upswing and again during the downswing of the pulse. This is done with a dual-trace oscilloscope by applying the input clock pulses to one of the channels and the output of the sixth inverter to the second channel. Set the time-base knob to the lowest time-per-division setting. The rise or fall time of the two pulses should appear on the screen. Divide the total delay by 6 to obtain an average propagation delay per inverter.

Universal NAND Gate.   Using a single 7400 IC, connect a circuit that produces:

(a) an inverter
(b) a 2-input AND
(c) a 2-input OR
(d) a 2-input NOR
(e) a 2-input XOR (see Fig. 4-21)

In each case, verify your circuit by checking its truth table.

NAND Circuit.   Using a single 7400 IC, construct a circuit with NAND gates that implements the Boolean function

$$F = AB + CD$$

1. Draw the circuit diagram.
2. Obtain the truth table for F as a function of the four inputs.
3. Connect the circuit and verify the truth table.
4. Record the patterns of 1's and 0's for F as inputs A, B, C, and D go from binary 0 to binary 15.
5. Connect the four outputs of the binary counter shown in Fig. 11-3 to the four in-

puts of the NAND circuit. Connect the input clock pulses from the counter to one channel and output F to the other channel of a dual-trace oscilloscope. Observe and record the 1's and 0's pattern of F after each clock pulse and compare it to the pattern recorded in Step 4.

11-3 SIMPLIFICATION OF BOOLEAN FUNCTIONS

This experiment demonstrates the relationship between a Boolean function and the corresponding logic diagram. The Boolean functions are simplified by using the map method, as discussed in Chapter 3. The logic diagrams are to be drawn using NAND gates, as explained in Section 3-6.

The gate ICs to be used for the logic diagrams must be those from Fig. 11-1 that contain NAND gates:

7400 2-input NAND

7404 Inverter (1-input NAND)

7410 3-input NAND

7420 4-input NAND

If an input to a NAND gate is not used, it should not be left open, but, instead, should be connected to another input that is used. For example, if the circuit needs an inverter and there is an extra 2-input gate available in a 7400 IC, then both inputs of the gate are to be connected together to form a single input for an inverter.

Logic Diagram.   This part of the experiment starts with a given logic diagram from which we proceed to apply simplification procedures to reduce the number of gates and possibly the number of ICs. The logic diagram shown in Fig. 11-6 requires two ICs, a 7400 and a 7410. Note that the inverters for inputs x, y, and z are obtained from the remaining three gates in the 7400 IC. If the inverters were taken from a 7404 IC, the circuit would have required three ICs. Also note that in drawing SSI circuits, the gates are not enclosed in blocks as is done with MSI circuits.

Assign pin numbers to all inputs and outputs of the gates and connect the circuit with the x, y, and z inputs going to three switches and the output F to an indicator lamp. Test the circuit by obtaining its truth table.

Obtain the Boolean function of the circuit and simplify it using the map method. Construct the simplified circuit without disconnecting the original circuit. Test both circuits by applying identical inputs to both and observing the separate outputs. Show that for each of the eight possible input combinations, the two circuits have identical outputs. This will prove that the simplified circuit behaves exactly as the original circuit.

Boolean Functions.   Given the two Boolean functions in sum of minterms:

$$F_1(A, B, C, D) = (0, 1, 4, 5, 8, 9, 10, 12, 13)$$

$$F_2(A, B, C, D) = (3, 5, 7, 8, 10, 11, 13, 15)$$

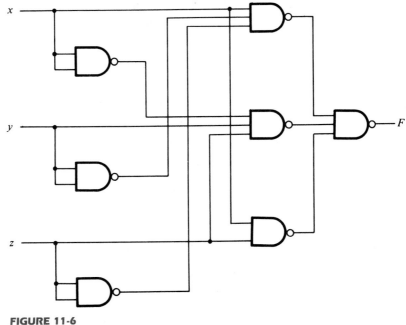

FIGURE 11-6
Logic diagram for Experiment 3

Simplify the two functions by means of maps. Obtain a composite logic diagram with four inputs, A, B, C, and D, and two outputs, F_1 and F_2. Implement the two functions together using a minimum number of NAND ICs. Do not duplicate the same gate if the corresponding term is needed for both functions. Use any extra gates in existing ICs for inverters when possible. Connect the circuit and check its operation. The truth table for F_1 and F_2 obtained from the circuit should conform with the minterms listed.

Complement.   Plot the following Boolean function in a map:

$$F = A'D + BD + B'C + AB'D$$

Combine the 1's in the map to obtain the simplified function for F in sum of products. Then combine the 0's in the map to obtain the simplified function for F' also in sum of products. Implement both F and F' using NAND gates and connect the two circuits to the same input switches, but to separate output indicator lamps. Obtain the truth table of each circuit in the laboratory and show that they are the complements of each other.

11-4   COMBINATIONAL CIRCUITS

In this experiment, you will design, construct, and test four combinational logic circuits. The first two circuits are to be constructed with NAND gates, the third with XOR gates, and the fourth with a decoder and NAND gates. Reference to a parity generator can be found in Section 4-9. Implementation with a decoder is discussed in Section 5-5.

Design Example.   Design a combinational circuit with four inputs, A, B, C, and D, and one output, F. F is to be equal to 1 when $A = 1$ provided that $B = 0$, or when $B = 1$ provided that either C or D is also equal to 1. Otherwise, the output is to be equal to 0.

1. Obtain the truth table of the circuit.
2. Simplify the output function.
3. Draw the logic diagram of the circuit using NAND gates with a minimum number of ICs.
4. Construct the circuit and test it for proper operation by verifying the conditions stated above.

Majority Logic.   A majority logic is a digital circuit whose output is equal to 1 if the majority of the inputs are 1's. The output is 0 otherwise. Design and test a 3-input majority circuit using NAND gates with a minimum number of ICs.

Parity Generator.   Design, construct, and test a circuit that generates an even parity bit from four message bits. Use XOR gates. Adding one more XOR gate, expand the circuit so it generates an odd parity bit also.

Decoder Implementation.   A combinational circuit has three inputs, x, y, and z, and three outputs, F_1, F_2, and F_3. The simplified Boolean functions for the circuit are as follows:

$$F_1 = xz + x'y'z'$$
$$F_2 = x'y + xy'z'$$
$$F_3 = xy + x'y'z$$

Implement and test the combinational circuit using a 74155 decoder IC and external NAND gates.

The block diagram of the decoder and its truth table are shown in Fig. 11-7. The 74155 can be connected as a dual 2×4 decoder or as a single 3×8 decoder. When a 3×8 decoder is desired, inputs $C1$ and $C2$ must be connected together as well as inputs $G1$ and $G2$, as shown in the block diagram. The function of the circuit is similar to the one shown in Fig. 5-10. G is the enable input and must be equal to 0 for proper operation. The eight outputs are labeled with symbols given in the data book. As in Fig. 5-10, the 74155 uses NAND gates, with the result that the selected output goes to 0 while all other outputs remain at 1. The implementation with the decoder is as shown in Fig. 5-9, except that the OR gates must be replaced with external NAND gates when the 74155 is used.

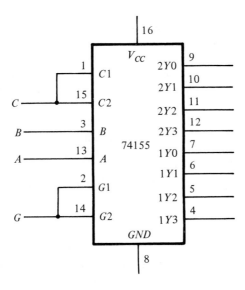

Truth table

Inputs				Outputs							
G	C	B	A	$2Y0$	$2Y1$	$2Y2$	$2Y3$	$1Y0$	$1Y1$	$1Y2$	$1Y3$
1	X	X	X	1	1	1	1	1	1	1	1
0	0	0	0	0	1	1	1	1	1	1	1
0	0	0	1	1	0	1	1	1	1	1	1
0	0	1	0	1	1	0	1	1	1	1	1
0	0	1	1	1	1	1	0	1	1	1	1
0	1	0	0	1	1	1	1	0	1	1	1
0	1	0	1	1	1	1	1	1	0	1	1
0	1	1	0	1	1	1	1	1	1	0	1
0	1	1	1	1	1	1	1	1	1	1	0

FIGURE 11-7
IC type 74155 connected as a 3 × 8 decoder

11-5   CODE CONVERTERS

The conversion from one binary code to another is common in digital systems. In this experiment, you will design and construct three combinational-circuit converters. Code conversion is discussed in Section 4-5.

Gray Code to Binary.   Design a combinational circuit with four inputs and four outputs that converts a four-bit Gray code number (Table 1-4) into the equivalent four-bit binary number. Implement the circuit with exclusive-OR gates. (This can be done with one 7486 IC.) Connect the circuit to four switches and four indicator lamps and check for proper operation.

9's Complementer.   Design a combinational circuit with four input lines that represent a decimal digit in BCD and four output lines that generate the 9's complement of the input digit. Provide a fifth output that detects an error in the input BCD number. This output should be equal to logic-1 when the four inputs have one of the unused combinations of the BCD code. Use any of the gates listed in Fig. 11-1, but minimize the total number of ICs used.

Seven-Segment Display.   A seven-segment indicator is used for displaying any one of the decimal digits 0 through 9. Usually, the decimal digit is available in BCD. A BCD-to-seven-segment decoder accepts a decimal digit in BCD and generates the corresponding seven-segment code. This is shown pictorially in Problem 4-16.

Figure 11-8 shows the connections necessary between the decoder and the display. The 7447 IC is a BCD-to-seven-segment decoder/driver. It has four inputs for the BCD digit. Input D is the most significant and input A the least significant. The 4-bit BCD digit is converted to a seven-segment code with outputs a through g. The outputs of the 7447 are applied to the inputs of the 7730 (or equivalent) seven-segment display. This

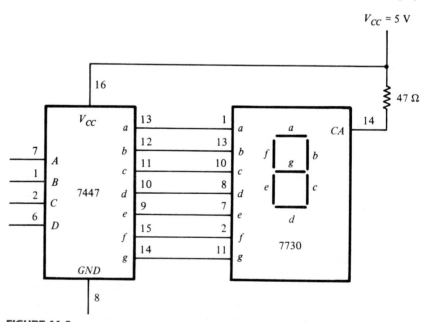

FIGURE 11-8
BCD-to-seven-segment decoder (7447) and seven-segment display (7730)

IC contains the seven LED (light-emitting diode) segments on top of the package. The input at pin 14 is the common anode (*CA*) for all the LEDs. A 47-Ω resistor to V_{CC} is needed in order to supply the proper current to the selected LED segments. Other equivalent seven-segment display ICs may have additional anode terminals and may require different resistor values.

Construct the circuit shown in Fig. 11-8. Apply the 4-bit BCD digits through four switches and observe the decimal display from 0 to 9. Inputs 1010 through 1111 have no meaning in BCD. Depending on the decoder, these values may cause either a blank or a meaningless pattern to be displayed. Observe and record the output displayed patterns of the six unused input combinations.

11-6 DESIGN WITH MULTIPLEXERS

In this experiment, you will design a combinational circuit and implement it with multiplexers, as explained in Section 5-6. The multiplexer to be used is IC type 74151, shown in Fig. 11-9. The internal construction of the 74151 is similar to the diagram shown in Fig. 5-16 except that there are eight inputs instead of four. The eight inputs are designated $D0$ through $D7$. The three selection lines, *C*, *B*, and *A*, select the particular input to be multiplexed and applied to the output. A strobe control, *S*, acts as an enable signal. The function table specifies the value of output *Y* as a function of the selection lines. Output *W* is the complement of *Y*. For proper operation, the strobe input *S* must be connected to ground.

Design Specifications.   A small corporation has 10 shares of stock, and each share entitles its owner to one vote at a stockholder's meeting. The 10 shares of stock are owned by four people as follows:

Mr. W: 1 share
Mr. X: 2 shares
Mr. Y: 3 shares
Mrs. Z: 4 shares

Each of these persons has a switch to close when voting yes and to open when voting no for his or her shares.

It is necessary to design a circuit that displays the total number of shares that vote yes for each measure. Use a seven-segment display and a decoder, as shown in Fig. 11-8, to display the required number. If all shares vote no for a measure, the display should be blank. (Note that binary input 15 into the 7447 blanks all seven segments.) If 10 shares vote yes for a measure, the display should show 0. Otherwise, the display shows a decimal number equal to the number of shares that vote yes. Use four 74151 multiplexers to design the combinational circuit that converts the inputs from the stock owners' switches into the BCD digit for the 7447. Do not use 5 V for logic-1. Use the output of an inverter whose input is grounded.

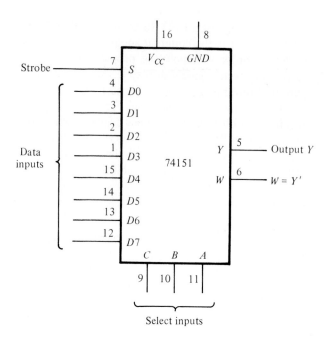

Function table

Strobe S	Select C	B	A	Output Y
1	X	X	X	0
0	0	0	0	D0
0	0	0	1	D1
0	0	1	0	D2
0	0	1	1	D3
0	1	0	0	D4
0	1	0	1	D5
0	1	1	0	D6
0	1	1	1	D7

FIGURE 11-9
IC type 74151 8 × 1 multiplexer

11-7 ADDERS AND SUBTRACTORS

In this experiment, you will construct and test various adder and subtractor circuits. The subtractor circuit is then used for comparing the relative magnitude of two numbers. Adders are discussed in Section 4-3. Subtraction with 2's complement is explained in Section 1-5. A 4-bit parallel adder is introduced in Section 5-2, and the comparison of two numbers is explained in Section 5-4.

Half-Adder.　Design, construct, and test a half-adder circuit using one XOR gate and two NAND gates.

Full-Adder.　Design, construct, and test a full-adder circuit using two ICs, 7486 and 7400.

Parallel Adder.　IC type 7483 is a 4-bit binary parallel adder. Its internal construction is similar to Fig. 5-5. The pin assignment is shown in Fig. 11-10. The two 4-bit input binary numbers are $A1$ through $A4$ and $B1$ through $B4$. The 4-bit sum is obtained from $S1$ through $S4$. $C0$ is the input carry and $C4$ the output carry.

Test the 4-bit binary adder 7483 by connecting the power supply and ground terminals. Then connect the four A inputs to a fixed binary number such as 1001 and the B inputs and the input carry to five toggle switches. The five outputs are applied to indicator lamps. Perform the addition of a few binary numbers and check that the output sum and output carry give the proper values. Show that when the input carry is equal to 1, it adds 1 to the output sum.

Adder-Subtractor.　The subtraction of two binary numbers can be done by taking the 2's complement of the subtrahend and adding it to the minuend. The 2's complement can be obtained by taking the 1's complement and adding 1. To perform $A - B$,

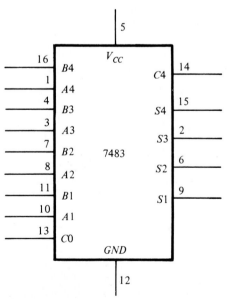

FIGURE 11-10
IC type 7483 4-bit binary adder

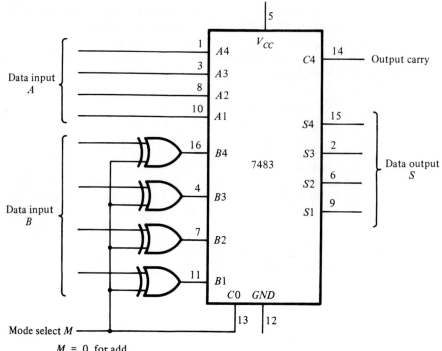

FIGURE 11-11
4-bit adder-subtractor

we complement the four bits of B, add them to the four bits of A, and add 1 through the input carry. This is done as shown in Fig. 11-11. The four XOR gates complement the bits of B when the mode select $M = 1$ (because $x \oplus 1 = x'$) and leave the bits of B unchanged when $M = 0$ (because $x \oplus 0 = x$). Thus, when the mode select M is equal to 1, the input carry $C0$ is equal to 1 and the sum output is A plus the 2's complement of B. When M is equal to 0, the input carry is equal to 0 and the sum generates $A + B$.

Connect the adder-subtractor circuit and test it for proper operation. Connect the four A inputs to a fixed binary number 1001 and the B inputs to switches. Perform the following operations and record the values of the output sum and the output carry $C4$.

$$9 + 5 \qquad 9 - 5$$
$$9 + 9 \qquad 9 - 9$$
$$9 + 15 \qquad 9 - 15$$

Show that during addition, the output carry is equal to 1 when the sum exceeds 15. Also show that when $A \geq B$, the subtraction operation gives the correct answer, $A - B$, and the output carry $C4$ is equal to 1. But when $A < B$, the subtraction gives the 2's complement of $B - A$ and the output carry is equal to 0.

Magnitude Comparator.   The comparison of two numbers is an operation that determines whether one number is greater than, equal to, or less than the other number. Two numbers, A and B, can be compared by first subtracting $A - B$ as done in Fig. 11-11. If the output in S is equal to zero, we know that $A = B$. The output carry from $C4$ determines the relative magnitude: when $C4 = 1$, we have $A \geq B$; when $C4 = 0$, we have $A < B$; and when $C4 = 1$ and $S \neq 0$, we have $A > B$.

It is necessary to supplement the subtractor circuit of Fig. 11-11 to provide the comparison logic. This is done with a combinational circuit that has five inputs, $S1$ through $S4$ and $C4$, and three outputs designated by x, y, and z, so that

$$x = 1 \quad \text{if} \quad A = B \quad (S = 0000)$$

$$y = 1 \quad \text{if} \quad A < B \quad (C4 = 0)$$

$$z = 1 \quad \text{if} \quad A > B \quad (C4 = 1 \text{ and } S \neq 0000)$$

The combinational circuit can be implemented with the two ICs, 7404 and 7408.

Construct the comparator circuit and test its operation. Use at least two sets of numbers for A and B to check each of the outputs x, y, and z.

11-8 FLIP-FLOPS

In this experiment, you will construct, test, and investigate the operation of various flip-flop circuits. The internal construction of the flip-flops can be found in Sections 6-2 and 6-3.

SR Latch.   The SR latch is a basic flip-flop made with two cross-coupled NAND gates. Construct a basic flip-flop circuit and connect the two inputs to switches and the two outputs to indicator lamps. Set the two switches to logic-1, then momentarily turn each switch separately to the logic-0 position and back to 1. Obtain the truth table of the circuit.

RS Flip-Flop.   Construct a clocked RS flip-flop with four NAND gates. Connect the S and R inputs to two switches and the clock input to a pulser. Verify the characteristic table of the flip-flop.

D Flip-Flop.   Construct a clocked D flip-flop with four NAND gates. This can be done by modifying the circuit shown in Fig. 6-5 as follows. Remove gate number 5. Connect the output of gate 3 to the input of gate 4, where gate 5 was originally connected. Show that the modified circuit is identical to the original circuit by deriving the Boolean function for the output of gate 4 in each case. Connect the modified D flip-flop circuit and verify its characteristic table.

Master–Slave Flip-Flop.   Construct a clocked master–slave JK flip-flop with one 7410 and two 7400 ICs. Connect the J and K inputs to logic-1 and the clock input to a pulser. Connect the normal output of the master flip-flop to one indicator lamp and the normal output of the slave flip-flop to another indicator lamp. Press the push button in the pulser and then release it to produce a single positive pulse. Observe that the master flip-flop changes when the pulse goes positive and the slave flip-flop follows the change when the pulse goes negative. Repeat a few times while observing the two indicator lamps. Explain the transfer sequence from input to master and from master to slave.

Disconnect the clock input from the pulser and connect it to a clock generator with a frequency of 10 kHz or higher. Using a dual-trace oscilloscope, observe the waveforms of the clock and the master and slave outputs. Verify that the delay between the master and slave outputs is equal to the pulse width. Obtain a timing diagram showing the relationship between the clock waveform and the master and slave flip-flop outputs.

Edge-Triggered D Flip-Flop.   Construct a *D*-type positive-edge-triggered flip-flop using six NAND gates. Connect the clock input to a pulser, the *D* input to a toggle switch, and the output *Q* to an indicator lamp. Set the value of *D* to the complement value of *Q*. Show that the flip-flop output changes only in response to a positive transi-

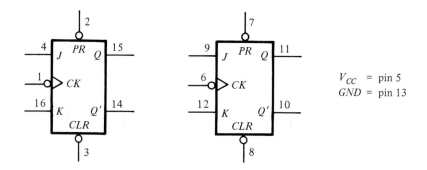

V_{CC} = pin 5
GND = pin 13

Function table

	Inputs				Outputs	
Preset	Clear	Clock	*J*	*K*	*Q*	*Q'*
0	1	X	X	X	1	0
1	0	X	X	X	0	1
0	0	X	X	X	1	1
1	1	⊓	0	0	No change	
1	1	⊓	0	1	0	1
1	1	⊓	1	0	1	0
1	1	⊓	1	1	Toggle	

FIGURE 11-12
IC type 7476 dual *JK* master–slave flip-flops

tion of the clock pulse. Verify that the output does not change when the clock input is logic-1, or when the clock goes through a negative transition, or when it is logic-0. Continue changing the D input to correspond to the complement of the Q output at all times.

Disconnect the input from the pulser and connect it to the clock generator. Connect the complement output Q' to the D input. This causes the output to complement with each positive transition of the clock pulse. Using a dual-trace oscilloscope, observe and record the timing relationship between the input clock and output Q. Show that the output changes in response to a positive edge transition.

IC Flip-Flops.   IC type 7476 consists of two JK master–slave flip-flops with preset and clear. The pin assignment for each flip-flop is shown in Fig. 11-12. The function table specifies the circuit operation. The first three entries in the table specify the operation of the asynchronous preset and clear inputs. These inputs behave like a NAND SR latch and are independent of the clock or the J and K inputs (the X's indicate don't-care conditions). The last four entries in the function table specify the clock operation with both the preset and clear inputs maintained at logic-1. The clock value is shown as a

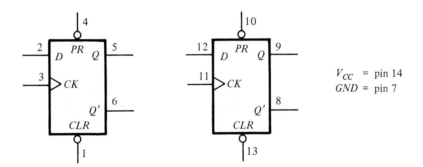

Function table

Inputs				Outputs	
Preset	Clear	Clock	D	Q	Q'
0	1	X	X	1	0
1	0	X	X	0	1
0	0	X	X	1	1
1	1	↑	0	0	1
1	1	↑	1	1	0
1	1	0	X	No change	

FIGURE 11-13
IC type 7474 dual D positive-edge-triggered flip-flops

single pulse. The positive transition of the pulse changes the master flip-flop, and the negative transition changes the slave flip-flop as well as the output of the circuit. With $J = K = 0$, the output does not change. The flip-flop toggles or complements when $J = K = 1$. Investigate the operation of one 7476 flip-flop and verify its function table.

IC type 7474 consists of two D positive-edge-triggered flip-flops with preset and clear. The pin assignment is shown in Fig. 11-13. The function table specifies the preset and clear operations and the clock operation. The clock is shown with an upward arrow to indicate that it is a positive-edge-triggered flip-flop. Investigate the operation of one of the flip-flops and verify its function table.

11-9 SEQUENTIAL CIRCUITS

In this experiment, you will design, construct, and test three synchronous sequential circuits. Use IC type 7476 JK flip-flops (Fig. 11-12) in all three designs. Choose any gate type that will minimize the total number of ICs. The design of synchronous sequential circuits is covered in Sections 6-7 and 6-8.

Up–Down Counter with Enable.   Design, construct, and test a 2-bit counter that counts up or down. An enable input E determines whether the counter is on or off. If $E = 0$, the counter is disabled and remains at its present count even though clock pulses are applied to the flip-flops. If $E = 1$, the counter is enabled and a second input, x, determines the count direction. If $x = 1$, the circuit counts up with the sequence 00, 01, 10, 11, and the count repeats. If $x = 0$, the circuit counts down with the sequence 11, 10, 01, 00, and the count repeats. Do not use E to disable the clock. Design the sequential circuit with E and x as inputs.

State Diagram.   Design, construct and test a sequential circuit whose state diagram is shown in Fig. 11-14. Designate the two flip-flops as A and B, the input as x, and the output as y.

Connect the output of the least significant flip-flop B to the input x and predict the sequence of states and output that will occur with the application of clock pulses. Verify the state transition and output by testing the circuit.

Design of Counter.   Design, construct, and test a counter that goes through the following sequence of binary states: 0, 1, 2, 3, 6, 7, 10, 11, 12, 13, 14, 15, and back to 0 to repeat. Note that binary states 4, 5, 8, and 9 are not used. The counter must be self-starting; that is, if the circuit starts from any one of the four invalid states, the count pulses must transfer the circuit to one of the valid states to continue the count correctly.

Check the circuit operation for the required count sequence. Verify that the counter is self-starting. This is done by initializing the circuit to each unused state by means of the preset and clear inputs and then applying pulses to see whether the counter reaches one of the valid states.

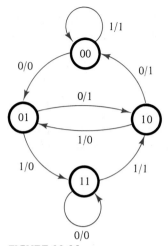

FIGURE 11-14
State diagram for Experiment 9

11-10 COUNTERS

In this experiment, you will construct and test various ripple and synchronous counter circuits. Ripple counters are discussed in Section 7-4, and synchronous counters are covered in Section 7-5.

Ripple Counter.   Construct a 4-bit binary ripple counter using two 7476 ICs (Fig. 11-12). Connect all asynchronous clear and preset inputs to logic-1. Connect the count-pulse input to a pulser and check the counter for proper operation.

Modify the counter so it will count down instead of up. Check that each input pulse decrements the counter by 1.

Synchronous Counter.   Construct a synchronous 4-bit binary counter and check its operation. Use two 7476 ICs and one 7408 IC.

Decimal Counter.   Design a synchronous BCD counter that counts from 0000 to 1001. Use two 7476 ICs and one 7408 IC. Test the counter for the proper sequence. Determine whether it is self-starting. This is done by initializing the counter to each of the six unused states by means of the preset and clear inputs. The application of pulses must transfer the counter to one of the valid states if the counter is self-starting.

Binary Counter with Parallel Load.   IC type 74161 is a 4-bit synchronous binary counter with parallel load and asynchronous clear. The internal logic is similar to the circuit shown in Fig. 7-19. The pin assignment to the inputs and outputs is shown in Fig. 11-15. When the load signal is enabled, the four data inputs are transferred into four internal flip-flops, QA through QD, with QD being the most significant bit. There

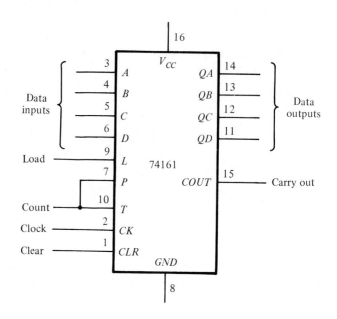

Function table

Clear	Clock	Load	Count	Function
0	X	X	X	Clear outputs to 0
1	↑	0	X	Load input data
1	↑	1	1	Count to next binary value
1	↑	1	0	No change in output

FIGURE 11-15
IC type 74161 binary counter with parallel load

are two count-enable inputs called P and T. Both must be equal to 1 for the counter to operate. The function table is similar to Table 7-6 with one exception: the load input in the 74161 is enabled when equal to 0. To load the input data, the clear input must be equal to 1 and the load input must be equal to 0. The two count inputs have don't-care conditions and may be equal to either 1 or 0. The internal flip-flops trigger on the positive transition of the clock pulse. The circuit functions as a counter when the load input is equal to 1 and both count inputs P and T are equal to 1. If either P or T goes to 0, the output does not change. The carry-out output is equal to 1 when all four data outputs are equal to 1. Perform an experiment to verify the operation of the 74161 IC according to the function table.

Show how the 74161 IC together with a 2-input NAND gate can be made to operate as a synchronous BCD counter that counts from 0000 to 1001. Do not use the clear input. Use the NAND gate to detect the count of 1001, which then causes all 0's to be loaded into the counter.

Connect each of the four circuits shown in Fig. 7-20 to achieve a mod-6 counter. Remember that the load input is enabled when equal to 0 and therefore a NAND gate will be needed in each case.

11-11   SHIFT REGISTERS

In this experiment, you will investigate the operation of shift registers. The IC to be used is the 74195 shift register with parallel load. Shift registers are explained in Section 7-3. Some applications are introduced in Section 7-6.

IC Shift Register.   IC type 74195 is a 4-bit shift register with parallel load and asynchronous clear. The internal logic is similar to Fig. 7-9 except that the 74195 is unidirectional, that is, it is capable of shifting only in one direction. The pin assignment to the inputs and outputs is shown in Fig. 11-16. The single control line labeled SH/LD (shift/load) determines the synchronous operation of the register. When $SH/LD = 0$, the control input is in the load mode and the four data inputs are transferred into the four internal flip-flops, QA through QD. When $SH/LD = 1$, the control input is in the shift mode and the information in the register is shifted right from QA toward QD. The serial input into QA during the shift is determined from the J and $\overline{K}$ inputs. The two inputs behave like the J and the complement of K of a JK flip-flop. When both J and $\overline{K}$ are equal to 0, flip-flop QA is cleared to 0 after the shift. If both inputs are equal to 1, QA is set to 1 after the shift. The other two conditions for the J and $\overline{K}$ inputs (not shown in the function table) provide a complement or no change in the output of flip-flop QA after the shift.

The function table for the 74195 shows the mode of operation of the register. When the clear input goes to 0, the four flip-flops clear to 0 asynchronously, that is, without the need of a clock. Synchronous operations are affected by a positive transition of the clock. To load the input data, the SH/LD must be equal to 0 and a positive clock-pulse transition must occur. To shift right, the SH/LD must be equal to 1. The J and $\overline{K}$ inputs must be connected together to form the serial input.

Perform an experiment that will verify the operation of the 74195 IC. Show that it performs all the operations listed in the function table. Include in your function table the two conditions for $\overline{JK} = 01$ and 10.

Ring Counter.   A ring counter is a circular shift register with the signal from the serial output QD going into the serial input. Connect the J and $\overline{K}$ input together to form the serial input. Use the load condition to preset the ring counter to an initial value of 1000. Rotate the single bit with the shift condition and check the state of the register after each clock pulse.

A switch-tail ring counter uses the complement output of QD for the serial input. Preset the switch-tail ring counter to 0000 and predict the sequence of states that will result from shifting. Verify your prediction by observing the state sequence after each shift.

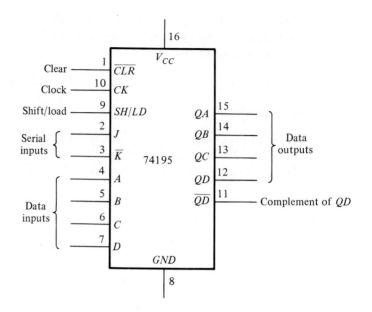

Function table

Clear	Shift/ load	Clock	J	$\overline{K}$	Serial input	Function
0	X	X	X	X	X	Asynchronous clear
1	X	0	X	X	X	No change in output
1	0	↑	X	X	X	Load input data
1	1	↑	0	0	0	Shift from QA toward QD, $QA = 0$
1	1	↑	1	1	1	Shift from QA toward QD, $QA = 1$

FIGURE 11-16
IC type 74195 shift register with parallel load

Feedback Shift Register.   A feedback shift register is a shift register whose serial input is connected to some function of selected register outputs. Connect a feedback shift register whose serial input is the exclusive-OR of outputs QC and QD. Predict the sequence of states of the register starting from state 1000. Verify your prediction by observing the state sequence after each clock pulse.

Bidirectional Shift Register.   The 74195 IC can shift only right from QA toward QD. It is possible to convert the register to a bidirectional shift register by using the load mode to obtain a shift left operation (from QD toward QA). This is accomplished by connecting the output of each flip-flop to the input of the flip-flop on its left and using the load mode of the SH/LD input as a shift-left control. Input D becomes the serial input for the shift-left operation.

Connect the 74195 as a bidirectional shift register (without parallel load). Connect the serial input for shift right to a toggle switch. Construct the shift left as a ring counter by connecting the serial output QA to the serial input D. Clear the register and then check its operation by shifting a single 1 from the serial input switch. Shift right three more times and insert 0's from the serial input switch. Then rotate left with the shift-left (load) control. The single 1 should remain visible while shifting.

Bidirectional Shift Register with Parallel Load.   The 74195 IC can be converted to a bidirectional shift register with parallel load in conjunction with a multiplexer circuit. We will use IC type 74157 for this purpose. This is a quadruple 2-to-1-line multiplexers

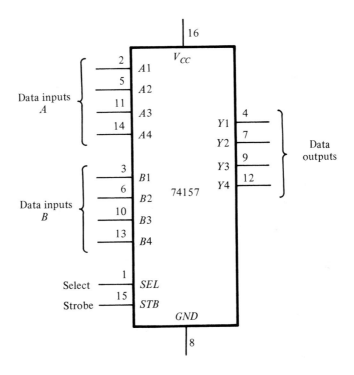

Function table

Strobe	Select	Data outputs Y
1	X	All 0's
0	0	Select data inputs A
0	1	Select data inputs B

FIGURE 11-17
IC type 74157 quadruple 2 × 1 multiplexers

whose internal logic is shown in Fig. 5-17. The pin assignment to the inputs and outputs of the 74157 is shown in Fig. 11-17. Note that the enable input is called a strobe in the 74157.

Construct a bidirectional shift register with parallel load using the 74195 register and the 74157 multiplexer. The circuit should be able to perform the following operations:

1. Asynchronous clear
2. Shift right
3. Shift left
4. Parallel load
5. Synchronous clear.

Derive a table for the five operations as a function of the clear, clock, and SH/LD inputs of the 74195 and the strobe and select inputs of the 74157. Connect the circuit and verify your function table. Use the parallel-load condition to provide an initial value into the register and connect the serial outputs to the serial inputs of both shifts in order not to lose the binary information while shifting.

11-12 SERIAL ADDITION

In this experiment, you will construct and test a serial adder-subtractor circuit. Serial addition of two binary numbers can be done by means of shift registers and a full adder, as explained in Section 7-3. Example 7-3 shows that the number of gates for the full-adder can be reduced if a *JK* flip-flop is used for storing the carry.

Serial Adder.   Starting from the diagram of Fig. 7-11, design and construct a 4-bit serial adder using the following ICs: 74195 (two), 7408, 7486, and 7476. Note that 74195 has a complement output for QD that is equivalent to the variables x' and y' in Fig. 7-11. Provide a facility for register B to accept parallel data from four toggle switches and connect its serial input to ground so that 0's are shifted into register B during the addition. Provide a toggle switch to clear the registers and the flip-flop. Another switch will be needed to specify whether register B is to accept parallel data or is to be shifted during the addition.

Testing the Adder.   To test your serial adder, perform the binary addition $5 + 6 + 15 = 26$. This is done by first clearing the registers and the carry flip-flop. Parallel load the binary value 0101 into register B. Apply four pulses to add B to A serially and check that the result in A is 0101. (Note that clock pulses for the 7476 must be as shown in Fig. 11-12.) Parallel load 0110 into B and add it to A serially. Check that A has the proper sum. Parallel load 1111 into B and add to A. Check that the value in A is 1010 and that the carry flip-flop is set.

Clear the registers and flip-flop and try a few other numbers to verify that your serial adder is functioning properly.

Serial Adder-Subtractor.   If we follow the procedure used in Example 7-3 to design a serial subtractor that subtracts $A - B$, we will find that the output difference is the same as the output sum, but that the input to the J and K of the borrow flip-flop needs the complement of x (see Problem 7-11 and its answer in the Appendix). Using the other two XOR gates from the 7486, convert the serial adder to a serial adder-subtractor with a mode control M. When $M = 0$, the circuit adds $A + B$. When $M = 1$, the circuit subtracts $A - B$ and the flip-flop holds the borrow instead of the carry.

Test the adder part of the circuit by repeating the operations recommended above to ensure that the modification did not change the operation. Test the serial subtractor part by performing the operations $15 - 4 - 5 - 13 = -7$. Binary 15 can be transferred to register A by first clearing it to 0 and adding 15 from B. Check the intermediate results during the subtraction. Note that -7 will appear as the 2's complement of 7 with a borrow of 1 in the flip-flop.

11-13 MEMORY UNIT

In this experiment, you will investigate the behavior of a random-access memory (RAM) unit and its storage capability. The RAM will be used to simulate a read-only memory (ROM). The ROM simulator will then be used to implement combinational circuits, as explained in Section 5-7. The memory unit is discussed in Sections 7-7 and 7-8.

IC RAM.   IC type 7489 is a 16×4 random-access memory. The internal logic is similar to the circuit shown in Fig. 7-27 for a 4×3 RAM. The pin assignment to the inputs and outputs is shown in Fig. 11-18. The four address inputs select one of 16 words in the memory. The least significant bit of the address is A, and the most significant is D. The memory enable (ME) input must be equal to 0 to enable the memory. If ME is equal to 1, the memory is disabled and all four outputs are at a logic-1 level. The write enable (WE) input determines the type of operation as indicated in the function table. The write operation is performed when $WE = 0$. This is a transfer of the binary number from the data inputs into the selected word in memory. The read operation is performed when $WE = 1$. This transfers the complement value stored in the selected word into the output data lines. The outputs are open-collector to allow external wired logic for memory expansion.

Testing the RAM.   An open-collector gate requires an external resistor for proper operation. However, an open-collector gate can be operated without an external resistor if its output is connected to the input of another gate. Since the outputs of the 7489 produce the complement values, we might as well insert four inverters to change the outputs to their normal value and, at the same time, avoid the need for external resistors. The RAM can be tested after making the following connections. Connect the address

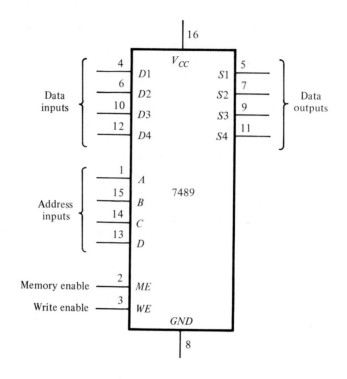

FIGURE 11-18
IC type 7489 16 × 4 RAM

Function table

ME	WE	Operation	Data outputs
0	0	Write	Complement of data inputs
0	1	Read	Complement of selected word
1	X	Disable	All 1's

inputs to a binary counter using the 7493 IC, as shown in Fig. 11-3. Connect the four data inputs to toggle switches and the data outputs to four 7404 inverters. Provide four indicator lamps for the address and four more for the outputs of the inverters. Connect input *ME* to ground and *WE* to a toggle switch (or a pulser that provides a negative pulse). Store a few words into the memory and then read them to verify that the write and read operations are functioning properly. You must be careful when using the *WE* switch. Always leave the *WE* input in the read mode, unless you want to write into memory. The proper way to write is first to set the address in the counter and the inputs in the four toggle switches. To store the word in memory, flip the *WE* switch to the write position and then return it to the read position. Be careful not to change the address or the inputs when *WE* is in the write mode.

ROM Simulator.   A ROM simulator is obtained from a RAM when operated in the read mode only. The pattern of 1's and 0's is first entered into the simulating RAM by placing the unit momentarily in the write mode. Simulation is achieved by placing the unit in the read mode and taking the address lines as inputs for the ROM. The ROM can then be used to implement any combinational circuit.

Implement a combinational circuit using the ROM simulator that converts a 4-bit binary number to its equivalent Gray code as defined in Table 1-4. This is done as follows. Obtain the truth table of the code converter. Store the truth table into the 7489 memory by setting the address inputs to the binary value and the data inputs to the corresponding Gray code value. After all 16 entries of the table are written in memory, the ROM simulator is set by connecting the *WE* line to logic-1 permanently. Check the code converter by applying the inputs to the address lines and verifying the correct outputs in the data output lines.

Memory Expansion.   Expand the memory unit to a 32×4 RAM using two 7489 ICs. Use the *ME* inputs to select between the two ICs. Note that since the data outputs are open-collector, you can tie pairs of terminals together to obtain a logic wired-OR operation in conjunction with the output inverter. Test your circuit by using it as a ROM simulator that adds a 3-bit number to a 2-bit number to produce a 4-bit sum. For example, if the input of the ROM is 10110, then the output is calculated to be $101 + 10 = 0111$. (The first three bits of the input represent 5, the last two bits represent 2, and the output sum is binary 7.) Use the counter to provide four bits of the address and a switch for the fifth bit of the address.

11-14 LAMP HANDBALL

In this experiment, you will construct an electronic game of handball using a single light to simulate the moving ball. This project demonstrates the application of a bidirectional shift register with parallel load. It also shows the operation of the asynchronous inputs of flip-flops. We will first introduce an IC that is needed for this experiment and then present the logic diagram of the simulated lamp handball game.

IC Type 74194.   This is a 4-bit bidirectional shift register with parallel load. The internal logic is similar to Fig. 7-9. The pin assignment to the inputs and outputs is shown in Fig. 11-19. The two mode-control inputs determine the type of operation as specified in the function table. The operation of the circuit is described in detail in Section 7-3 in conjunction with Fig. 7-9.

Logic Diagram.   The logic diagram of the electronic lamp handball is shown in Fig. 11-20. It consists of two 74194 ICs, a dual *D* flip-flop 7474 IC, and three gate ICs: 7400, 7404, and 7408. The ball is simulated by a moving light that is shifted left or right through the bidirectional shift register. The rate at which the light moves is determined by the frequency of the clock. The circuit is first initialized with the *reset*

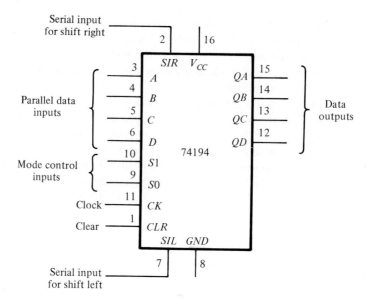

Function table

| Clear | Clock | Mode | | Function |
		S1	S0	
0	X	X	X	Clear outputs to 0
1	↑	0	0	No change in output
1	↑	0	1	Shift right in the direction from QA to QD. SIR to QA
1	↑	1	0	Shift left in the direction from QD to QA. SIL to QD
1	↑	1	1	Parallel-load input data

FIGURE 11-19
IC type 74194 bidirectional shift register with parallel load

switch. The *start* switch starts the game by placing the ball (an indicator lamp) at the extreme right. The player must press the pulser push button to start the ball moving to the left. The single light shifts to the left until it reaches the leftmost position (the wall), at which time the ball returns to the player by reversing the direction of shift of the moving light. When the light is again at the rightmost position, the player must press the pulser again to reverse the direction of shift. If the player presses the pulser too soon or too late, the ball disappears and the light goes off. The game can be restarted by turning the start switch on and then off. The start switch must be open (logic-1) during the game.

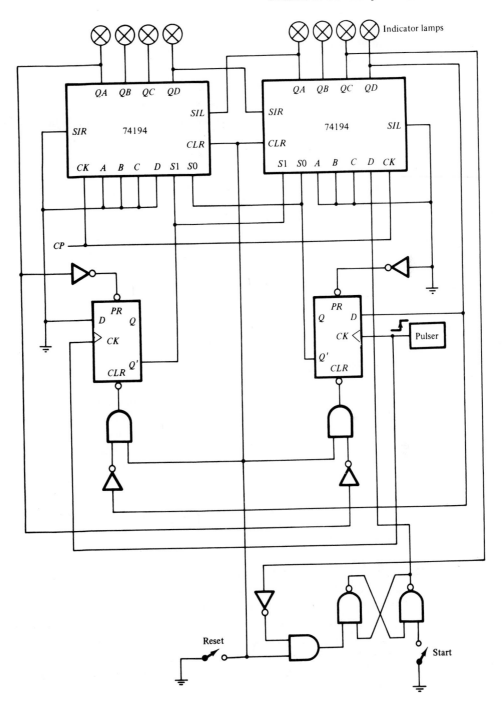

FIGURE 11-20

Lamp handball logic diagram

Circuit Analysis.   Prior to connecting the circuit, analyze the logic diagram to ensure that you understand how the circuit operates. In particular, try to answer the following questions:

1. What is the function of the reset switch?

2. Explain how the light in the rightmost position comes on when the start switch is grounded. Why is it necessary to place the start switch in the logic-1 position before the game starts?

3. What happens to the two mode-control inputs, $S1$ and $S0$, once the ball is set in motion?

4. What happens to the mode-control inputs and to the ball if the pulser is pressed while the ball is moving to the left? What happens if it is moving to the right but has not reached the rightmost position yet?

5. Suppose that the ball returned to the rightmost position, but the pulser has not been pressed yet; what is the state of the mode-control inputs if the pulser is pressed? What happens if it is not pressed?

Playing the Game.   Wire the circuit of Fig. 11-20. Test the circuit for proper operation by playing the game. Note that the pulser must provide a positive-edge transition and that both the reset and start switches must be open (be in the logic-1 state) during the game. Start with a low clock rate and increase the clock frequency to make the handball game more challenging.

Counting the Number of Losses.   Design a circuit that keeps score of the number of times the player loses while playing the game. Use a BCD-to-seven-segment decoder and a seven-segment display as in Fig. 11-8 to display the count from 0 through 9. Counting is done with a decimal counter using either the 7493 as a ripple decimal counter or the 74161 and a NAND gate as a synchronous decimal counter. The display should show 0 when the circuit is reset. Every time the ball disappears and the light goes off, the display should increase by 1. If the light stays on during the play, the number in the display should not change. The final design should be an automatic scoring circuit, with the decimal display incremented automatically each time the player loses when the light disappears.

Lamp Ping-Pong.   Modify the circuit of Fig. 11-20 so as to obtain a lamp Ping-Pong game. Two players can participate in this game, with each player having his own pulser. The player with the right pulser returns the ball when in the extreme right position, and the player with the left pulser returns the ball when in the extreme left position. The only modification required for the Ping-Pong game is a second pulser and a change of few wires.

   With a second start circuit, the game can be made to start (serve) by either one of the two players. This addition is optional.

11-15   CLOCK-PULSE GENERATOR

In this experiment, you will use an IC timer unit and connect it to produce clock pulses at a given frequency. The circuit requires the connection of two external resistors and two external capacitors. The cathode-ray oscilloscope is used to observe the waveforms and measure the frequency.

IC Timer.   IC type 72555 (or 555) is a precision timer circuit whose internal logic is shown in Fig. 11-21. (The resistors, R_A and R_B, and the two capacitors are not part of the IC.) It consists of two voltage comparators, a flip-flop, and an internal transistor.

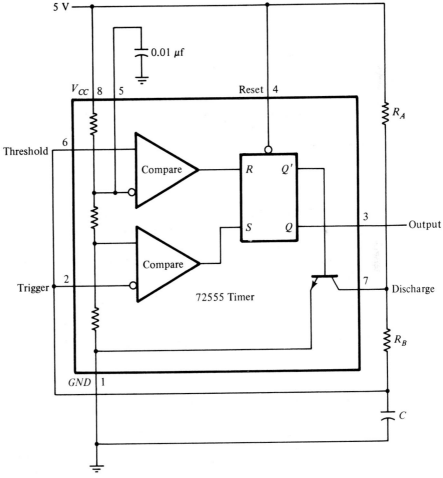

FIGURE 11-21
IC type 72555 timer connected as a clock-pulse generator

The voltage division from $V_{CC} = 5$ V through the three internal resistors to ground produce $\frac{2}{3}$ and $\frac{1}{3}$ of V_{CC} (3.3 V and 1.7 V) into the fixed inputs of the comparators. When the threshold input at pin 6 goes above 3.3 V, the upper comparator resets the flip-flop and the output goes low to about 0 V. When the trigger input at pin 2 goes below 1.7 V, the lower comparator sets the flip-flop and the output goes high to about 5 V. When the output is low, Q' is high and the base–emitter junction of the transistor is forward-biased. When the output is high, Q' is low and the transistor is cut off (see Section 10-2). The timer circuit is capable of producing accurate time delays controlled by an external RC circuit. In this experiment the IC timer will be operated in the astable mode to produce clock pulses.

Circuit Operation.   Figure 11-21 shows the external connections for the astable operation. The capacitor C charges through resistors R_A and R_B when the transistor is cut off and discharges through R_B when the transistor is forward-biased and conducting. When the charging voltage across capacitor C reaches 3.3 V, the threshold input at pin 6 causes the flip-flop to reset and the transistor turns on. When the discharging voltage reaches 1.7 V, the trigger input at pin 2 causes the flip-flop to set and the transistor turns off. Thus, the output continually alternates between two voltage levels at the output of the flip-flop. The output remains high for a duration equal to the charge time. This duration is determined from the equation

$$t_H = 0.693(R_A + R_B)C$$

The output remains low for a duration equal to the discharge time. This duration is determined from the equation

$$t_L = 0.693R_B C$$

Clock-Pulse Generator.   Starting with a capacitor C of 0.001 μF, calculate values for R_A and R_B to produce clock pulses, as shown in Fig. 11-22. The pulse width is 1 μs in the low level, and it is repeating at a frequency rate of 100 kHz (every 10 μs). Connect the circuit and check the output in the oscilloscope.

Observe the output across the capacitor C and record its two levels to verify that they are between the trigger and threshold values.

Observe the waveform in the collector of the transistor at pin 7 and record all pertinent information. Explain the waveform by analyzing the circuit action.

Connect a variable resistor (potentiometer) in series with R_A to produce a variable-

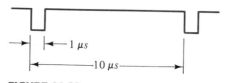

FIGURE 11-22
Output waveform for clock generator

frequency pulse generator. The low-level duration remains at 1 μs. The frequency should range from 20 to 100 kHz.

Change the low-level pulses to high-level pulses with a 7404 inverter. This will produce positive pulses of 1 μs with a variable-frequency range.

11-16 PARALLEL ADDER

In this experiment, you will construct a 4-bit parallel adder whose sum can be loaded into a register. The numbers to be added will be stored in a random-access memory. A set of binary numbers will be selected from memory and their sum will be accumulated in the register.

Block Diagram.   Use the RAM circuit from the memory experiment of Section 11-13, a 4-bit parallel adder, a 4-bit shift register with parallel load, a carry flip-flop, and a multiplexer to construct the circuit. The block diagram and the ICs to be used are shown in Fig. 11-23. Information can be written into RAM from data in four switches or from the 4-bit data available in the outputs of the register. The selection is done by means of a multiplexer. The data in RAM can be added to the contents of the register and the sum transferred back to the register.

Control of Register.   Provide toggle switches to control the 74194 register and the 7476 carry flip-flop as follows:

(a) A LOAD condition to transfer the sum to the register and the output carry to the flip-flop upon application of a clock pulse.
(b) A SHIFT condition to shift the register right with the carry from the carry flip-flop transferred into the leftmost position of the register upon application of a clock pulse. The value in the carry flip-flop should not change during the shift.
(c) A NO-CHANGE condition that leaves the contents of the register and flip-flop unchanged even when clock pulses are applied.

Carry Circuit.   In order to conform with the above specifications, it is necessary to provide a circuit between the output carry from the adder and the J and K inputs of the 7476 flip-flop so that the output carry is transferred into the flip-flop (whether it is equal to 0 or 1) only when the LOAD condition is activated and a pulse is applied to the clock input of the flip-flop. The carry flip-flop should not change if the LOAD condition is disabled or the SHIFT condition is enabled.

Detailed Circuit.   Draw a detailed diagram showing all the wiring between the ICs. Connect the circuit and provide indicator lamps for the outputs of the register and carry flip-flop and for the address and output data of the RAM.

Checking the Circuit.   Store the following numbers in RAM and then add them to

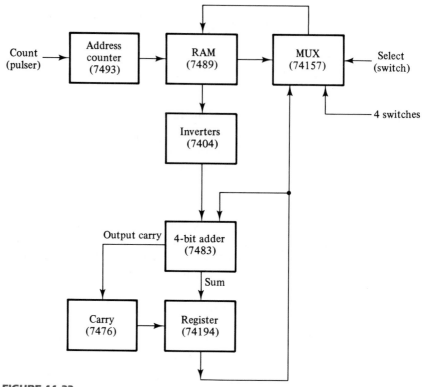

FIGURE 11-23

Block diagram of a parallel adder for Experiment 16

the register one at a time. Start with a cleared register and flip-flop. Predict the values in the output of the register and carry after each addition and verify your results.

$$0110 + 1110 + 1101 + 0101 + 0011$$

Circuit Operation.   Clear the register and the carry flip-flop to zero and store the following 4-bit numbers in RAM in the indicated addresses.

Address	Content
0	0110
3	1110
6	1101
9	0101
12	0011

Now perform the following four operations:

1. Add the contents of address 0 to the contents of the register using the LOAD condition.
2. Store the sum from the register into RAM at address 1.
3. Shift right the contents of the register and carry with the SHIFT condition.
4. Store the shifted contents of the register at address 2 of RAM.

Check that the contents of the first three locations in RAM are as follows:

Address	Content
0	0110
1	0110
2	0011

Repeat the above four operations for each of the other four binary numbers stored in RAM. Use addresses 4, 7, 10, and 13 to store the sum from the register in step 2. Use addresses 5, 8, 11, and 14 to store the shifted value from the register in step 4. Predict what the contents of RAM at addresses 0 through 14 would be and check to verify your results.

11-17   BINARY MULTIPLIER

In this experiment, you will design and construct a circuit that multiplies two 4-bit unsigned numbers to produce an 8-bit product. An algorithm for multiplying two binary numbers is presented in Section 8-6.

Block Diagram.   The block diagram of the binary multiplier with the recommended ICs to be used is shown in Fig. 11-24(a). This is similar to the block diagram of Fig. 8-20 with some minor variations. The multiplicand B is available from four switches instead of a register. The multiplier Q is obtained from another set of four switches. The product is displayed with eight indicator lamps. Counter P is initialized to 0 and then incremented after each partial product is formed. When the counter reaches the count of four, output P_c becomes 1 and the multiplication operation terminates.

Control of Registers.   The ASM chart for the binary multiplier in Fig. 8-21 shows that the three registers and the carry flip-flop are controlled with signals T_1, T_2, and T_3. An additional control signal that depends on Q_1 loads the sum into register A and the output carry into flip-flop E. Q_1 is the least significant bit of register Q. The control-state diagram and the operations to be performed in each state are listed in Fig. 11-24(b). $T_2 Q_1$ is generated with an AND gate whose inputs are T_2 and Q_1. Note that carry flip-flop E can be cleared to 0 with every clock pulse except when the output carry is transferred to it.

Multiplication Example.   Before connecting the circuit, make sure that you under-

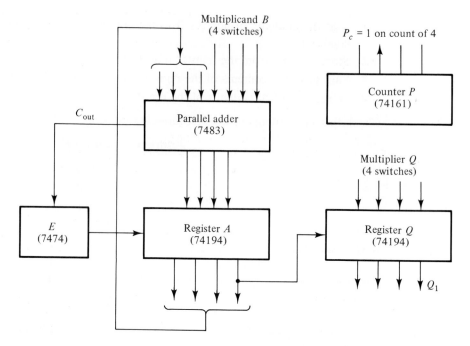

(a) Data-processor block diagram

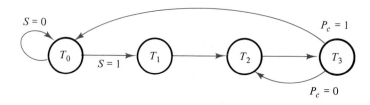

T_1: $A \leftarrow 0,\ E \leftarrow 0,\ P \leftarrow 0,\ Q \leftarrow$ Multiplier
T_2: $P \leftarrow P + 1$
$T_2 Q_1$: $A \leftarrow A + B,\ E \leftarrow C_{out}$
T_3:   Shift right $EAQ,\ E \leftarrow 0$

(b) Control state diagram

FIGURE 11-24
Binary multiplier circuit

stand the operation of the multiplier. To do this, construct a table similar to the one shown in Fig. 8-22, but with $B = 1111$ for the multiplicand and $Q = 1011$ for the multiplier. Along each comment listed on the left side of the table, specify which one of the state variables, T_1 or T_2 or T_3, is enabled in each case. (The states should start with T_1 and then repeat T_2 and T_3 four times.)

Data-Processor Design.   Draw a detailed diagram of the data-processor part of the multiplier, showing all IC pin connections. Generate the control signals, T_1, T_2, and T_3, with three switches and use them to provide the required control operations for the various registers. Connect the circuit and check that each component is functioning properly. With the three control variables at 0, set the multiplicand switches to 1111 and the multiplier switches to 1011. Sequence the control variables manually by means of the control switches as specified by the state diagram of Fig. 11-24(b). Apply a single pulse while in each control state and observe the outputs of registers A and Q and the values in E and P_c. Compare with the numbers in your numerical example to verify that the circuit is functioning properly. Note that IC type 74161 has master–slave flip-flops. To operate it manually, it is necessary that the single clock pulse be a negative pulse.

Design of Control.   Design the control circuit specified by the state diagram. You can use any method of control implementation discussed in Section 8-4 (see also Problem 8-18). Choose the method that minimizes the number of ICs. Verify the operation of the control circuit prior to its connection to the data processor.

Checking the Multiplier.   Connect the outputs of the control circuit to the data processor and verify the total circuit operation by repeating the steps of multiplying 1111 by 1011. The single clock pulses should now sequence the control states as well (remove the manual switches). The start signal S can be generated with a switch that is on while the control is in state T_0.

Generate the start signal S with a pulser or any other short pulse and operate the multiplier with continuous clock pulses from a clock generator. Pressing the S pulser should initiate the multiplication operation and upon completion, the product should be displayed in the A and Q registers. Note that the multiplication will be repeated as long as signal S is enabled. Make sure that S goes back to 0, then set the switches to two other 4-bit numbers and press S again. The new product should appear at the outputs. Repeat the multiplication of a few numbers to verify the operation of the circuit.

11-18 ASYNCHRONOUS SEQUENTIAL CIRCUITS

In this experiment, you will analyze and design asynchronous sequential circuits. These type of circuits are presented in Chapter 9.

Analysis Example.   The analysis of asynchronous sequential circuits with SR latches is outlined in Section 9-3. Analyze the circuit of Fig. P9-9 (shown with Problem 9-9) by deriving the transition table and output map of the circuit. From the transition table and output map, determine: (a) what happens to output Q when input x_1 is a 1 irrespective of the value of input x_2; (b) what happens to output Q when input x_2 is a 1 and x_1 is equal to 0; and (c) what happens to output Q when both inputs go back to 0? Connect the circuit and show that it operates according to the way you analyzed it.

Design Example.   The circuit of a positive-edge-triggered D-type flip-flop is shown in Fig. 6-12. The circuit of a negative-edge T-type flip-flop is shown in Fig. 9-46. Using the six-step procedure recommended in Section 9-8, design, construct, and test a D-type flip-flop that triggers on both the positive and negative transitions of the clock. The circuit has two inputs, D and C, and a single output, Q. The value of D at the time C changes from 0 to 1 becomes the flip-flop output, Q. The output remains unchanged irrespective of the value of D as long as C remains at 1. On the next clock transition, the output is again updated to the value of D when C changes from 1 to 0. The output then remains unchanged as long as C remains at 0.

Standard Graphic Symbols

12-1 RECTANGULAR-SHAPE SYMBOLS

Digital components such as gates, decoders, multiplexers, and registers are available commercially in integrated circuits and are classified as SSI or MSI circuits. Standard graphic symbols have been developed for these and other components so that the user can recognize each function from the unique graphic symbol assigned to it. This standard, known as ANSI/IEEE Std. 91–1984, has been approved by industry, government, and professional organizations and is consistent with international standards.

The standard uses a rectangular-shape outline to represent each particular logic function. Within the outline, there is a general qualifying symbol denoting the logical operation performed by the unit. For example, the general qualifying symbol for a multiplexer is MUX. The size of the outline is arbitrary and can be either a square or a rectangular shape with an arbitrary length–width ratio. Input lines are placed on the left and output lines are placed on the right. If the direction of signal flow is reversed, it must be indicated by arrows.

The rectangular-shape graphic symbols for SSI gates are shown in Fig. 12-1. The qualifying symbol for the AND gate is the ampersand (&). The OR gate has the qualifying symbol that designates greater than or equal to 1, indicating that at least one input must be active for the output to be active. The symbol for the buffer gate is 1, showing that only one input is present. The exclusive-OR symbol designates the fact that only one input must be active for the output to be active. The inclusion of the logic negation small circle in the output converts the gates to their complement values. Although the

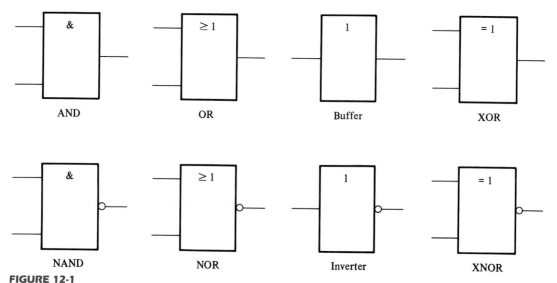

FIGURE 12-1

Rectangular-shape graphic symbols for gates

rectangular-shape symbols for the gates are recommended, the standard also recognizes the distinctive-shape symbols for the gates shown in Fig. 2-5.

An example of an MSI standard graphic symbol is the 4-bit parallel adder shown in Fig. 12-2. The qualifying symbol for an adder is the Greek letter Σ. The preferred letters for the arithmetic operands are P and Q. The bit-grouping symbols in the two types of inputs and the sum output are the decimal equivalents of the weights of the bits to the power of 2. Thus, the input labeled 3 corresponds to the value of $2^3 = 8$. The input carry is designated by CI and the output carry by CO. When the digital component rep-

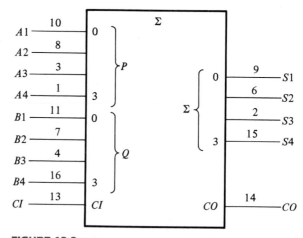

FIGURE 12-2

Standard graphic symbol for a 4-bit parallel adder, IC type 7483

resented by the outline is also a commercial integrated circuit, it is customary to write the IC pin number along each input and output. Thus, IC type 7483 is a 4-bit adder with look-ahead carry. It is enclosed in a package with 16 pins. The pin numbers for the nine inputs and five outputs are shown in Fig. 12-2. The other two pins are for the power supply. Internally, the adder is equivalent to the circuit shown in Fig. 5-5.

Before introducing the graphic symbols of other components, it is necessary to review some terminology. As mentioned in Section 2-8, a positive-logic system defines the more positive of two signal levels (designated by H) as logic-1 and the more negative signal level (designated by L) as logic-0. Negative logic assumes the opposite assignment. A third alternative is to employ a mixed-logic convention, where the signals are considered entirely in terms of their H and L values. At any point in the circuit, the user is allowed to define the logic polarity by assigning logic-1 to either the H or L signal. The mixed-logic notation uses a small right-angle-triangle graphic symbol to designate a negative-logic polarity at any input or output terminal. (See Fig. 2-11(f).)

Integrated-circuit manufacturers specify the operation of integrated circuits in terms of H and L signals. When an input or output is considered in terms of positive logic, it is defined as *active-high*. When it is considered in terms of negative logic, it is defined as *active-low*. Active-low inputs or outputs are recognized by the presence of the small-triangle polarity-indicator symbol. When positive logic is used exclusively throughout the entire system, the small-triangle polarity symbol is equivalent to the small circle that designates negation. In this book, we have assumed positive logic throughout and employed the small circle when drawing logic diagrams. When an input or output line does not include the small circle, we define it to be active if it is logic-1. An input or output that includes the small-circle symbol is considered active if it is in the logic-0 state. However, we will use the small-triangle polarity symbol to indicate active-low assignment in all drawings that represent standard diagrams. This will conform with integrated-circuit data books, where the polarity symbol is usually employed. Note that the bottom four gates in Fig. 12-1 could have been drawn with a small triangle in the output lines instead of a small circle.

Another example of a graphic symbol for an MSI circuit is shown in Fig. 12-3. This is a 2-to-4-line decoder representing one-half of IC type 74155. Inputs are on the left

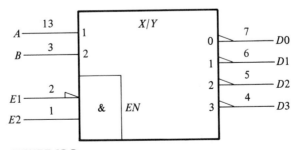

FIGURE 12-3

Standard graphic symbol for a 2-to-4-line decoder (one-half of IC type 74155)

and outputs on the right. The identifying symbol X/Y indicates that the circuit converts from code X to code Y. Data inputs A and B are assigned binary weights 1 and 2 equivalent to 2^0 and 2^1, respectively. The outputs are assigned numbers from 0 to 3, corresponding to outputs D_0 through D_3, respectively. The decoder has one active-low input E_1 and one active-high input E_2. These two inputs go through an internal AND gate to enable the decoder. The output of the AND gate is labeled EN (enable) and is activated when E_1 is at a low-level state and E_2 at a high-level state. For positive-logic assignment, the inputs and outputs are as specified in the truth table of Fig. 5-10, with $EN = E = E_1' E_2$.

12-2   QUALIFYING SYMBOLS

The IEEE standard graphic symbols for logic functions provides a list of qualifying symbols to be used in conjunction with the outline. A qualifying symbol is added to the basic outline to designate the overall logic characteristics of the element or the physical characteristics of an input or output. Table 12-1 lists some of the general qualifying symbols specified in the standard. A general qualifying symbol defines the basic function performed by the device represented in the diagram. It is placed near the top cen-

TABLE 12-1
General Qualifying Symbols

Symbol	Description
&	AND gate or function
≥ 1	OR gate or function
1	Buffer gate or inverter
$= 1$	Exclusive-OR gate or function
2k	Even function or even parity element
2k + 1	Odd function or odd parity element
X/Y	Coder, decoder, or code converter
MUX	Multiplexer
DMUX	Demultiplexer
Σ	Adder
Π	Multiplier
COMP	Magnitude comparator
ALU	Arithmetic logic unit
SRG	Shift register
CTR	Counter
RCTR	Ripple counter
ROM	Read-only memory
RAM	Random-access memory

Symbol	Description

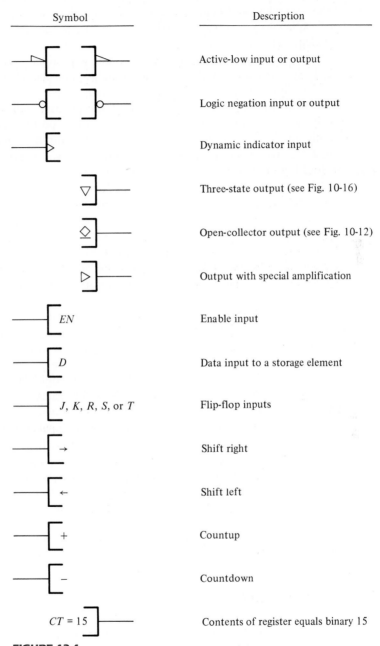

Active-low input or output

Logic negation input or output

Dynamic indicator input

Three-state output (see Fig. 10-16)

Open-collector output (see Fig. 10-12)

Output with special amplification

EN Enable input

D Data input to a storage element

J, K, R, S, or *T* Flip-flop inputs

→ Shift right

← Shift left

+ Countup

– Countdown

CT = 15 Contents of register equals binary 15

FIGURE 12-4
Qualifying symbols associated with inputs and outputs

ter position of the rectangular-shape outline. The general qualifying symbols for the gates, decoder, and adder were shown in previous diagrams. The other symbols are self-explanatory and will be used later in diagrams representing the corresponding digital elements.

Some of the qualifying symbols associated with inputs and outputs are shown in Fig. 12-4. Symbols associated with inputs are placed on the left side of the column labeled *symbol*. Symbols associated with outputs are placed on the right side of the column. The active-low input or output symbol is the polarity indicator. As mentioned previously, it is equivalent to the logic negation when positive logic is assumed. The dynamic input is associated with the clock input in flip-flop circuits. It indicates that the input is active on a transition from a low-to-high-level signal. The three-state output has a third high-impedance state, which has no logic significance. When the circuit is enabled, the output is in the normal 0 or 1 logic state, but when the circuit is disabled, the three-state output is in a high-impedance state. This state is equivalent to an open circuit.

The open-collector output has one state that exhibits a high-impedance condition. An externally connected resistor is sometimes required in order to produce the proper logic level. The diamond-shape symbol may have a bar on top (for high type) or on the bottom (for low type). The high or low type specifies the logic level when the output is not in the high-impedance state. For example, TTL-type integrated circuits have special outputs called open-collector outputs. These outputs are recognized by a diamond-shape symbol with a bar under it. This indicates that the output can be either in a high-impedance state or in a low-level state. When used as part of a distribution function, two or more open-collector NAND gates when connected to a common resistor perform a positive-logic AND function or a negative-logic OR function (see Fig. 10-12).

The output with special amplification is used in gates that provide special driving capabilities. Such gates are employed in components such as clock drivers or bus-oriented transmitters. The *EN* symbol designates an enable input. It has the effect of enabling all outputs when it is active. When the input marked with *EN* is inactive, all outputs are disabled. The symbols for flip-flop inputs have the usual meaning. The *D* input is also associated with other storage elements such as memory input.

The symbols for shift right and shift left are arrows pointing to the right or the left, respectively. The symbols for count-up and count-down counters are the plus and minus symbols, respectively. An output designated by $CT = 15$ will be active when the contents of the register reach the binary count of 15. When nonstandard information is shown inside the outline, it is enclosed in square brackets [like this].

12-3 DEPENDENCY NOTATION

The most important aspect of the standard logic symbols is the dependency notation. Dependency notation is used to provide the means of denoting the relationship between different inputs or outputs without actually showing all the elements and interconnections between them. We will first demonstrate the dependency notation with an example of the AND dependency and then define all the other symbols associated with this notation.

The AND dependency is represented with the letter G followed by a number. Any input or output in a diagram that is labeled with the number associated with G is considered to be ANDed with it. For example, if one input in the diagram has the label $G1$ and another input is labeled with the number 1, then the two inputs labeled $G1$ and 1 are considered to be ANDed together internally.

An example of AND dependency is shown in Fig. 12-5. In (a), we have a portion of a graphic symbol with two AND dependency labels, $G1$ and $G2$. There are two inputs labeled with the number 1 and one input labeled with the number 2. The equivalent interpretation is shown in part (b) of the figure. Input X associated with $G1$ is considered to be ANDed with inputs A and B, which are labeled with a 1. Similarly, input Y is ANDed with input C to conform with the dependency between $G2$ and 2.

The standard defines ten other dependencies. Each dependency is denoted by a letter symbol (except EN). The letter appears at the input or output and is followed by a number. Each input or output affected by that dependency is labeled with that same number. The eleven dependencies and their corresponding letter designation are as follows:

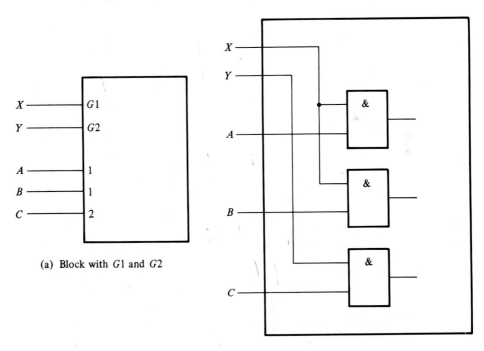

(a) Block with $G1$ and $G2$

(b) Equivalent interpretation

FIGURE 12-5

Example of G (AND) dependency

G	Denotes an AND (gate) relationship
V	Denotes an OR relationship
N	Denotes a negate (exclusive-OR) relationship
EN	Specifies an enable action
C	Identifies a control dependency
S	Specifies a setting action
R	Specifies a resetting action
M	Identifies a mode dependency
A	Identifies an address dependency
Z	Indicates an internal interconnection
X	Indicates a controlled transmission

The *V* and *N* dependencies are used to denote the Boolean relationships of OR and exclusive-OR similar to the *G* that denotes the Boolean AND. The *EN* dependency is similar to the qualifying symbol *EN* except that a number follows it (for example, *EN* 2). Only the outputs marked with that number are disabled when the input associated with *EN* is active.

The control dependency *C* is used to identify a clock input in a sequential element and to indicate which input is controlled by it. The set *S* and reset *R* dependencies are used to specify internal logic states of an *SR* flip-flop. The *C*, *S*, and *R* dependencies are explained in Section 12-5 in conjunction with the flip-flop circuit. The mode *M* dependency is used to identify inputs that select the mode of operation of the unit. The mode dependency is presented in Section 12-6 in conjunction with registers and counters. The address *A* dependency is used to identify the address input of a memory. It is introduced in Section 12-8 in conjunction with the memory unit.

The *Z* dependency is used to indicate interconnections inside the unit. It signifies the existence of internal logic connections between inputs, outputs, internal inputs, and internal outputs, in any combination. The *X* dependency is used to indicate the controlled transmission path in a CMOS transmission gate similar to the one presented in Section 10-23.

12-4 SYMBOLS FOR COMBINATIONAL ELEMENTS

The examples in this section and the rest of this chapter illustrate the use of the standard in representing various digital components with graphic symbols. The examples demonstrate actual commercial integrated circuits with the pin numbers included in the inputs and outputs. Most of the ICs presented in this chapter are included with the suggested experiments outlined in Chapter 11.

The graphic symbols for the adder and decoder were shown in Section 12-2. IC type 74155 can be connected as a 3 × 8 decoder, as shown in Fig. 12-6. (The truth table of this decoder is shown in Fig. 11-7.) There are two *C* and two *G* inputs in the IC. Each pair must be connected together as shown in the diagram. The enable input is active

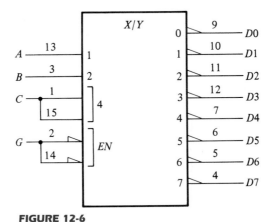

FIGURE 12-6

IC type 74155 connected as a 3 × 8 decoder

when in the low-level state. The outputs are all active-low. The inputs are assigned binary weights 1, 2, and 4, equivalent to 2^0, 2^1 and 2^2, respectively. The outputs are assigned numbers from 0 to 7. The sum of the weights of the inputs determines the output that is active. Thus, if the two input lines with weights 1 and 4 are activated, the total weight is $1 + 4 = 5$ and output 5 is activated. Of course, the *EN* input must be activated for any output to be active.

The decoder is a special case of a more general component referred to as a *coder*. A coder is a device that receives an input binary code on a number of inputs and produces a different binary code on a number of outputs. Instead of using the qualifying symbol X/Y, the coder can be specified by the code name. For example, the 3-to-8-line decoder of Fig. 12-6 can be symbolized with the name *BIN/OCT* since the circuit converts a 3-bit binary number into 8 octal values, 0 through 7.

Before showing the graphic symbol for the multiplexer, it is necessary to show a variation of the AND dependency. The AND dependency is sometimes represented by a shorthand notation like $G\frac{0}{7}$. This symbol stands for eight AND dependency symbols from 0 to 7 as follows:

$$G0, G1, G2, G3, G4, G5, G6, G7$$

At any given time, only one out of the eight AND gates can be active. The active AND gate is determined from the inputs associated with the G symbol. These inputs are marked with weights equal to the powers of 2. For the eight AND gates just listed, the weights are 0, 1, and 2, corresponding to the numbers 2^0, 2^1, and 2^2, respectively. The AND gate that is active at any given time is determined from the sum of the weights of the active inputs. Thus, if inputs 0 and 2 are active, then the AND gate that is active has the number $2^0 + 2^2 = 5$. This makes $G5$ active and the other seven AND gates inactive.

The standard graphic symbol for a 8 × 1 multiplexer is shown in Fig. 12-7(a). The qualifying symbol MUX identifies the device as a multiplexer. The symbols inside the

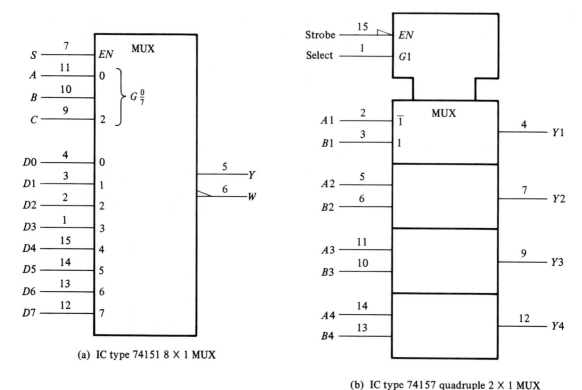

(a) IC type 74151 8 × 1 MUX

(b) IC type 74157 quadruple 2 × 1 MUX

FIGURE 12-7

Graphic symbols for multiplexers

block are part of the standard notation, but the symbols marked outside are user-defined symbols. The function table of the 74151 IC can be found in Fig. 11-9. The AND dependency is marked with $G\frac{0}{7}$ and is associated with the inputs enclosed in brackets. These inputs have weights of 0, 1, and 2. They are actually what we have called the selection inputs. The eight data inputs are marked with numbers from 0 to 7. The net weight of the active inputs associated with the G symbol specifies the number in the data input that is active. For example, if selection inputs $CBA = 110$, then inputs 1 and 2 associated with G are active. This gives a numerical value for the AND dependency of $2^2 + 2^1 = 6$, which makes $G6$ active. Since $G6$ is ANDed with data input number 6, it makes this input active. Thus, the output will be equal to data input D_6 provided that the enable input is active.

Figure 12-7(b) represents the quadruple 2 × 1 multiplexer IC type 74157 whose function table 1s listed in Fig. 11-17. The enable and selection inputs are common to all four multiplexers. This is indicated in the standard notation by the indented box at the top of the diagram, which represents a *common control block*. The inputs to a common control block control all lower sections of the diagram. The common enable input *EN* is active when in the low-level state. The AND dependency, *G*1, determines which input is active in each multiplexer section. When *G*1 = 0, the *A* inputs marked with $\overline{1}$ are active. When *G*1 = 1, the *B* inputs marked with 1 are active. The active inputs are applied to the corresponding outputs if *EN* is active. Note that the input symbols $\overline{1}$ and 1 are marked in the upper section only instead of repeating them in each section.

12-5 SYMBOLS FOR FLIP-FLOPS

The standard graphic symbols for different types of flip-flops are shown in Fig. 12-8. A flip-flop is represented by a rectangular-shape block with inputs on the left and outputs on the right. One output designates the normal state of the flip-flop and the other output

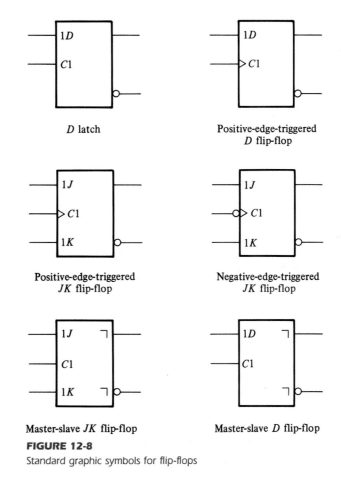

D latch

Positive-edge-triggered
D flip-flop

Positive-edge-triggered
JK flip-flop

Negative-edge-triggered
JK flip-flop

Master-slave *JK* flip-flop

Master-slave *D* flip-flop

FIGURE 12-8
Standard graphic symbols for flip-flops

with a small-circle negation symbol (or polarity indicator) designates the complement output. The graphic symbols distinguish between three types of flip-flops: the D latch, whose internal construction is shown in Fig. 6-5; the master–slave flip-flop, shown in Fig. 6-9; and the edge-triggered flip-flop, introduced in Fig. 6-12. The graphic symbol for the D latch or D flip-flop has inputs D and C indicated inside the block. The graphic symbol for the JK flip-flop has inputs J, K, and C inside. The notation $C1$, $1D$, $1J$, and $1K$ are examples of control dependency. The input in $C1$ controls input $1D$ in a D flip-flop and inputs $1J$ and $1K$ in a JK flip-flop.

The D latch has no other symbols besides the $1D$ and $C1$ inputs. The edge-triggered flip-flop has an arrowhead-shape symbol in front of the control dependency $C1$ to designate a dynamic input. The dynamic indicator symbol denotes that the flip-flop responds to the positive-edge transition of the input clock pulses. A small circle outside the block along the dynamic indicator designates a negative-edge transition for triggering the flip-flop. The master–slave is considered to be a pulse-triggered flip-flop and is indicated as such with an upside-down L symbol in front of the outputs. This is to show that the output signal changes on the falling edge of the pulse. Note that the master–slave flip-flop is drawn without the dynamic indicator.

Flip-flops available in integrated-circuit packages provide special inputs for setting and resetting the flip-flop asynchronously. These inputs are usually called direct set and direct reset. They affect the output on the negative level of the signal without the need of a clock. The graphic symbol of a master–slave JK flip-flop with direct set and reset is shown in Fig. 12-9(a). The notations $C1$, $1J$, and $1K$ represent control dependency, showing that the clock input at $C1$ controls inputs $1J$ and $1K$. S and R have no 1 in front of the letters and, therefore, they are not controlled by the clock at $C1$. The S and R inputs have a small circle along the input lines to indicate that they are active when in the logic-0 level. The function table for the 7476 flip-flop is shown in Fig. 11-12.

The graphic symbol for a positive-edge-triggered D flip-flop with direct set and reset is shown in Fig. 12-9(b). The positive-edge transition of the clock at input $C1$ controls input $1D$. The S and R inputs are independent of the clock. This is IC type 7474, whose function table is listed in Fig. 11-13.

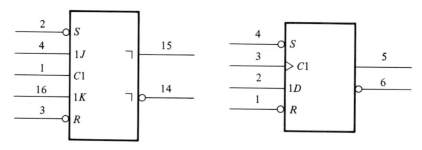

    (a) One-half 7476 JK flip-flop   (b) One-half 7474 D flip-flop

FIGURE 12-9

IC flip-flops with direct set and reset

12-6   SYMBOLS FOR REGISTERS

The standard graphic symbol for a register is equivalent to the symbol used for a group of flip-flops with a common clock input. Figure 12-10 shows the standard graphic symbol of IC type 74175, consisting of four D flip-flops with common clock and clear inputs. The clock input $C1$ and the clear input R appear in the common control block. The inputs to the common control block are connected to each of the elements in the lower sections of the diagram. The notation $C1$ is the control dependency that controls all the $1D$ inputs. Thus, each flip-flop is triggered by the common clock input. The dynamic input symbol associated with $C1$ indicates that the flip-flops are triggered on the positive edge of the input clock. The common R input resets all flip-flops when its input is at a low-level state. The $1D$ symbol is placed only once in the upper section instead of repeating it in each section. The complement outputs of the flip-flops in this diagram are marked with the polarity symbol rather than the negation symbol.

The standard graphic symbol for a shift register with parallel load is shown in Fig. 12-11. This is IC type 74195, whose function table can be found in Fig. 11-16. The qualifying symbol for a shift register is SRG followed by a number that designates the number of stages. Thus, $SRG4$ denotes a 4-bit shift register. The common control

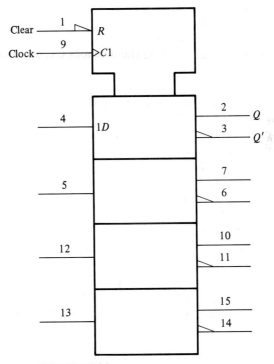

FIGURE 12-10
Graphic symbol for a 4-bit register, IC type 74175

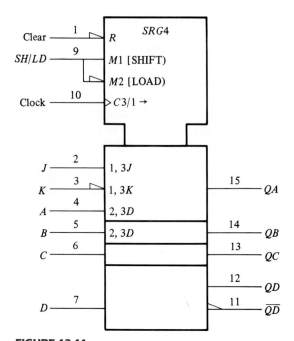

FIGURE 12-11
Graphic symbol for a shift register with parallel load, IC type 74195

block has two mode dependencies, $M1$ and $M2$, for the shift and load operations, respectively. Note that the IC has a single input labeled SH/LD (shift/load), which is split into two lines to show the two modes. $M1$ is active when the SH/LD input is high and $M2$ is active when the SH/LD input is low. $M2$ is recognized as active-low from the polarity indicator along its input line. Note the convention in this symbology: we must recognize that a single input actually exists in pin 9, but it is split into two parts in order to assign to it the two modes, $M1$ and $M2$. The control dependency $C3$ is for the clock input. The dynamic symbol along the $C3$ input indicates that the flip-flops trigger on the positive edge of the clock. The symbol $/1 \rightarrow$ following $C3$ indicates that the register shifts to the right or in the downward direction when mode $M1$ is active.

The four sections below the common control block represent the four flip-flops. Flip-flop QA has three inputs: two are associated with the serial (shift) operation and one with the parallel (load) operation. The serial input label 1, $3J$ indicates that the J input of flip-flop QA is active when $M1$ (shift) is active and $C3$ goes through a positive clock transition. The other serial input with label 1, $3K$ has a polarity symbol in its input line corresponding to the complement of input K in a JK flip-flop. The third input of QA and the inputs of the other flip-flops are for the parallel input data. Each input is denoted by the label 2, $3D$. The 2 is for $M2$ (load), and 3 is for the clock $C3$. If the input in pin number 9 is in the low level, $M1$ is active and a positive transition of the clock at $C3$ causes a parallel transfer from the four inputs, A through D, into the four flip-flops,

QA through QD. Note that the parallel input is labeled only in the first and second sections. It is assumed to be in the other two sections below.

Figure 12-12 shows the graphic symbol for the bidirectional shift register with parallel load, IC type 74194. The function table for this IC is listed in Fig. 11-19. The common control block shows an R input for resetting all flip-flops to 0 asynchronously. The mode select has two inputs and the mode dependency M may take binary values from 0 to 3. This is indicated by the symbol $M\frac{0}{3}$, which stands for $M0$, $M1$, $M2$, $M3$, and is similar to the notation for the G dependency in multiplexers. The symbol associated with the clock is

$$C4/1 \rightarrow \quad /2 \leftarrow$$

$C4$ is the control dependency for the clock. The $/1 \rightarrow$ symbol indicates that the register shifts right (down in this case) when the mode is $M1$ ($S_1 S_0 = 01$). The $/2 \leftarrow$ symbol indicates that the register shifts left (up in this case) when the mode is M_2 ($S_1 S_0 = 10$). The right and left directions are obtained when the page is turned 90 degrees counterclockwise.

The sections below the common control block represent the four flip-flops. The first flip-flop has a serial input for shift right, denoted by 1, $4D$ (mode $M1$, clock $C4$, input D). The last flip-flop has a serial input for shift left, denoted by 2, $4D$ (mode $M2$, clock $C4$, input D). All four flip-flops have a parallel input denoted by the label 3, $4D$ (mode $M3$, clock $C4$, input D). Thus, $M3$ ($S_1 S_0 = 11$) is for parallel load. The remaining

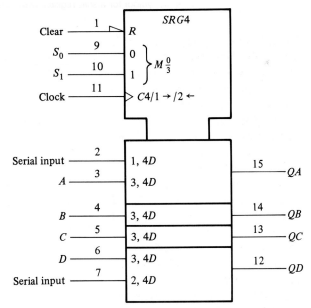

FIGURE 12-12

Graphic symbol for a bidirectional shift register with parallel load, IC type 74194 mode $M0$ ($S_1 S_0 = 00$) has no effect on the outputs because it is not included in the input labels.

12-7 SYMBOLS FOR COUNTERS

The standard graphic symbol of a binary ripple counter is shown in Fig. 12-13. The qualifying symbol for a ripple counter is *RCTR*. The designation *DIV* 2 stands for the divide-by-2 circuit that is obtained from the single flip-flop *QA*. The *DIV* 8 designation is for the divide-by-8 counter obtained from the other three flip-flops. The diagram represents IC type 7493, whose internal circuit diagram is shown in Fig. 11-2. The common control block has an internal AND gate with inputs $R1$ and $R2$. When both of these inputs are equal to 1, the content of the counter goes to zero. This is indicated by the symbol $CT = 0$. Since the count input does not go to the clock inputs of all flip-flops, it has no $C1$ label and, instead, the symbol $+$ is used to indicate a count-up operation. The dynamic symbol next to the $+$ together with the polarity symbol along the input line signify that the count is affected with a negative-edge transition of the input signal. The bit grouping from 0 to 2 in the output represents values for the weights to the power of 2. Thus, 0 represents the value of $2^0 = 1$ and 2 represents the value $2^2 = 4$.

The standard graphic symbol for the 4-bit counter with parallel load, IC type 74161, is shown in Fig. 12-14. The qualifying symbol for a synchronous counter is *CTR* followed by the symbol *DIV* 16 (divide by 16), which gives the cycle length of the counter. There is a single load input at pin 9 that is split into the two modes, $M1$ and $M2$.

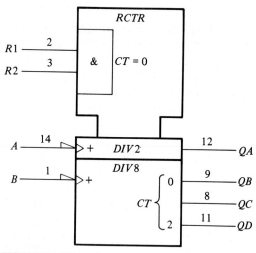

FIGURE 12-13
Graphic symbol for ripple counter, IC type 7493

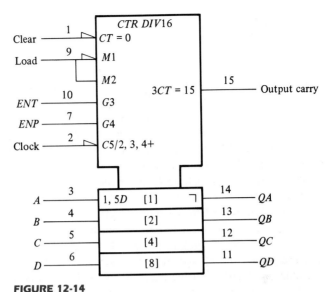

FIGURE 12-14

Graphic symbol for 4-bit binary counter with parallel load, IC type 74161

$M1$ is active when the load input at pin 9 is low and $M2$ is active when the load input at pin 9 is high. $M1$ is recognized as active-low from the polarity indicator along its input line. The count-enable inputs use the G dependencies. $G3$ is associated with the T input and $G4$ with the P input of the count enable. The label associated with the clock is

$$C5/2, 3, 4 +$$

This means that the circuit counts up (the + symbol) when $M2$, $G3$, and $G4$ are active (load = 1, $ENT = 1$, and $ENP = 1$) and the clock in $C5$ goes through a positive transition. This condition is specified in the function table of the 74161 listed in Fig. 11-15. The parallel inputs have the label 1, $5D$, meaning that the D inputs are active when $M1$ is active (load = 0) and the clock goes through a positive transition. The output carry is designated by the label

$$3CT = 15$$

This is interpreted to mean that the output carry is active (equal to 1) if $G3$ is active ($ENT = 1$) and the content (CT) of the counter is 15 (binary 1111). Note that the outputs have an inverted L symbol, indicating that all the flip-flops are of the master–slave type. The polarity symbol in the $C5$ input designates an inverted pulse for the input clock. This means that the master is triggered on the negative transition of the clock pulse and the slave changes state on the positive transition. Thus, the output changes on the positive transition of the clock pulse. It should be noted that IC type 74LS161 (low-power Schottky version) has positive-edge-triggered flip-flops.

12-8 SYMBOL FOR RAM

The standard graphic symbol for the random-access memory (RAM) 7489 is shown in Fig. 12-15. The numbers 16 × 4 that follow the qualifying symbol RAM designate the number of words and the number of bits per word. The common control block is shown with four address lines and two control inputs. Each bit of the word is shown in a separate section with an input and output data line. The address dependency A is used to identify the address inputs of the memory. Data inputs and outputs affected by the address are labeled with the letter A. The bit grouping from 0 through 3 provides the binary address that ranges from $A0$ through $A15$. The diamond symbol with a bar under it signifies an open-collector output. The polarity symbol specifies the inversion of the outputs.

The operation of the memory is specified by means of the dependency notation. The RAM graphic symbol uses four dependencies: A (address), G (AND), EN (enable), and C (control). Input $G1$ is to be considered ANDed with $1EN$ and $1C2$ because $G1$ has a 1 after the letter G and the other two have a 1 in their label. The EN dependency is used to identify an enable input that controls the data outputs. The dependency $C2$ controls the inputs as indicated by the $2D$ label. Thus, for a write operation, we have the $G1$ and $1C2$ dependency $(ME = 0)$, the $C2$ and $2D$ dependency $(WE = 0)$, and the A dependency, which specifies the binary address in the four address inputs. For a read operation, we have the $G1$ and $1EN$ dependencies $(ME = 0, WE = 1)$ and the A dependency for the outputs. The interpretation of these dependencies results in the operation of the memory as listed in the function table of Fig. 11-18.

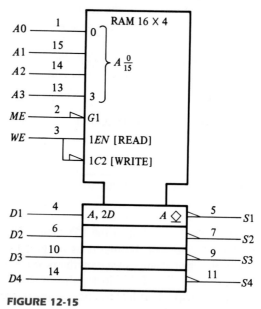

FIGURE 12-15

Graphic symbol for 16 × 4 RAM, IC type 7489

REFERENCES

1. *IEEE Standard Graphic Symbols for Logic Functions* (ANSI/IEEE Std. 91-1984). New York: Institute of Electrical and Electronics Engineers, 1984.

2. KAMPEL, I., *A Practical Introduction to the New Logic Symbols*. Boston: Butterworth, 1985.

3. MANN, F. A., *Explanation of New Logic Symbols*. Dallas: Texas Instruments, 1984.

4. *The TTL Data Book,* Volume 1. Dallas: Texas Instruments, 1985.

PROBLEMS

12-1 Figure 11-1 shows various small-scale integration circuits with pin assignment. Using this information, draw the rectangular-shape graphic symbols for the 7400, 7404, and 7486 ICs.

12-2 Define the following in your own words:
(a) Positive and negative logic.
(b) Active-high and active-low.
(c) Polarity indicator.
(d) Dynamic indicator.
(e) Dependency notation.

12-3 Show an example of a graphic symbol that has the three Boolean dependencies, G, V, and N. Draw the equivalent interpretation.

12-4 Draw the graphic symbol of a BCD-to-decimal decoder. This is similar to a decoder with four inputs and ten outputs.

12-5 Draw the graphic symbol for a binary-to-octal decoder with three enable inputs, $E1$, $E2$, and $E3$. The circuit is enabled if $E1 = 1$, $E2 = 0$, and $E3 = 0$ (assuming positive logic).

12-6 Draw the graphic symbol of a dual 4-to-1-line multiplexers with common selection inputs and a separate enable input for each multiplexer.

12-7 Draw the graphic symbol for the following flip-flops:
(a) Negative-edge-triggered D flip-flop.
(b) Master–slave RS flip-flop.
(c) Positive-edge-triggered T flip-flop.

12-8 Explain the function of the common control block when used with the standard graphic symbols.

12-9 Draw the graphic symbol of a 4-bit register with parallel load using the label $M1$ for the load input and $C2$ for the clock.

12-10 Explain all the symbols used in the standard graphic diagram of Fig. 12-12.

12-11 Draw the graphic symbol of an up–down synchronous binary counter with mode input (for up or down) and count-enable input with G dependency. Show the output carries for the up count and the down count.

12-12 Draw the graphic symbol of a 256×1 RAM. Include the symbol for three-state outputs.

APPENDIX
Answers
to
Selected Problems

Chapter 1

1-1. 0, 1, 2, 3, 4, 5, 6, 7, 8, 9, A, B, 10, 11, 12, 13

1-2. 65,535

1-3. 46; 117.75; 436

1-4. 151; 580; 35; 260

1-5. 10011001111; 1010100001.0011101; 10011100010000; 11111001110

1-6. (a) 16612.34631...
(b) 792.41CAC...
(c) 10101111.001011...

1-7. $(111100111010011111000010)_2 = (74723702)_8$

1-8.

Decimal	Binary	Octal	Hexadecimal
225	11100001	341	E1
215	11010111	327	D7
403	110010011	623	193
10949	10101011000101	25305	2AC5

1-9. (a) 1304; 336313
(b) 206; E4F9
(c) 1101011; 101100101110

1-10. 110011 (255/5 = 51)

1-11. $x = 7$

1-12. $(73642815)_9$

1-13. 87650123; 99019899; 09990048; 99999999

1-14. 876100; 909343; 900000; 000000

1-15.

Number	1's complement	2's complement
10101110	01010001	01010010
10000001	01111110	01111111
10000000	01111111	10000000
00000001	11111110	11111111
00000000	11111111	00000000

1-17. (a) 01010; (b) 01101; (c) −101100; (d) 0000000

1-19. (a) 100011 (−29); (b) 000000; (c) 101111 (−17); (d) 000101

1-20. 0001 0011 0101 1001 0111
1001 0011 0010 1000 0110
1001 1001 1000 1000 0000

1-21.

	7421
0	0000
1	0001
2	0010
3	0011
4	0100
5	0101
6	0110
7	0111 (or 1000)
8	1001
9	1010

1-22. (b) 012345
(c) 000, 001, 010, 101, 110, 111; for digits 0, 1, 2, 3, 4, 5, respectively

1-23. (a) 1000 0110 0010 0000
(b) 1011 1001 0101 0011
(c) 1110 1100 0010 0000
(d) 10000110101100

1-24. 3864: 0011 1110 1100 0100
6135: 1100 0001 0011 1011

1-28. John Doe

1-29. (a) 100100111
(b) 001010010101
(c) 011001001110010110101

1-30. 94 printing characters; 32 special characters

1-31. (a) 597 in BCD
(b) 264 in excess-3 code
(c) Not valid for the 2, 4, 2, 1 code of Table 1-2

1-32. 0100000001 + 1000000010 = 1100000011

1-34. $L = (A + B) \cdot C$

Chapter 2

2-2. (a) $x' + y$
(b) x

(c) 1

(d) $x' + y + z'$

(e) $xy' + x'z'$

2-3. (a) B

(b) $z(x + y)$

(c) $x'y'$

(d) $x(y + w)$

(e) 0

2-4. (a) $AB + C'$

(b) $x + y + z$

(c) B

(d) $A'(B + C'D)$

2-6. (a) $xy + x'y'$

(b) $(A' + B + D)(C' + D)E'$

(d) $x'yz' + xz + x'y'$

2-7. (a) $F = (x + y)' + (x + z')' + (y + z')'$

(b) $F = [(y + z')' + (x + y)' + (y' + z)']'$

2-8. (a) $F = [(x'y')'(x'z)'(y'z)']'$

(b) $F = (y'z)'(x'y)'(yz')'$

2-9. (a) $\Sigma(3, 5, 6, 7) = \Pi(0, 1, 2, 4)$

(b) $\Sigma(0, 1, 3, 7) = \Pi(2, 4, 5, 6)$

2-10. (a) $F = \Sigma(2, 3, 6, 7)$

(b) $F' = \Sigma(0, 1, 4, 5)$

(d) $F = y$

2-11. (c) $F = y'z + y(w + x)$

2-12. (a) $\Sigma(1, 3, 5, 7, 9, 11, 13, 15) = \Pi(0, 2, 4, 6, 8, 10, 12, 14)$

(b) $\Sigma(3, 5, 6, 7) = \Pi(0, 1, 2, 4)$

2-16. (a) $AB + BC; (A + C)B$

(b) $x' + y + z'$

2-18. $F' = (x'y + xy')' = xy + x'y'$ Dual of F is $(x' + y)(x + y') = xy + x'y'$

2-21. (a) 4; (b) 3; (c) 2; (d) 2; (e) 1

Chapter 3

3-1. (a) $x'y' + xz$

(b) $y + x'z$

(c) $xy + xz + yz$

(d) $A'B + C'$

3-2. (a) $xy + x'z'$

(b) $x' + yz$

(c) $C' + A'B$

3-3. (a) $BCD + A'BD'$

(b) $wx + w'x'y$

(c) $ABD + ABC + CD$

3-4. (a) $xz' + w'y'z + wxy$

(b) $A'C' + A'B'D' + ACD + A'BD$ (or BCD)

(c) $wx + x'y$

(d) $BD + B'D' + A'B$ (or $A'D'$)

3-5. (a) $x'y + z$

(b) $BC' + B'D + AB'C$

(c) $AC + B'D' + A'BD + B'C$ (or CD)

(d) $xz + wy + x'y$

3-6.   (a) $F(x, y, z) = \Sigma(3, 5, 6, 7)$

(b) $F(A, B, C, D) = \Sigma(1, 3, 5, 9, 12, 13, 14)$

(c) $F(w, x, y, z) = \Sigma(0, 2, 5, 7, 8, 10, 14, 15)$

3-7.   (a) The essential prime implicants are xz and $x'z'$;

$\quad\quad F = xz + x'z' + w'x$ (or $w'z'$)

(b) The essential prime implicants are AC, $B'D'$ and $A'BD$;

$\quad\quad F = AC + B'D' + A'BD + CD$ (or $B'C$)

(c) The essential prime implicants are BC' and AC;

$\quad\quad F = BC' + AC + A'B'D$

3-8.   (a) $A'B'D' + AD'E + B'C'D'$

(b) $DE + A'B'C + B'C'E'$

(c) $A'B'D' + B'D'E + B'CD' + CDE' + BDE'$

3-9.   (a) $(w' + x')(x + z')(x' + y + z)$

(b) $(A + D')(B' + D')$

(c) y

(d) $(B + C')(A + B)(A + C + D)$

3-10.   (a) $xy + z' = (x + z')(y + z')$

(b) $AC' + CD + B'D = (A + D)(C' + D)(A + B' + C)$

(c) $B'D' + AD' + A'C' = (A' + D')(C' + D')(A + B' + C')$

3-11. $F = B'D' + A'BD + A'BC = (A' + B')(B + D')(B' + C + D)$

3-12.   (a) $A + BC' + C'D'$

(b) $BD + BC + AB'C'D'$

3-13. $F' = BD + BC + AC$

3-15.   (a) $F = (w + z')(x' + z')(w' + x' + y')$

(b) $F = (w + x)(w' + x')(y + z)(w' + z')$

3-18. $F = B'D'(A' + C) + BD(A' + C')$

$\quad\quad = [B' + D(A' + C')][B + D'(A' + C)]$

$\quad\quad = [D' + B(A' + C')][D + B'(A' + C)]$

3-21.

$AND-AND = AND$ $\quad\quad\quad$ $OR-OR = OR$

$AND-NAND = NAND$ $\quad\quad$ $OR-NOR = NOR$

$NOR-NAND = OR$ $\quad\quad\quad$ $NAND-NOR = AND$

$NOR-AND = NOR$ $\quad\quad\quad$ $NAND-OR = NAND$

3.22.   (a) $F = 1 = \Sigma(0, 1, 2, 3, 4, 5, 6, 7)$

(b) $F = B'D' + CD' + ABC'D = \Sigma(0, 2, 6, 8, 10, 13, 14)$

(c) $F = A'D + BD + C'D = \Sigma(1, 3, 5, 7, 9, 13, 15)$

3-23.   (a) $F = x'z' + w'z = (x' + z)(w' + z')$

3-24. $F = C + AD'$

3-27.   (a) $A'CEF'G'$

(b) $ABCDEFG + A'CEF'G' + BC'D'EF$

(c) $A'B'C'DEF' + A'BC'D'E + CE'F + A'BD'EF$ (or $A'BCD'F$)

Chapter 4

4-1. $F = x'y' + x'z'$

4-2. $A = xy + xz + yz$; $B = x \oplus y \oplus z$; $C = z'$

4-3. $F = xy + xz + yz$

4-7.

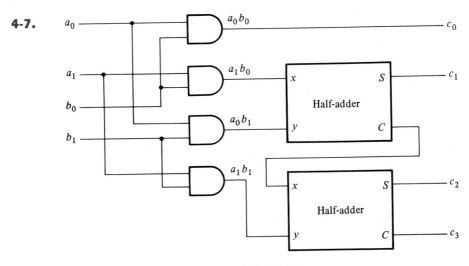

4-9. Inputs: x, y, z;   outputs: A, B, C, D, E, F

$A = xy$ $D = yz'$
$B = xy' + xz$ $E = 0$
$C = z(x \oplus y)$ $F = z$

4-10. Inputs: A, B, C, D;   outputs: $w, x, y, z; d = \Sigma(10, 11, 12, 13, 14, 15)$;

$w = A'B'C'$
$x = B \oplus C$
$y = C$
$z = D'$

4-11. Inputs: A, B, C, D;   outputs: w, x, y, z;

$w = A'B + A'C + A'D + AB'C'D'$
$x = B'C + B'D + BC'D'$
$y = CD' + C'D$
$z = D$

4-12. Inputs: A, B, C, D;   output: $E = AB + AC$
4-13. Inputs: A, B, C, D;   outputs: $w, x, y, z; d = \Sigma(1, 2, 3, 12, 13, 14)$;

$w = AB + AC'D'$
$x = B'C + B'D + BC'D'$
$y = CD' + C'D$
$z = D$

4-14. Inputs: A, B, C, D;   outputs: $w, x, y, z; d = \Sigma(5, 6, 7, 8, 9, 10)$;

$w = A$
$x = A'C + A'B + A'D + BCD$
$y = ACD + AC'D' + A'C'D + A'CD'$
$z = D$

4-15. Inputs: w, x, y, z;   outputs: E, A, B, C, D;

$E = wx + wy$
$A = wx'y'$
$B = w'x + xy$
$C = w'y + wxy'$
$D = z$

4-16. Inputs: A, B, C, D;   outputs: a, b, c, d, e, f, g;

$a = A'C + A'BD + B'C'D' + AB'C'$
$b = A'B' + A'C'D' + A'CD + AB'C'$
$c = A'B + A'D + B'C'D' + AB'C'$
$d = A'CD' + A'B'C + B'C'D' + AB'C' + A'BC'D$
$e = A'CD' + B'C'D'$
$f = A'BC' + A'C'D' + A'BD' + AB'C'$
$g = A'CD' + A'B'C + A'BC' + AB'C'$

4-17. Full-adder circuit

4-21. $F = ABC + A'D$
$G = ABC + A'D'$

4-28. Inputs: A, B, C, D;   outputs: w, x, y, z

$w = A$
$x = A \oplus B$
$y = A \oplus B \oplus C = x \oplus C$
$z = A \oplus B \oplus C \oplus D = y \oplus D$

4-30. $(A \oplus B)(C \oplus D)$

Chapter 5

5-3. (a) $A = 0110; B = 1001; M = 1$
(b) $S = 1101$ (2's complement of 0011); $C_5 = 0$ (because $A < B$)

5-4.

Sum	C_4	
(a) 1101	0	$7 + 6 = 13$
(b) 0001	1	$8 + 9 = 16 + 1$
(c) 0100	1	$12 - 8 = 4$
(d) 1011	0	$5 - 10 = -5$ (in 2's complement)
(e) 1111	0	$0 - 1 = -1$ (in 2's complement)

5-5. (b) $C_4 = (G_3'P_3' + G_3'G_2'P_2' + G_3'G_2'G_1'P_1' + G_3'G_2'G_1'C_1')'$

5-6. (c) $C_4 = (P_3' + G_3'P_2' + G_3'G_2'P_1' + G_3'G_2'G_1'C_1')'$

5-7. 60 ns

5-8. $C_5 = G_4 + P_4G_3 + P_4P_3G_2 + P_4P_3P_2G_1 + P_4P_3P_2P_1C_1$

5-9. 312

5-10. See the answer to Problem 4-10.

5-13. $x = (A_0 \odot B_0)(A_1 \odot B_1)(A_2 \odot B_2)(A_3 \odot B_3)$

5-15. $F_1(x, y, z) = \Sigma(0, 5, 7)$
$F_2(x, y, z) = \Sigma(2, 3, 4)$
$F_3(x, y, z) = \Sigma(1, 6, 7)$

5-16. Use NAND gates for F_1 and F_2; AND gate for F_3.

5-20. $x = D_0'D_1'$
$y = D_0'D_1' + D_0'D_2'$
$V = D_0 + D_1 + D_2 + D_3$

5-21. For inputs $D_5 = D_3 = 1$, the outputs are $xyz = 101$; $V = 1$.

5-27. $F(A, B, C, D) = \Sigma(1, 6, 7, 9, 10, 11, 12)$

5-28. When $AB = 00$, $F = D$
When $AB = 01$, $F = (C + D)'$ (use a NOR gate)
When $AB = 10$, $F = CD$ (use an AND gate)
When $AB = 11$, $F = 1$

5-30. 24 pins

5-31. (a) 256×8; (b) 512×5; (c) 1024×4; (d) 32×7

5-33. Six product terms: yz', xz', $x'y'z$, xy', $x'y$, z

5-37. $A = yz' + xz' + x'y'z$
$B = x'y' + xy + yz$
$C = A + xyz$
$D = z + x'y$

Chapter 6

6-6. (c)

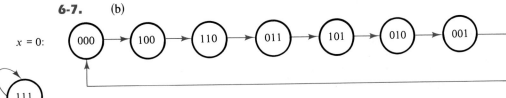

6-7. (b)

$x = 0$:

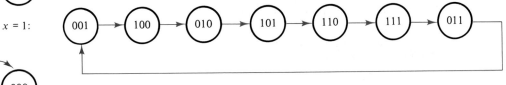

$x = 1$:

6-8.

Present State	Inputs		Next State	Output
Q	x	y	Q	S
0	0	0	0	0
0	0	1	0	1
0	1	0	0	1
0	1	1	1	0
1	0	0	0	1
1	0	1	1	0
1	1	0	1	0
1	1	1	1	1

6-9.   A counter with a repeated sequence of 00, 01, 10.

6-10.   (a)

J	N	$Q(t+1)$
0	0	0
0	1	$Q(t)$
1	0	$Q'(t)$
1	1	1

(b)

$Q(t)$	$Q(t+1)$	J	N
0	0	0	X
0	1	1	X
1	0	X	0
1	1	X	1

6-11.

Present State		Input	Next State		Output
A	B	x	A	B	y
0	0	0	0	1	0
0	0	1	0	0	1
0	1	0	1	0	1
0	1	1	1	1	0
1	0	0	0	0	1
1	0	1	0	1	0
1	1	0	1	1	0
1	1	1	1	0	1

6-12.   (c) $A(t+1) = xB + x'A + yA + y'A'B'$
$B(t+1) = xA'B' + (x' + y')A'B$

6-13. Present state: 00 00 01 00 01 11 00 01 11 10 00 01 11 10 10
     Input: 0 1 0 1 1 0 1 1 1 0 1 1 1 1 0
    Output: 0 0 1 0 0 1 0 0 0 1 0 0 0 0 1
   Next state: 00 01 00 01 11 00 01 11 10 00 01 11 10 10 00

6-14.

	Next State		Output	
Present State	0	1	0	1
a	f	b	0	0
b	d	a	0	0
d	g	a	1	0
f	f	b	1	1
g	g	d	0	1

6-15.   State: $a\ f\ b\ c\ e\ d\ g\ h\ g\ g\ h\ a$
   Input: 0 1 1 1 0 0 1 0 0 1 1
   Output: 0 1 0 0 0 1 1 1 0 1 0

6-16.   State: $a\ f\ b\ a\ b\ d\ g\ d\ g\ g\ d\ a$
   Input: 0 1 1 1 0 0 1 0 0 1 1
   Output: 0 1 0 0 0 1 1 1 0 1 0

6-19. $DQ = Q'J + QK'$

6-20. $DA = Ax' + Bx$
   $DB = A'x + Bx'$

6-21. $JA = KA = (Bx + B'x')E$
   $JB = KB = E$

6-22. (a) $DA = A'B'x$
    $DB = A + C'x' + BCx$
    $DC = Cx' + Ax + A'B'x'$
    $y = A'x$
   (b) $JA = B'x$ $KA = 1$
    $JB = A + C'x'$ $KB = C'x + Cx'$
    $JC = Ax + A'B'x'$ $KC = x$
    $y = A'x$

6-23. $SA = BX'$ $RA = BX$
   $SB = B'x$ $RB = A'x' + ABx$

6-24. $TA = ABx + A'Bx'$
   $TB = ABx + A'Bx' + B'x$

6-25. (a) $JA = BC$ $KA = B$
    $JB = C$ $KB = A + C$
    $JC = A' + B'$ $KC = 1$
   (b) $DA = A \oplus B$
    $DB = AB' + C$
    $DC = A'B'C'$

Chapter 9

9-2. Sequence of $Y_1 Y_2$: 00, 00, 01, 11, 11, 01, 00.

9-3. (d) When the input is 01, the output is 0. When the input is 10, the output is 1. Whenever the input assumes one of the other two combinations, the output retains its previous value.

9-4. (c)

	00	01	11	10
a	ⓐ, 0	b, 1	c, 1	d, 0
b	a, 0	ⓑ, 1	c, 1	ⓑ, 0
c	ⓒ, 1	b, 1	ⓒ, 1	d, 1
d	c, 1	b, 1	c, 1	ⓓ, 1

9-5. (c) $Y_1 = x_1'x_2 + x_2y_1$
$Y_2 = x_2 + x_1y_2$
$z = x_1x_2y_1' + x_1y_2'$

9-10. $S = x_1x_2'$
$R = x_1'x_2$

9-13. (b) Two possible transition tables:

	00	01	11	10
a	ⓐ, 0	b, −	−, −	e, −
b	ⓑ, 1	ⓑ, 1	−, −	d, −
d	a, −	ⓓ, 1	−, −	ⓓ, 1
e	ⓔ, 1	d, −	−, −	ⓔ, 1

	00	01	11	10
a	ⓐ, 0	b, −	−, −	b, −
b	c, −	ⓑ, 1	−, −	ⓑ, 0
c	ⓒ, 1	d, −	−, −	d, −
d	a, −	ⓓ, 1	−, −	ⓓ, 1

9-18. 3a: $(a, b)(c, d)(e, f, g, h)$
3b: $(a, e, f)(b, j)(c, d)(g, h)(k)$

9-20. Add states g and h to binary assignment.

	00	01	11	10
0	a	g	b	f
1	c	h	d	e

9-22. $F = A'D' + AC'D' + A'BC + A'CD'$

9-23. $Y = (x_1 + x_2')(x_2 + x_3)(x_1 + x_3)$

Chapter 10

10-1. Fan-out = 10; power dissipation = 18.75 mW; propagation delay = 3 ns; noise margin = 0.3 V

10-2. (a) 1.058 V (b) 0.82 V (c) 0.238 V

10-3. $I_B = 0.44$ mA, $I_{CS} = 2.4$ mA

10-4. (a) 2.4 mA (b) 0.82 mA (c) $2.4 + 0.82N$ (d) 7.8 (e) 7

10-5. (b) 3.53 (c) 2.585 mA (d) 16 mA (e) 300 Ω

10-9. (a) 4.62 mA (b) 4 mA

10-10. 0.3 V

Index